QUANTUM MECHANICS OF MANY DEGREES OF FREEDOM

QUANTUM MECHANICS OF MANY DEGREES OF FREEDOM

DANIEL S. KOLTUN
University of Rochester
Rochester, New York, U.S.A.

JUDAH M. EISENBERG
Tel Aviv University
Tel Aviv, Israel

WILEY

A Wiley-Interscience Publication
JOHN WILEY & SONS
New York Chichester Brisbane Toronto Singapore

Library of Congress Cataloging in Publication Data:

Koltun, Daniel S.
    Quantum mechanics of many degrees of freedom / Daniel S. Koltun,
Judah M. Eisenberg.
        p.   cm.
    "A Wiley-Interscience publication."
    Includes index.
    ISBN 0-471-88842-7
    1. Degree of freedom.   2. Quantum field theory.   3. Quantum
theory.   4. Many-body problem.   I. Eisenberg, Judah M.   II. Title.
QC174.52.D43K65 1988
530.1'2—dc19
                                                                87-28579
                                                                CIP

Printed in the United States of America

10 9 8 7 6 5 4 3 2

To our parents,
Charlotte S. and Samuel B. Koltun,
Rose L. Eisenberg,
and to the memory of Azriel L. Eisenberg

Preface

This book was written because its authors came to feel that there is a clear need in the standard physics graduate curriculum for a particular course in advanced quantum theory to follow the standard one-year course in quantum mechanics that has become almost universal in physics education. This feeling, in turn, was generated by the realization that each of us was teaching such a course at his own institution, and that our two syllabuses strongly resembled each other even though they had been independently constructed at far geographical remove and in response to the dynamics of two rather different physics curricula. When we checked further with colleagues in Europe and North America it became clear that a similar course was becoming standard in a large number of graduate physics programs.

This course covers material having to do with the quantum theory of systems possessing many degrees of freedom, either because the systems contain many particles, or because of the possibility of particle creation and annihilation, or both. This subject matter is pertinent in a great number of areas of modern physics research, and—apart from considerations of efficiency in not repeating the introductory aspects of the material in each specialized curriculum in atomic physics, condensed-matter physics, elementary-particle physics, nuclear physics, and so forth—there is a clear educational gain in stressing the commonality of the phenomena and methods involved. A quick perusal of the table of contents shows that the physical examples considered in detail include the many-boson aspects of the electromagnetic field, single-particle approximations in atoms and nuclei, the pairing interaction and superconductivity in metals, hard-sphere gases, and the electron gas. Theoretical techniques developed in the text begin with second quantization and the elements of quantum field theory, eventually leading to diagrammatic methods for the many-body problem and for scattering theory.

The text includes a number of exercises at the end of every chapter, intended both to give concrete examples of some of the techniques, and in some cases to carry some topics further than they have been developed in the main text. A small number of appendices are included, mostly to fill in details of proofs of some of the main results used in the text. This is to keep the book material essentially self-contained.

The organization of material in more detail is as follows: The book begins with a discussion of second quantization. Since this material is usually touched upon in the one-year introductory course in quantum mechanics, it appears here partly for purposes of review. On the other hand, there is rarely time in the first year of quantum theory to show the usefulness and power of methods based on second quantization—which is our main purpose here—so that it is important to provide this crucial link between the introductory and the more advanced material. We then develop in Chapter 2 the rudiments of quantum field theory, starting from a given Lagrangian and proceeding through the stage of second quantization; our purpose here is not to give a thorough introduction to field theory, but rather to teach just enough of the first stages of this subject to exhibit the universality of the physics of second quantization. In Chapter 3, we apply second-quantization methods to the electromagnetic field in some detail. This represents a topic of central importance for almost all physics subdisciplines, and so we take time to explore differences between photon statistics and random statistics, laser applications, the effect of Hanbury Brown and Twiss, and the uses of the coherent state in generating a classical limit.

We focus on systems of many fermions beginning in Chapter 4, where Hartree–Fock theory is developed and some of its ramifications, extensions, and applications are discussed. Here again much weight is given to this topic because it serves as a major point of departure—conceptual or computational—in so many specialities of physics. For similar reasons, and also because it is paradigmatic for a situation where single-particle methods fail utterly, we devote Chapter 5 to an extensive discussion of pairing based primarily on the methods of Bardeen, Cooper, and Schrieffer and of Bogolyubov. To illustrate the abstract theory in direct, physical terms we make application to the tunneling between a superconductor and a normal metal and to the Josephson effects.

Two more examples of many-fermion systems are introduced in Chapter 6, which serve as illustrations of two important and rather different kinds of physical behavior. The hard-sphere gas is chosen to characterize systems dominated by strongly repulsive potentials at short distance, such as liquid helium (^{3}He or ^{4}He) or nuclear matter. The electron gas represents a fundamental model of electrons in metals, in which the long range of the Coulomb potential is responsible for the characteristic behavior of the system. An elementary treatment in perturbation theory shows that more powerful theoretical tools will be needed. Unlike the treatments in Chapters 4 and 5, we do not immediately provide methods of solution of the particular problems raised in this chapter, but motivate development of diagrammatic perturbation theory. This is then presented systematically in the next three chapters, following which we return to specific applications to the hard-sphere and electron gases in Chapter 10.

Chapter 7 provides general background on the handling of time development for many-particle systems, and Chapter 8 works out the expansion of the time-development operator in terms of diagrammatic perturbation theory in both the Feynman and the Goldstone approaches. The full exploitation of these methods for calculating properties of the ground state is given in Chapter 9, which centers around the linked-cluster expansion. In Chapter 10 applications are made to a number of Fermi-gas problems. Further applications of diagrammatic methods to particle propagation and scattering problems are given in Chapter 11.

In presenting this material in Chapters 6 through 10—and indeed in the material of Chapters 1 through 5—we are purposely trying to achieve a more gradual rise in level from the base of the first-year course in quantum mechanics than exists in the many excellent books already available on the subject of diagrammatic techniques in many-body theory. We do not attempt to be nearly as encyclopedic as are those works, but we hope for a presentation which is more easily accessible to students who are still at an early stage of developing their formal skills and powers of abstraction. As a consequence, Chapters 6 through 10 contain a good deal of assessment and explanation of what the theory is trying to do and how it goes about doing it, as well as extensive comparison between different lines of approach, in particular as between Feynman and Goldstone diagrams. Given the pedagogical emphasis of the material, we do not refer to the original literature when the presentation is self-contained, particularly for subjects which have been developed repeatedly in books, articles, and lectures. We do give references for matters that are only touched on lightly in the text.

In the nature of the introductory scope of the course we have in mind, most of this book is restricted to ground-state properties of Fermi gases. There are, however, a number of places that could serve as natural points of departure for the discussion of other material, leading in the direction of more specialized areas almost at the research-seminar level. Thus there are potential ligatures to the topics of excitations in systems of fermions, to nonzero temperatures, to systems of bosons, and to the application of the methods of the course to scattering problems. This last subject is treated at somewhat greater length in Chapter 11 in order to ease the transition to the study of diagrammatic methods, especially in elementary-particle and nuclear physics, where the scattering situation is often of paramount importance. In those more specialized cases, the basic methods of diagrammatic perturbation theory are often somewhat obscured by the needs for a covariant theory and for the embodiment of sometimes rather complicated internal symmetries in the theory. We therefore choose, in Chapter 11, to restrict ourselves to the simplest illustrative cases, which we see—in common with much of the rest of this course—as preparing the way for advanced study in the pertinent subdiscipline of physics.

Over the years, and more especially during the preparation of this book, we have benefited enormously from discussions with a very large number of colleagues on the topics treated here. It is a pleasure to express here our thanks and indebtednesss to all of them, and in particular to B. S. Deaver, W. Greiner, Y. Imry, H. P. Kelly, E. Levin, and J. Shapir. We are indeed grateful to S. Okubo for his astute comments on the manuscript at an earlier stage.

We are most appreciative of the excellent typing of all versions of the manuscript by J. Mack and J. Gorsky, with additional help from C. Jones, and administrative assistance from Z. Oron.

One of us (D.S.K.) benefited greatly during the work on this book from the hospitality of two institutions where he was a visitor during a year of academic leave. He would like to express his gratitude for support from the Center for Theoretical Physics of M.I.T. during fall 1984, and from the Lady Davis Fellowship Trust of Jerusalem, and the Racah Institute of the Hebrew University, during spring 1985.

Last, we acknowledge the assistance of the U.S. Department of Energy through their support of research in nuclear theory at the University of Rochester, and of the Yuval Ne'eman Chair in Theoretical Nuclear Physics at Tel Aviv University in the preparation of this book.

DANIEL S. KOLTUN
JUDAH M. EISENBERG

Rochester, New York, U.S.A.
Tel Aviv, Israel
January 1988

Contents

QUANTUM MECHANICS
OF MANY DEGREES
OF FREEDOM

CHAPTER 1

Number Representation

This book deals with physical systems that have many degrees of freedom. This comes about for these systems for one or more of several reasons: The physical problem may intrinsically involve the continued presence of a number of particles, as for example the many electrons in a complex atom or in a solid, or the many nucleons in a nucleus. As a second possibility, the system may be such that in the course of its time development it creates many particles, even though initially there were only a few particles present. Such is the case for example in particle-producing collisions, such as the multiphoton bremsstrahlung of an electron in an external Coulomb field provided by a nucleus, or the hadronic reaction $N + N \rightarrow N + N + \pi + \cdots$. Indeed, even when the energy is insufficient to produce these particles as real particles, the uncertainty principle allows their involvement in the problem as virtual particles, which appear so briefly on the scene that their presence does not violate energy conservation. Last, the many degrees of freedom in question may arise because other channels are coupled to the system of interest and so must be taken into account—again either as real or as virtual participants. For instance, one may have pion–nucleon scattering which for the elastic case is $\pi + N \rightarrow \pi + N$ but in other channels may lead to $\pi + N \rightarrow K + \Lambda$ or $\pi + N \rightarrow K + \Sigma$ and so on; above a total center-of-mass system energy equal to the combined rest-mass energies of the kaon and the hyperon (Λ, Σ, and other baryons carrying strangeness) these particles may be produced as real particles, while below that value they will be virtual. Another example of many degrees of freedom arising even though initially and finally only two "bodies" are present occurs in the case of nuclear rearrangement collisions, such as $d + \alpha \rightarrow d + \alpha$ or $\rightarrow {}^6\text{Li} + \gamma$ or $\rightarrow p + {}^5\text{He}$, and so on, where again the different combinations will be real or virtual depending on whether the total center-of-mass system energy is sufficient to reach the combined rest-mass energies of the final channel.

As even these few examples show, systems with many degrees of freedom arise in almost every field of modern physics. In fact in a quantum theory the uncertainty principle offers the intrinsic possibility of introducing many degrees of freedom in a problem that seemed only to involve a few particles (even two, or one, or for that matter none—since empty space may dissociate virtually into antiparticle pairs such as $e^+ e^-$ or $\pi^+ \pi^-$ or $\bar{p}p$, say). It is our purpose here to study the numerous common aspects of the description of systems with many degrees of freedom. In the case of systems with a fixed but large number of particles we must be able to deal correctly with the boson or fermion statistics of the particles. For situations where particles are produced or annihilated, we must have a way of keeping track of these changes in the number of particles. In both cases we wish to develop special techniques for handling the complex dynamics of the system by a calculational device that keeps track of the number of particles of each variety that are present at each moment in time.

1.1 SINGLE-PARTICLE CREATION AND ANNIHILATION

We consider a rewriting of the usual description of a single particle by means of a wave function in terms of an operator that will depict the state of the particle at the moment of time in question. In fact before we can do that usefully we must first define the vacuum state $|0\rangle$, a state with no physical particles present; in our notation here, the zero inside the ket indicates the number of physical particles in the vacuum state. Suppose we are interested in a single particle occupying a state characterized by a complete set of quantum numbers that we shall designate by ρ. This set might include, for example, the momentum of the particle $\mathbf{k}$, its spin and spin projection, its isospin and isospin projection, and other internal quantum numbers for the particle. We characterize this single-particle state by the action of a creation operator $a_\rho^\dagger$ on the vacuum to produce a ket with one particle in the state ρ, namely

$$|1_\rho\rangle = a_\rho^\dagger |0\rangle . \tag{1}$$

To spell out a little more explicitly what our notation means in terms of a basis of single-particle states designated by the (sets of) quantum numbers $\xi, \rho, \sigma, \tau, \ldots$, the vacuum really refers to no occupation of any of these states,

$$|0\rangle = |\ldots, 0_\xi, 0_\rho, 0_\sigma, 0_\tau, \ldots\rangle , \tag{2}$$

and the single-particle ket to the occupation of only one of them (and that by one particle),

$$|1_\rho\rangle = |\ldots, 0_\xi, 1_\rho, 0_\sigma, 0_\tau, \ldots\rangle . \qquad (3)$$

The projection of this single-particle ket onto configuration space is then what we mean by a single-particle wave function,

$$u_\rho(\mathbf{r}) = \langle \mathbf{r}|1_\rho\rangle = \langle \mathbf{r}|a_\rho^\dagger|0\rangle . \qquad (4)$$

For the annihilation or destruction of a particle we use the Hermitian conjugate of the creation operator; thus

$$a_\rho|1_\rho\rangle = |0\rangle . \qquad (5)$$

If there are no particles present to be annihilated in the single-particle state in question, then the operation leads to a null result, e.g.

$$a_\rho|1_\sigma\rangle = 0 , \quad \rho \neq \sigma , \qquad (6)$$

or, taking these together,

$$a_\rho|1_\sigma\rangle = a_\rho a_\sigma^\dagger|0\rangle = \delta_{\rho\sigma}|0\rangle . \qquad (7)$$

Since the vacuum state is unoccupied for any single-particle label,

$$a_\rho|0\rangle = 0 . \qquad (8)$$

For our present purpose we assume that the vacuum state is unique, and thus the scheme for building up occupied states should also be unique.

1.2 MORE THAN ONE PARTICLE: COMMUTATION AND ANTICOMMUTATION, STATISTICS

Suppose we have two particles present in the system. These may be in different single-particle states,

$$|1_\rho 1_\sigma\rangle \sim a_\rho^\dagger a_\sigma^\dagger|0\rangle , \qquad (9)$$

or in the same single-particle modes,

$$|2_\rho\rangle \sim a_\rho^\dagger a_\rho^\dagger|0\rangle . \qquad (10)$$

In these expressions we have assumed that acting twice with creation operators on the vacuum will produce two-particles states, though these need not necessarily be normalized even if the vacuum and single-particle states are. Generalizations for states with more particles are obvious: we merely list the number of particles in each single-particle mode; for instance, for three particles, two in the single-particle state ρ and one in σ,

$$|2_\rho 1_\sigma\rangle \sim a_\rho^\dagger a_\rho^\dagger a_\sigma^\dagger |0\rangle \, , \tag{11}$$

and so forth. From the left-hand side of this equation, it is clear why this is called the *number representation*; for most of the manipulations we consider the right-hand form will be more useful. The space in which these kets exist is called *Fock space*, after the name of the physicist who introduced its use. And the whole procedure is often referred to as *second quantization*, since one views the first quantization as that which gives discrete levels as embodied in the single-particle Schrödinger equation, and this one as yielding a description in terms of discrete particles.

Now in what order should we write the creation operators when more than one particle is present? For two identical particles, say, it is clear that

$$|1_\rho 1_\sigma\rangle \sim a_\rho^\dagger a_\sigma^\dagger |0\rangle \tag{12a}$$

is equivalent to (i.e., refers to the same state as)

$$|1_\sigma 1_\rho\rangle \sim a_\sigma^\dagger a_\rho^\dagger |0\rangle \, , \tag{12b}$$

and so these should be related by a phase. We are free to choose this phase as we wish at a given moment in time, and we take it here as real. In fact we are easily guided in our procedure at this point because we know that two identical bosons are to be described by symmetrical states, and two identical fermions by antisymmetrical ones. Within the selection of the real phase this motivates, for *bosons*,

$$|1_\rho 1_\sigma\rangle = |1_\sigma 1_\rho\rangle \tag{13a}$$

and the *commutation relations*

$$a_\rho^\dagger a_\sigma^\dagger = a_\sigma^\dagger a_\rho^\dagger \, , \tag{13b}$$

while for *fermions* we have

$$|1_\rho 1_\sigma\rangle = -|1_\sigma 1_\rho\rangle \tag{14a}$$

and *anticommutation relations*

$$a_\rho^\dagger a_\sigma^\dagger = -a_\sigma^\dagger a_\rho^\dagger \, . \tag{14b}$$

For the fermion case we immediately see

$$a_\rho^\dagger a_\rho^\dagger = -a_\rho^\dagger a_\rho^\dagger = 0 , \qquad (14c)$$

so that two fermions cannot occupy the same single-particle state and the Pauli exclusion principle is immediately satisfied. For bosons, of course, one may have an arbitrary number, $0, 1, 2, 3, \ldots$ of particles in a given single-particle state. If this state ρ has energy ω_ρ, each additional particle adds energy ω_ρ to the system, just as if we had a harmonic oscillator of energy ω_ρ that we excited to successively higher levels of excitation; each such excitation adds energy ω_ρ. In fact we can exploit this analogy quite completely by associating a corresponding oscillator with each single-particle mode and using its degree of excitation to "count" the number of particles in the system and their contribution to the total system energy. As is developed in Exercise 1.1 at the end of this chapter, the creation and annihilation operators then have the formal analogy of the ladder operators for the harmonic oscillator. These have the explicit representation

$$a^\dagger = \frac{1}{\sqrt{2m\omega}} (p + im\omega x) , \qquad a = \frac{1}{\sqrt{2m\omega}} (p - im\omega x) , \qquad (15a, b)$$

for each mode $\rho, \sigma, \ldots$, where ω is the mode frequency or energy* and m is the oscillator mass, the Hamiltonian being given by

$$H = \frac{1}{2m} p^2 + \tfrac{1}{2} m\omega^2 x^2 = \omega(a^\dagger a + \tfrac{1}{2}) . \qquad (15c)$$

Let us explore further the description of boson systems, returning later to fermions. The Hermitian conjugate of the commutation relation (13b) immediately yields also

$$a_\sigma a_\rho = a_\rho a_\sigma . \qquad (16)$$

From Eqs. 7 and 8 we are led to consider

$$(a_\rho a_\sigma^\dagger - a_\sigma^\dagger a_\rho)|0\rangle = \delta_{\rho\sigma}|0\rangle , \qquad (17)$$

which suggests the further extension of the commutation relations to**

$$a_\rho a_\sigma^\dagger - a_\sigma^\dagger a_\rho = \delta_{\rho\sigma} . \qquad (18)$$

---

*Here and throughout this book we use units such that $\hbar = 1$. We also take the velocity of light to be unity, $c = 1$. To transcribe units, it is useful to remember that $\hbar c = 197.33$ MeV-fm $= 1973.3$ eV-Å.

**This equation may be taken as a definition for a system of elementary bosons.

Thus altogether we expect to get a description of boson properties that has the symmetry properties expected by taking for our creation and annihilation operators the commutation relations

$$[a_\rho, a_\sigma] = 0 , \qquad [a_\rho^\dagger, a_\sigma^\dagger] = 0 , \qquad [a_\rho, a_\sigma^\dagger] = \delta_{\rho\sigma} . \qquad (19a\text{–}c)$$

We now define the *number operator* for a given mode ρ,

$$N_\rho = a_\rho^\dagger a_\rho , \qquad (20)$$

or for the total system,

$$N_{\text{tot}} \equiv \sum_\rho N_\rho = \sum_\rho a_\rho^\dagger a_\rho . \qquad (21)$$

This operator simply counts the number of particles in each mode. From the commutation relations 19 and the relationships

$$[A, BC] = B[A, C] + [A, B]C \qquad (22a)$$

and

$$[AB, C] = A[B, C] + [A, C]B , \qquad (22b)$$

we have

$$[N_\rho, a_\sigma] = [a_\rho^\dagger a_\rho, a_\sigma]$$
$$= a_\rho^\dagger [a_\rho, a_\sigma] + [a_\rho^\dagger, a_\sigma] a_\rho = -\delta_{\rho\sigma} a_\rho , \qquad (23a)$$

and similarly

$$[N_\rho, a_\sigma^\dagger] = \delta_{\rho\sigma} a_\rho^\dagger , \qquad (23b)$$

while

$$[N_\rho, N_\sigma] = [N_\rho, a_\sigma^\dagger a_\sigma] = a_\sigma^\dagger [N_\rho, a_\sigma] +]N_\rho, a_\sigma^\dagger] a_\sigma$$
$$= \delta_{\rho\sigma}(-a_\sigma^\dagger a_\rho + a_\rho^\dagger a_\sigma) = 0 . \qquad (23c)$$

Thus the numbers of particles in the different modes are diagonal or "sharp" in this representation, as one might expect if the counting of the occupations of the different single-particle states is to be independent and well defined.

The counting of the number of particles in a given mode now proceeds as follows (we suppress the mode index ρ for a while): We populate the mode by constructing

$$(a^\dagger)^n|0\rangle = \underbrace{a^\dagger a^\dagger \cdots a^\dagger}_{n \text{ operators}}|0\rangle \tag{24a}$$

and to verify the number of particles in this state we act on it with the number operator $N = a^\dagger a$ for that mode

$$N(a^\dagger)^n|0\rangle = N\underbrace{a^\dagger a^\dagger \cdots a^\dagger}_{n}|0\rangle = a^\dagger(1+N)\underbrace{a^\dagger a^\dagger \cdots a^\dagger}_{n-1}|0\rangle$$
$$= a^\dagger a^\dagger(2+N)\underbrace{a^\dagger a^\dagger \cdots a^\dagger}_{n-2}|0\rangle$$
$$= \cdots = \underbrace{a^\dagger a^\dagger \cdots a^\dagger}_{n}(n+N)|0\rangle$$
$$= (a^\dagger)^n(n+a^\dagger a)|0\rangle = n(a^\dagger)^n|0\rangle , \tag{24b}$$

where we have used Eq. 23b and $a|0\rangle = 0$. Thus the state constructed in 24a is an eigenstate of the number operator N_ρ for the mode in question, with eigenvalue equal to the number of particles $n = n_\rho$ placed in that mode. We label the states by the occupation number, $|n_\rho\rangle \sim (a_\rho^\dagger)^{n_\rho}|0\rangle$.

These states have not yet been normalized. For that purpose we consider

$$a_\rho^\dagger|n_\rho\rangle = c_+(n_\rho)|n_\rho + 1\rangle , \qquad a_\rho|n_\rho\rangle = c_-(n_\rho)|n_\rho - 1\rangle , \tag{25a, b}$$

where $c_\pm(n_\rho)$ are the normalization coefficients which we must determine; we anticipate that they may depend on the present occupation number. Then for the normalization condition $\langle n_\rho \pm 1|n_\rho \pm 1\rangle = 1$ we have (as in the harmonic-oscillator problem)

$$|c_-(n_\rho)|^2 = \langle n_\rho|a_\rho^\dagger a_\rho|n_\rho\rangle = n_\rho \tag{26a}$$

and

$$|c_+(n_\rho)|^2 = \langle n_\rho|a_\rho a_\rho^\dagger|n_\rho\rangle = n_\rho + 1 . \tag{26b}$$

Again choosing a real phase, we have

$$a_\rho^\dagger|n_\rho\rangle = \sqrt{n_\rho + 1}|n_\rho + 1\rangle , \qquad a_\rho|n_\rho\rangle = \sqrt{n_\rho}|n_\rho - 1\rangle . \tag{27a, b}$$

Since we shall encounter these relationships time and again in what follows, it is useful to note the mnemonic suggested by Feynman. The argument in the radical is always the larger of the two occupation numbers involved in the relation. Of course a simple test case is Eq. 8, which is immediately embodied in 27b here.

As we have generated the eigenstates of the number operator here it is clear by construction that they must involve an integral number of particles. We now also check that these are the only admissible form of number eigenstates. First consider

$$n_\rho = \langle n_\rho | N_\rho | n_\rho \rangle = \langle n_\rho | a_\rho^\dagger a_\rho | n_\rho \rangle$$

$$= (a_\rho | n_\rho \rangle)^\dagger (a_\rho | n_\rho \rangle) \geq 0, \tag{28}$$

showing that the number-operator eigenvalues must be positive. Next we note that n_ρ must be an integer, since otherwise the repeated action of a_ρ on $|n_\rho\rangle$, as in $(a_\rho)^{n_\rho+1}|n_\rho\rangle$, would eventually yield a negative number label, which is impossible by virtue of Eq. 28. If n_ρ is an integer we are spared this by Eq. 8 or 27b, since we eventually reach $a_\rho|0_\rho\rangle = 0$. (We assume implicitly here that the vacuum state $|0_\rho\rangle$ is unique.) Thus all the eigenstates of N_ρ are those reached by the successive action of $a_\rho^\dagger$ operators on the vacuum, namely

$$a_\rho^\dagger|0\rangle = |1_\rho\rangle, \qquad a_\rho^\dagger|1_\rho\rangle = \sqrt{2}|2_\rho\rangle, \dots. \tag{29}$$

An explicit representation for the creation and annihilation operators, and hence for the number operator and its eigenstates, is immediately provided by the harmonic-oscillator analogy of Eqs. 15 and Exercise 1.1, or by direct verification of the commutation relations 19 and definition 20. We have, for the mode ρ,

$$a_\rho = \begin{bmatrix} 0 & \sqrt{1} & 0 & 0 & 0 & \cdots \\ 0 & 0 & \sqrt{2} & 0 & 0 & \cdots \\ 0 & 0 & 0 & \sqrt{3} & 0 & \cdots \\ 0 & 0 & 0 & 0 & \sqrt{4} & \cdots \\ \vdots & \vdots & \vdots & \vdots & & \end{bmatrix}_\rho ,$$

$$a_\rho^\dagger = \begin{bmatrix} 0 & 0 & 0 & 0 & 0 & \cdots \\ \sqrt{1} & 0 & 0 & 0 & 0 & \cdots \\ 0 & \sqrt{2} & 0 & 0 & 0 & \cdots \\ 0 & 0 & \sqrt{3} & 0 & 0 & \cdots \\ \vdots & \vdots & \vdots & \vdots & & \end{bmatrix}_\rho ,$$

$$N_\rho = \begin{bmatrix} 0 & 0 & 0 & 0 & 0 & \cdots \\ 0 & 1 & 0 & 0 & 0 & \cdots \\ 0 & 0 & 2 & 0 & 0 & \cdots \\ 0 & 0 & 0 & 3 & 0 & \cdots \\ \vdots & \vdots & \vdots & \vdots & & \end{bmatrix}_\rho ,$$

and

$$|0_\rho\rangle = \begin{bmatrix} 1 \\ 0 \\ 0 \\ 0 \\ \vdots \end{bmatrix}_\rho , \qquad |1_\rho\rangle = \begin{bmatrix} 0 \\ 1 \\ 0 \\ 0 \\ \vdots \end{bmatrix}_\rho , \qquad |2_\rho\rangle = \begin{bmatrix} 0 \\ 0 \\ 1 \\ 0 \\ \vdots \end{bmatrix}_\rho , \dots. \tag{30}$$

The analogous results for fermions are rather simpler because the exclusion principle guarantees that single-particle occupation is binary in character: a given mode is either occupied or unoccupied. As we saw in Eqs. 14, the antisymmetry of fermion states implies anticommutators, and, instead of Eqs. 19, we expect*

$$\{a_\rho, a_\sigma\} \equiv a_\rho a_\sigma + a_\sigma a_\rho = 0, \qquad \{a_\rho^\dagger, a_\sigma^\dagger\} = 0, \qquad \text{(31a, b)}$$

$$\{a_\rho, a_\sigma^\dagger\} = \delta_{\rho\sigma}, \qquad \text{(31c)}$$

with the immediate corollaries

$$a_\rho^\dagger a_\rho^\dagger = 0, \qquad a_\rho a_\rho = 0. \qquad \text{(32a, b)}$$

The number operator is defined as before,

$$N_\rho = a_\rho^\dagger a_\rho, \qquad \text{(33)}$$

and we immediately see that

$$N_\rho^2 = a_\rho^\dagger a_\rho a_\rho^\dagger a_\rho = a_\rho^\dagger (1 - a_\rho^\dagger a_\rho) a_\rho = a_\rho^\dagger a_\rho = N_\rho. \qquad \text{(34)}$$

Thus the number eigenvalues must satisfy $n_\rho^2 = n_\rho$, or $n_\rho = 0$ or 1, exactly as we would expect on the basis of the exclusion principle. Thus all we need consider for a given mode ρ are the states $|0_\rho\rangle$ and $|1_\rho\rangle$ with

$$|1_\rho\rangle = a_\rho^\dagger |0_\rho\rangle, \qquad a_\rho |1_\rho\rangle = |0_\rho\rangle. \qquad \text{(35)}$$

An explicit representation for the operators and states in the case of fermions is

$$a_\rho = \begin{pmatrix} 0 & 1 \\ 0 & 0 \end{pmatrix}_\rho, \qquad a_\rho^\dagger = \begin{pmatrix} 0 & 0 \\ 1 & 0 \end{pmatrix}_\rho,$$

$$N_\rho = \begin{pmatrix} 0 & 0 \\ 0 & 1 \end{pmatrix}_\rho, \qquad \text{(36)}$$

and

$$|0_\rho\rangle = \begin{pmatrix} 1 \\ 0 \end{pmatrix}_\rho, \qquad |1_\rho\rangle = \begin{pmatrix} 0 \\ 1 \end{pmatrix}_\rho.$$

---

*Equation 31c may be taken as a definition for a system of elementary fermions; see Eq. 18 and footnote.

As we shall see in the many applications of the coming chapters, the use of fermion creation and annihilation operators greatly simplifies the treatment of antisymmetrization as opposed to the construction of explicitly antisymmetric wave functions or Slater determinants. Of course it is quite crucial to keep careful track of the minus signs introduced by the anticommutation relations of Eqs. 31a–c.

In order to complete the picture of the method based on creation and annihilation operators, we show the explicit mapping of two-particle states in Fock space into conventional symmetrized or antisymmetrized states for bosons or fermions, respectively. For particles in two different single-particle states these are

$$|1_\rho 1_\sigma\rangle = \frac{1}{\sqrt{1!1!}}\, a_\rho^\dagger a_\sigma^\dagger |0\rangle$$

$$\rightarrow \frac{1}{\sqrt{2}}\left[u_\rho(\mathbf{r})u_\sigma(\mathbf{r}') \pm u_\sigma(\mathbf{r})u_\rho(\mathbf{r}')\right], \qquad \rho \neq \sigma, \qquad (37a)$$

while for particles in the same mode

$$|2_\rho\rangle = \frac{1}{\sqrt{2!}}\, a_\rho^\dagger a_\rho^\dagger |0\rangle$$

$$\rightarrow \tfrac{1}{2}\left[u_\rho(\mathbf{r})u_\rho(\mathbf{r}') \pm u_\rho(\mathbf{r})u_\rho(\mathbf{r}')\right]$$

$$= \begin{cases} u_\rho(\mathbf{r})u_\rho(\mathbf{r}'), & \text{bosons}, \\ 0, & \text{fermions}; \end{cases} \qquad (37b)$$

the upper signs apply, obviously, to bosons and the lower signs to fermions. For two particles in the same single-particle states, the product wave function is immediately symmetrical for the boson case, while it vanishes for fermions. We have shown the mapping of Fock states to configuration-space wave functions explicitly for the situation of two particles because that is the most transparent case in which particle statistics are nontrivial. Naturally the scheme generalizes immediately to the construction of completely symmetrized or antisymmetrized states for bosons or fermions, respectively, when more than two particles of a given species are present.

1.3　CONSTRUCTION OF ONE- AND TWO-PARTICLE OPERATORS

In treating systems with many degrees of freedom we shall constantly encounter situations in which we must count the contribution of the particles present to various quantities that characterize the system. These quantities may be of one-particle nature, that is, each particle present makes its separate contribution to the total system value, without reference to other particles. Examples of one-particle quantities are the number operator of

Eqs. 20 and 21 itself, the kinetic energy in the system, the angular momentum, and so on. It is also very common to encounter two-particle quantities, that is, features of the system whose characterization requires knowledge concerning two particles together, as for example the interaction energy arising from a potential that acts between two particles. One may also require a description of three-, four-, or many-particle aspects of the system, which arise, for instance, in nuclear systems where two-particle interactions may not exhaust the nature of the potentials. We shall not, however, deal much with these many-particle situations. This is mainly because their thorough treatment generally requires a prior detailed knowledge of the nature and consequences of the two-particle forces, which often have not yet been adequately mastered. It is also partly because at least the early stages of the generalization to three-particle forces, or still more complex situations, follow as a straightforward generalization of the two-particle case.

To make clear the nature of one-particle operators, consider, as an example, a collection of noninteracting bosons; these could be, for instance, the photons in a blackbody cavity, which to an excellent approximation are noninteracting. The energy operator or Hamiltonian for this system is, in Fock space,

$$H = \sum_{ks} \omega_k a_{ks}^\dagger a_{ks} , \tag{38}$$

where $\mathbf{k}$ is the particle momentum and s is the particle spin. We have assumed that the energy ω_k is independent of the spin (and of the direction of $\mathbf{k}$), though that is not at all a necessary assumption; for the photon example, $\omega_k = |\mathbf{k}| = k$ in our units. The operator in Eq. 38 is so constructed that the number operator $a_{ks}^\dagger a_{ks}$ for each single-particle mode $\mathbf{k}, s$ counts how many particles are in the mode and weights that number with the energy $\omega_\mathbf{k}$ of the mode, summing over the modes. If we take the expectation value of this one-particle operator for a state having definite occupation numbers for the single-particle modes, we get

$$\langle n_{\mathbf{k}_1 s_1}, n_{\mathbf{k}_2 s_2}, n_{\mathbf{k}_3 s_3}, \ldots | H | n_{\mathbf{k}_1 s_1}, n_{\mathbf{k}_2 s_2}, n_{\mathbf{k}_3 s_3}, \ldots \rangle$$
$$= n_{\mathbf{k}_1 s_1} \omega_{k_1} + n_{\mathbf{k}_2 s_2} \omega_{k_2} + n_{\mathbf{k}_3 s_3} \omega_{k_3} + \cdots . \tag{39}$$

For more general situations, where we may not always be interested in diagonal matrix elements, we must extend the construction of the one-particle operator slightly. The kinetic energy for a system—also a one-particle operator—can thus be written in Fock space as

$$T = \sum_{\alpha\beta} \langle \alpha | t | \beta \rangle a_\alpha^\dagger a_\beta , \tag{40}$$

where t is the kinetic-energy operator for an individual particle, and α, β refer to the single-particle basis to which the number representation is being applied. For example, if this basis referred to configuration space, the matrix element might be evaluated as

$$\langle \alpha|t|\beta \rangle = \int u_\alpha^\dagger(\mathbf{r}) \left[-\frac{1}{2m} \nabla^2 \right] u_\beta(\mathbf{r}) \, d\mathbf{r} \,, \tag{41}$$

for particle mass m and wave functions $u_{\alpha,\beta}$, while in a momentum-space basis one would have

$$\langle \alpha|t|\beta \rangle = \int \tilde{u}_\alpha^\dagger(\mathbf{p}) \frac{p^2}{2m} \tilde{u}_\beta(\mathbf{p}) \frac{d\mathbf{p}}{(2\pi)^3} \,, \tag{42}$$

where $\mathbf{p}_\alpha$ is the momentum eigenvalue of the state $|\alpha\rangle$. We have here taken nonrelativistic forms in the explicit examples.

The Fock-space matrix elements of the one-particle kinetic-energy operator are, for one particle present,

$$\langle 1_\rho|T|1_\sigma \rangle = \langle 0|a_\rho T a_\sigma^\dagger|0 \rangle$$

$$= \sum_{\alpha\beta} \langle \alpha|t|\beta \rangle \langle 0|a_\rho a_\alpha^\dagger a_\beta a_\sigma^\dagger|0 \rangle$$

$$= \sum_{\alpha\beta} \langle \alpha|t|\beta \rangle \delta_{\alpha\rho} \delta_{\beta\sigma} = \langle \rho|t|\sigma \rangle \,, \tag{43}$$

for bosons or for fermions. If the single-particle basis refers to plane-wave states, this last result is diagonal,

$$\langle 1_\rho|T|1_\sigma \rangle = \langle \rho|t|\sigma \rangle = \frac{p_\sigma^2}{2m} \delta_{\rho\sigma} \,, \tag{44}$$

where $\mathbf{p}_\sigma$ is the momentum of the plane-wave state $|\sigma\rangle$. But this need not be the case, of course; one may expand in any complete set of single-particle states. For the structure of Eq. 40 it was, however, crucial that T be of one-particle character, which would evidence itself in configuration space, say, through the appearance of a sum of operators each referring to only one particle variable,

$$T = -\frac{1}{2m} (\nabla_1^2 + \nabla_2^2 + \nabla_3^2 + \cdots) \,, \tag{45}$$

where we have also used the fact that the identical particles in the system will obviously all have the same mass m.

For a two-particle operator, such as a two-particle potential (again acting between identical particles),

$$V = \sum_{i<j} v(i, j) = \frac{1}{2} \sum_{i \neq j} v(i, j),$$ (46)

the corresponding Fock-space two-particle operator will be

$$V = \frac{1}{2} \sum_{\alpha\beta\gamma\delta} \langle \alpha\beta | v | \gamma\delta \rangle a_\alpha^\dagger a_\beta^\dagger a_\delta a_\gamma,$$ (47a)

where we again have in mind expansion on *any* complete, single-particle basis. The matrix element here is

$$\langle \alpha\beta | v | \gamma\delta \rangle = \int u_\alpha^\dagger(\mathbf{r}) u_\beta^\dagger(\mathbf{r}') v(\mathbf{r}, \mathbf{r}') u_\gamma(\mathbf{r}) u_\delta(\mathbf{r}') \, d\mathbf{r} \, d\mathbf{r}'.$$ (47b)

The change from alphabetical order in the factor $a_\delta a_\gamma$ of Eq. 47a is immaterial for bosons, where a_δ and a_γ commute, but flips a sign of matrix elements for fermions, where they anticommute. We can verify the validity of Eq. 47, and in particular of this sign reversal, by evaluating matrix elements of V when two identical particles are present,

$$\langle 1_\rho 1_\sigma | V | 1_\tau 1_\nu \rangle = \frac{1}{2} \sum_{\alpha\beta\gamma\delta} \langle \alpha\beta | v | \gamma\delta \rangle \langle 0 | a_\sigma a_\rho a_\alpha^\dagger a_\beta^\dagger a_\delta a_\gamma a_\tau^\dagger a_\nu^\dagger | 0 \rangle$$

$$= \frac{1}{2} \sum_{\alpha\beta\gamma\delta} \langle \alpha\beta | v | \gamma\delta \rangle [(\delta_{\alpha\rho}\delta_{\beta\sigma} \pm \delta_{\alpha\sigma}\delta_{\beta\rho})(\delta_{\gamma\tau}\delta_{\delta\nu} \pm \delta_{\gamma\nu}\delta_{\delta\tau})]$$

$$= \tfrac{1}{2}[\langle \rho\sigma | v | \tau\nu \rangle + \langle \sigma\rho | v | \nu\tau \rangle \pm (\langle \sigma\rho | v | \tau\nu \rangle + \langle \rho\sigma | v | \nu\tau \rangle)]$$

$$= \langle \rho\sigma | v | \tau\nu \rangle \pm \langle \rho\sigma | v | \nu\tau \rangle,$$ (48)

where, again, the upper sign refers to bosons and the lower sign to fermions. In deriving Eq. 48 it is convenient to note that just as $a_\xi | 0 \rangle = 0$, one also has $\langle 0 | a_\xi^\dagger = 0$. In addition we have made use of the natural symmetry of the matrix element for the mutual interaction between two identical particles,

$$\langle \beta\alpha | v | \delta\gamma \rangle = \int u_\beta^\dagger(\mathbf{r}) u_\alpha^\dagger(\mathbf{r}') v(\mathbf{r}, \mathbf{r}') u_\delta(\mathbf{r}) u_\gamma(\mathbf{r}') \, d\mathbf{r} \, d\mathbf{r}'$$

$$= \langle \alpha\beta | v | \gamma\delta \rangle,$$ (49)

here written explicitly in configuration space, as an example.

The final result of Eq. 48 gives clear expression to the symmetry or antisymmetry that we expect for bosons or fermions. Indeed, it is convenient to incorporate this into the notation by introducing a subscript to indicate the status of the matrix element, symmetric or antisymmetric as the case may be:

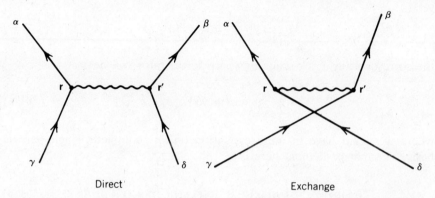

FIGURE 1.1   Direct and exchange graphs for two-particle interaction matrix elements.

$$\langle 1_\rho 1_\sigma | V | 1_\tau 1_\nu \rangle = \langle \rho\sigma | v | \tau\nu \rangle \pm \langle \rho\sigma | v | \nu\tau \rangle$$

$$= \begin{cases} \langle \rho\sigma | v | \tau\nu \rangle_S , & \text{bosons } (+) , \\ \langle \rho\sigma | v | \tau\nu \rangle_A , & \text{fermions } (-) . \end{cases} \qquad (50)$$

In view of this inevitable symmetry feature it is also useful on occasion to rewrite Eq. 47 in a completely equivalent way as

$$V = \frac{1}{4} \sum_{\alpha\beta\gamma\delta} \langle \alpha\beta | v | \gamma\delta \rangle_{A \text{ or } S} a_\alpha^\dagger a_\beta^\dagger a_\delta a_\gamma . \qquad (51)$$

Last, we note that it is frequently convenient to draw diagrams that represent the interaction matrix elements of Eqs. 48 and 50. The case for two particles is shown in Figure 1.1.

1.4   THE FERMI GAS IN FOCK SPACE

Partly as an exercise in the techniques of second quantization for the fermion case, and partly because of its enormous usefulness as a zero-order approximation in the treatments of the many-fermion system that follow, we develop here the basic features of a noninteracting gas of N nonrelativistic fermions with mass m. The N fermions in question are viewed as existing in a cubic box of volume Ω, eventually taken very large, with periodic boundary conditions applied. The single-particle basis then refers to definite momentum $\mathbf{k}$ and spin projection s, taken here for spin $\frac{1}{2}$ to have values $s = \pm\frac{1}{2}$; one could also consider other internal degrees of freedom, such as isospin for nuclear applications. The wave functions in question are

$$u_\alpha(\mathbf{r}) = \frac{1}{\sqrt{\Omega}} e^{i\mathbf{k}\cdot\mathbf{r}} \chi_s , \qquad \alpha = \{\mathbf{k}, s\} , \qquad (52)$$

with energy $\epsilon_k = k^2/2m$, degenerate in spin $s = \pm\frac{1}{2}$.

To construct the ground state of the Fermi gas we populate the lowest available single-particle levels with one fermion each, in accordance with Pauli's exclusion principle. This population is by pairs in the sense that, since the single-particle energies do not depend on spin, for each ϵ_k we have one fermion with spin up and one with spin down. The filled levels are referred to as the Fermi sea, and the last level filled is called the Fermi surface (in momentum or energy space) and designated by ϵ_F or k_F, the Fermi energy or momentum (see Figure 1.2). The ground state then has the structure

$$|\Phi_0\rangle = \prod_{\substack{k \leq k_F \\ s}} a_{ks}^\dagger |0\rangle . \tag{53}$$

The number of particles in this state (which by construction is to be N) is easily determined using the number operator $\sum_{ks} a_{ks}^\dagger a_{ks}$; the particle density ρ_0 is then that number divided by the volume Ω. Thus

$$\rho_0 = \frac{1}{\Omega} \langle \Phi_0 | \sum_{ks} a_{ks}^\dagger a_{ks} | \Phi_0 \rangle$$

$$= \frac{1}{\Omega} \sum_{\substack{s=\pm 1/2 \\ \mathbf{k}, k \leq k_F}} 1 = \frac{1}{\Omega} \sum_{ks} \theta(k_F - k)$$

$$\xrightarrow[\Omega \to \infty]{} \frac{1}{\Omega} \sum_{s=\pm 1/2} 1 \cdot \int \frac{\Omega \, d\mathbf{k}}{(2\pi)^3} \theta(k_F - k)$$

$$= 2 \int_0^{k_F} \frac{k^2 \, dk}{(2\pi)^3} \int d\hat{\mathbf{k}} = 2 \times 4\pi \times \frac{k_F^3}{3(2\pi)^3} = \frac{2k_F^3}{6\pi^2} . \tag{54}$$

The factor 2 in the numerator is simply the spin degeneracy factor $\sum_{s=\pm 1/2} 1$ and in general would be replaced by the degeneracy factor $\sum_s 1$ dictated by the number of (degenerate) substates for the internal quantum numbers in question, e.g. 4 for nuclei with spin up and down and isospin up and down (protons and neutrons). Thus the more general version of Eq. 54 is

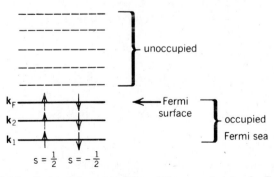

FIGURE 1.2 A pictorial representation of the ground state of a Fermi gas.

$$\rho_0 = \frac{gk_F^3}{6\pi^2},\tag{55}$$

where g is the degeneracy factor arising from internal degrees of freedom for the fermion, and $g = 2$ for collections of spin-$\frac{1}{2}$ fermions having only spin as an internal degree of freedom.

The energy of the Fermi gas is the kinetic energy of its fermions and thus is given by

$$\mathcal{E}_0 \equiv \langle \Phi_0 | H | \Phi_0 \rangle = \langle \Phi_0 | \sum_{ks} \frac{k^2}{2m} a_{ks}^\dagger a_{ks} | \Phi_0 \rangle = \sum_{\substack{s = \pm 1/2 \\ \mathbf{k},\, k \leq k_F}} \frac{k^2}{2m}$$

$$\xrightarrow[\Omega \to \infty]{} \sum_s 1 \cdot \int \frac{\Omega\, d\mathbf{k}}{(2\pi)^3} \frac{k^2}{2m} \theta(k_F - k)$$

$$= g\Omega \frac{1}{2m} \int_0^{k_F} \frac{k^4\, dk}{(2\pi)^3} \int d\hat{\mathbf{k}} = g\Omega \frac{1}{2m} \frac{k_F^5}{5} \frac{4\pi}{(2\pi)^3} = \rho_0 \Omega \frac{3}{5} \frac{k_F^2}{2m},\tag{56}$$

so that the energy density is

$$\frac{\mathcal{E}_0}{\Omega} = \tfrac{3}{5} \rho_0 \epsilon_F .\tag{57}$$

Note that all the physical quantities are defined as densities, or—equivalently—per particle, so that their meaning is preserved in the limit of a large system at fixed density (called the thermodynamic limit).

Let us now consider the possibility of a potential acting between the particles. We suppose that it is sufficiently weak that perturbation techniques may be adequate. It is then easy to write down the corrections to the system arising from this interaction. To lowest order the energy density will be

$$\frac{\mathcal{E}^{(1)}}{\Omega} = \frac{1}{\Omega} \langle \Phi_0 | V | \Phi_0 \rangle$$

$$= \frac{1}{\Omega} \cdot \frac{1}{2} \sum_{\substack{\mathbf{k}_1 s_1, \mathbf{k}_2 s_2 \\ \mathbf{k}_3 s_3, \mathbf{k}_4 s_4}} \langle \mathbf{k}_1 s_1, \mathbf{k}_2 s_2 | v | \mathbf{k}_3 s_3, \mathbf{k}_4 s_4 \rangle$$

$$\times \langle \Phi_0 | a_{\mathbf{k}_1 s_1}^\dagger a_{\mathbf{k}_2 s_2}^\dagger a_{\mathbf{k}_4 s_4} a_{\mathbf{k}_3 s_3} | \Phi_0 \rangle$$

$$= \frac{1}{2\Omega} \sum_{\substack{\mathbf{k}_1 s_1, \mathbf{k}_2 s_2 \\ k_1 \leq k_F, k_2 \leq k_F}} \langle \mathbf{k}_1 s_1, \mathbf{k}_2 s_2 | v | \mathbf{k}_1 s_1, \mathbf{k}_2 s_2 \rangle_A .\tag{58}$$

Now the antisymmetrized matrix element here is

$$\langle \mathbf{k}_1 s_1, \mathbf{k}_2 s_2 | v | \mathbf{k}_1 s_1, \mathbf{k}_2 s_2 \rangle_A = \frac{1}{\Omega^2} \int d\mathbf{r} \, d\mathbf{r}' \, v(\mathbf{r}, \mathbf{r}')$$

$$\times [1 - e^{i(\mathbf{k}_2 - \mathbf{k}_1)\cdot(\mathbf{r} - \mathbf{r}')}(\chi_{s_1}, \chi_{s_2})(\chi_{s_2}, \chi_{s_1})], \quad (59)$$

where we have assumed that the interaction is independent of spin. If we further take the interaction to be a function only of the particle separation vector $\boldsymbol{\rho} \equiv \mathbf{r} - \mathbf{r}'$, and introduce the center-of-mass variable $\mathbf{R} \equiv \frac{1}{2}(\mathbf{r}_1 + \mathbf{r}_2)$, such that $d\mathbf{r} \, d\mathbf{r}' = d\boldsymbol{\rho} \, d\mathbf{R}$ while $\int d\mathbf{R} = \Omega$, we have

$$\langle \mathbf{k}_1 s_1, \mathbf{k}_2 s_2 | v | \mathbf{k}_1 s_1, \mathbf{k}_2 s_2 \rangle_A = \frac{1}{\Omega} \int d\boldsymbol{\rho} \, v(\boldsymbol{\rho})[1 - e^{i(\mathbf{k}_2 - \mathbf{k}_1)\cdot\boldsymbol{\rho}} \delta_{s_1 s_2}], \quad (60)$$

where the first term is the direct term and the second is exchange. This expression, together with Eq. 58, gives us a simple, crude estimate of the energy of a very large system of (weakly) interacting fermions. After we have used the methods of second quantization presented in this chapter in order to develop the concept of a field—and in particular the electromagnetic field—we shall return to more refined techniques for handling many-fermion systems in the presence of interactions.

EXERCISES

1.1 Consider the Hamiltonian for a harmonic oscillator,

$$H = \frac{1}{2m} \, p^2 + \frac{1}{2}kx^2, \qquad k = m\omega^2.$$

(a) Defining ladder operators a and $a^\dagger$ as in Eq. 15, find the commutation relations for these operators using the commutators of p and x.

(b) Show that

$$H = \omega(a^\dagger a + \tfrac{1}{2}).$$

(c) Show that if, for a state $|0\rangle$, $H|0\rangle = \frac{1}{2}\omega|0\rangle$, then

$$H|n\rangle = H(a^\dagger)^n|0\rangle = (n + \tfrac{1}{2})\omega|n\rangle.$$

(d) Show that the state $|0\rangle$ of part (c), possessing the property $a|0\rangle = 0$, must exist.

(e) Construct the configuration-space representation of $|0\rangle$, and explain how to construct wave functions for excited states from it.

(f) Explain how to construct the states of part (e) in momentum space.

(g) Calculate the overlaps $\langle m|n\rangle$.

(h) Calculate $(\Delta x)^2 = \langle x^2\rangle - \langle x\rangle^2$ and $(\Delta p)^2 = \langle p^2\rangle - \langle p\rangle^2$ with respect to expectation values in the state $|n\rangle$, and calculate $(\Delta x)(\Delta p)$.

(i) How are these considerations changed if H, instead of being a one-dimensional operator, is an oscillator in d dimensions?

1.2 Calculate the commutators, for both bosons and fermions,

$$[N_{\text{tot}}, (a_\rho^\dagger)^n], \qquad [N_{\text{tot}}, a_\rho^n], \qquad [N_{\text{tot}}, A_{\rho\sigma}^\dagger],$$

$$[A_{\rho\sigma}^\dagger, A_{\tau\nu}^\dagger], \qquad [A_{\rho\sigma}^\dagger, A_{\tau\nu}], \qquad [A_{\rho\sigma}, A_{\tau\nu}],$$

where N_{tot} is given by Eq. 21, $A_{\rho\sigma}^\dagger \equiv a_\rho^\dagger a_\sigma^\dagger$, and $A_{\rho\sigma} = (A_{\rho\sigma}^\dagger)^\dagger$.

1.3 (a) Write a state in Fock space for a pair of identical bosons of spin one coupled to spin $J = 0, 1, 2$. What are the consequences of boson symmetry here?

(b) For two identical spin-$\frac{1}{2}$ fermions write states with total spin $J = 0$ or 1. How are the consequences of the Pauli principle expressed here?

1.4 We expect that one-particle operators in configuration space go over to Fock-space operators as

$$\Omega = \sum_{i=1}^{N} \omega(r_i) \to \hat{\Omega}_1 = \sum_{\alpha\beta} \langle\alpha|\omega|\beta\rangle a_\alpha^\dagger a_\beta,$$

while two-particle operators follow

$$\Omega_1 = \frac{1}{2}\sum_{i\neq j} \omega(r_i, r_j) \to \hat{\Omega}_2 = \frac{1}{2}\sum_{\alpha\beta\gamma\delta} \langle\alpha\beta|\omega|\gamma\delta\rangle a_\alpha^\dagger a_\beta^\dagger a_\delta a_\gamma.$$

Establish the validity of these by showing the equivalence of their matrix elements in a space of N particles.

CHAPTER 2

Fields

Second quantization has given us a tool for handling conveniently the statistics features of an assemblage of identical particles and for dealing with the creation or annihilation of a particle in the system. The main device for this purpose was a method based on counting the number of particles present in each possible single-particle state or mode. We now wish further to develop the treatment of the dynamics of such systems, and towards this end we introduce the notion of a *field*, which embodies an expansion over the infinite degrees of freedom that may be available in the system,

$$\psi(\mathbf{r}, t) = \sum_{\alpha} a_{\alpha}(t) u_{\alpha}(\mathbf{r}) . \tag{1}$$

In this expansion for the field ψ there appear annihilation operators $a_{\alpha}(t)$ for each single-particle state α with wave function $u_{\alpha}(\mathbf{r})$; though we only indicate explicitly dependence on $\mathbf{r}$, it is of course intended that α refer to a complete set of commuting quantum numbers for the modes and that u_{α} exist in a space that takes into account the necessary internal degrees of freedom of the particles, for example the u_{α} may contain spinors to describe intrinsic spin. The time dependence of the annihilation operators in Eq. 1 will eventually permit the appropriate dynamical treatment for the development of the system in time; the considerations of Chapter 1 all pertained to a given instant in time. [Note that the field $\psi(\mathbf{r}, t)$, as an operator carrying time dependence, might refer, say, to a description of the physical system in the Heisenberg picture or in the interaction picture—see Chapter 6—but not in the Schrödinger picture, where time dependence is to be found in the state vectors and not in the operators.] In general terms the concept of the field is the central entity in second-quantized theories for the description of the dynamics of the system. It takes over from the wave function of first quantization because it has the further advantage of describing the creation and annihilation of the particles in question.

It is obvious from the definition of Eq. 1 and from its Hermitian conjugate

$$\psi^\dagger(\mathbf{r}, t) = \sum_\alpha a_\alpha^\dagger(t) u_\alpha^\dagger(\mathbf{r}) \tag{2}$$

that the field operators do not necessarily commute and so care is required in their manipulation. The field $\psi(\mathbf{r}, t)$ is closely related to our usual expectations for a wave function in the sense that if one single-particle state ρ is populated by one particle then

$$\langle 0|\psi(\mathbf{r}, t)|1_\rho\rangle = \langle 0|\psi(\mathbf{r}, t)a_\rho^\dagger|0\rangle = c_\rho(t)u_\rho(\mathbf{r}) , \tag{3}$$

where $c_\rho(t)$ takes into account the time dependence of the annihilation operator. (For a noninteracting system, for example, this is given by $c_\rho(t) = e^{-i\omega t}$, as discussed in Eq. 7.11 below.) Correspondingly, $\psi^\dagger$ relates to the conjugate wave function.

In order to develop the dynamics of the quantized system in Fock space, we shall follow standard procedure: We construct a Lagrangian L embodying within it whatever symmetries we know the physical system to possess, and guessing—sometimes confining ourselves to a range of constructs bilinear in $\psi, \psi^\dagger$ for which we can hope to obtain a relatively simple solution—the rest of the features of L. The Lagrangian will here be a functional of the field variables $\psi, \psi^\dagger$. We determine the system dynamics by a least-action, variational principle which yields the Euler–Lagrange equations for the fields. From L we can determine the variables conjugate to the fields $\psi, \psi^\dagger$ and the system Hamiltonian. Then we construct the classical Poisson brackets for these quantities and go over to the quantum situation by replacing these by commutation relations; at that point the commutation features of the creation and annihilation operators enter and the second quantization is accomplished. The ultimate test of our guess for the Lagrangian is, of course, the degree to which the derived dynamics agree with experiment.

We shall now illustrate this procedure in some detail in the case of nonrelativistic scalar fields—also known as Schrödinger fields, since they correspond to the conventional solution of the Schrödinger wave equation. Schrödinger fields will suffice for the vast majority of the systems treated in this book. In particular we wish here to distinguish as clearly as possible those aspects of the theory that arise because of the existence of many degrees of freedom and the need to deal with them formally, as opposed to features that relate to relativistic behavior and covariant description or to more complex internal degrees of freedom and the group-theory methods they require. The formalism developed in the bulk of this text can be extended to handle these situations, but for the purpose of understanding the broad elements of systems with many degrees of freedom those extensions are secondary.

2.1 LAGRANGIAN FORMULATION OF THE NONRELATIVISTIC SCALAR FIELD

We assume that the Lagrangian L for the nonrelativistic, scalar field has explicit functional dependence on the field itself, $\psi(\mathbf{r}, t)$; on the spatial and time derivatives of the field, $\nabla\psi$ and $\dot{\psi}$; and, possibly, on the time as well:

$$L = L[\psi(\mathbf{r}, t), \nabla\psi(\mathbf{r}, t); \dot{\psi}(\mathbf{r}, t); t] ; \qquad (4)$$

we shall not usually treat explicit dependence of L on t, however. (In principle one may imagine an immediate extension to a situation in which ψ has several components—in which case it is not a scalar field in at least one of the relevant spaces. The following manipulations then pertain successively for each component.) As part of the consequences of the Lagrangian formulation of the problem we shall find a method to induce the second quantization of the fields ψ; thus at this stage we do not yet think of them as operators but as ordinary functions.

The action integral for the physical system is the time integral over the interval between t_1 and t_2,

$$S = \int_{t_1}^{t_2} L \, dt = \int_{t_1}^{t_2} dt \int_\Omega d\mathbf{r} \, \mathcal{L}[\psi, \nabla\psi; \dot{\psi}; t] , \qquad (5)$$

where we have introduced the Lagrangian density in terms of which the Lagrangian L is given by an integration over the volume Ω occupied by the physical system.

The action principle now requires that S be an extremum, that is, that the variations in S vanish ($\delta S = 0$) for arbitrary changes $\delta\psi$ in the field variables subject only to the restriction that these changes vanish at the endpoints (in time) of the system trajectory:

$$\delta\psi(\mathbf{r}, t_1) = \delta\psi(\mathbf{r}, t_2) = 0 . \qquad (6)$$

Thus

$$\delta S = \int_{t_1}^{t_2} dt \int_{\Omega} d\mathbf{r} \left[\frac{\partial \mathscr{L}}{\partial \psi} \delta \psi + \frac{\partial \mathscr{L}}{\partial (\nabla \psi)} \cdot \delta (\nabla \psi) + \frac{\partial \mathscr{L}}{\partial \dot{\psi}} \delta \dot{\psi} \right]$$

$$= \int_{t_1}^{t_2} dt \int_{\Omega} d\mathbf{r} \left[\frac{\partial \mathscr{L}}{\partial \psi} \delta \psi + \frac{\partial \mathscr{L}}{\partial (\nabla \psi)} \cdot \nabla \delta \psi + \frac{\partial \mathscr{L}}{\partial \dot{\psi}} \frac{\partial}{\partial t} \delta \psi \right]$$

$$= \int_{t_1}^{t_2} dt \int_{\Omega} d\mathbf{r} \left[\frac{\partial \mathscr{L}}{\partial \psi} - \nabla \cdot \frac{\partial \mathscr{L}}{\partial (\nabla \psi)} - \frac{\partial}{\partial t} \frac{\partial \mathscr{L}}{\partial \dot{\psi}} \right] \delta \psi$$

$$+ \int_{t_1}^{t_2} dt \int_{\Sigma} d\mathbf{a} \cdot \frac{\partial \mathscr{L}}{\partial (\nabla \psi)} \delta \psi + \int_{\Omega} d\mathbf{r} \frac{\partial \mathscr{L}}{\partial \dot{\psi}} \delta \psi \bigg|_{t_1}^{t_2} = 0 , \qquad (7)$$

where the first equality represents the consequences of varying ψ, $\nabla \psi$, and $\dot{\psi}$; the second equality exploits the assumption that the variation of the derivative of the field is equal to the derivative of the variation; and the last form follows from an integration by parts. In the surface-integral term over the surface Σ enclosing the system volume Ω we anticipate a sufficiently remote surface that the fields, and also $\partial \mathscr{L} / \partial (\nabla \psi)$, will be negligible on Σ so that the integral may be dropped. For the very last integral of Eq. 7, over the volume Ω but restricted to the starting and ending times, t_1 and t_2, we get no contribution because of the restriction in Eq. 6. Now since the variation $\delta \psi$ is arbitrary, its coefficient—the square bracket in the last version in Eq. 7—must vanish, and thus we have the *Euler–Lagrange equation* that determines the system dynamics,

$$\frac{\partial \mathscr{L}}{\partial \psi} - \nabla \cdot \frac{\partial \mathscr{L}}{\partial (\nabla \psi)} - \frac{\partial}{\partial t} \frac{\partial \mathscr{L}}{\partial \dot{\psi}} = 0 . \qquad (8)$$

In the further handling of the fields we shall often wish to base ourselves on methods appropriate to ordinary, discrete generalized coordinates for a physical system $q_s(t)$, where s ranges over the available degrees of freedom for the system. We can apply these methods to our field problem by imagining that the system volume Ω is divided into a (finite) number of cells labeled by s, located at $\mathbf{r}_s$, and of volume $\delta \mathbf{r}_s$ sufficiently small so that $\psi(\mathbf{r}, t)$ does not vary appreciably for $\mathbf{r}$ within the volume $\delta \mathbf{r}_s$ at $\mathbf{r}_s$. We then identify the generalized coordinates with the value of the field in each cell,

$$q_s(t) = \psi(\mathbf{r}_s, t) . \qquad (9)$$

In these terms, the Lagrangian is obtained from the Lagrangian density through a sum over the cells,

$$L = \sum_s \mathscr{L}(q_s, \dot{q}_s) \delta \mathbf{r}_s , \qquad (10)$$

in place of the integral of Eq. 5.

Note that in the Lagrangian density we have chosen to address directly the dependence of $\mathscr{L}$ on $\nabla\psi$, which is a common feature of the Lagrangians with which we shall work. From the point of view of the dynamical treatment we take $\mathscr{L}$ as a function of ψ (and of course $\dot{\psi}$), the treatment of variations in $\nabla\psi$ then being induced from the consequences of changes in ψ. This will lead to the recurring combination encountered, for example, in Eq. 19.

We are now able to define the *canonical conjugate momentum* as

$$p_s(t) \equiv \frac{\partial L}{\partial \dot{q}_s(t)} = \frac{\partial \mathscr{L}}{\partial \dot{q}_s} \, \delta\mathbf{r}_s \, , \tag{11}$$

and from this we can define an appropriate *conjugate field variable*

$$\pi_s(t) \equiv \frac{p_s(t)}{\delta\mathbf{r}_s} = \frac{\partial \mathscr{L}}{\partial \dot{q}_s} \, , \tag{12}$$

or, in the limit of small cell volumes $\delta\mathbf{r}_s \to 0$,

$$\pi(\mathbf{r}, t) \equiv \frac{\partial \mathscr{L}}{\partial \dot{\psi}(\mathbf{r}, t)} \, . \tag{13}$$

In terms of these variables the Hamiltonian is immediately defined in the usual way,

$$H = \sum_s p_s \dot{q}_s - L = \sum_s [\pi_s \dot{q}_s - \mathscr{L}(q_s, \dot{q}_s)] \, \delta\mathbf{r}_s$$

$$= \int_\Omega (\pi\dot{\psi} - \mathscr{L}) \, d\mathbf{r} \, . \tag{14}$$

The Hamiltonian density is then introduced in such a way that its integral over the volume yields H, namely

$$H = \int_\Omega \mathscr{H}(\psi, \nabla\psi; \pi, \nabla\pi) \, d\mathbf{r} \, , \tag{15}$$

where we have already anticipated in our notation the crucial point that H and $\mathscr{H}$ must have functional dependence only on ψ and π and their spatial derivatives, and not on $\dot{\psi}$. To verify that this is indeed the case we consider changes in H arising from all of its possible dependences,

$$\delta H = \int_\Omega d\mathbf{r} \left[\pi \, \delta\dot{\psi} + \delta\pi \, \dot{\psi} - \frac{\partial \mathscr{L}}{\partial \psi} \, \delta\psi \right.$$

$$\left. - \frac{\partial \mathscr{L}}{\partial(\nabla\psi)} \cdot \delta(\nabla\psi) - \frac{\partial \mathscr{L}}{\partial \dot{\psi}} \, \delta\dot{\psi} - \frac{\partial \mathscr{L}}{\partial t} \, dt \right) \, , \tag{16}$$

where in the last term we again (briefly) permit explicit time dependence in the problem. Now, by the definition in Eq. 13, the first and fifth terms cancel. The third and fourth terms, through the same manipulations as for these terms in Eq. 7 and using the Euler–Lagrange equation 8, are replaced by $-(\partial/\partial t)(\partial \mathscr{L}/\partial \dot{\psi})\,\delta \psi = -\dot{\pi}\,\delta \psi$. Thus we find

$$\delta H = \int_\Omega d\mathbf{r}\left[\delta \pi\,\dot{\psi} - \dot{\pi}\,\delta \psi - \frac{\partial \mathscr{L}}{\partial t}\,dt\right], \tag{17}$$

which verifies that there is no true dependence of H on $\dot{\psi}$ (i.e., no term in δH involving $\delta \dot{\psi}$). Further, for the dependence noted in Eq. 15 we must have

$$\delta H = \int_\Omega d\mathbf{r}\left[\frac{\partial \mathscr{H}}{\partial \psi}\,\delta \psi + \frac{\partial \mathscr{H}}{\partial(\boldsymbol{\nabla}\psi)} \cdot \delta(\boldsymbol{\nabla}\psi)\right.$$

$$\left. + \frac{\partial \mathscr{H}}{\partial \pi}\,\delta \pi + \frac{\partial \mathscr{H}}{\partial(\boldsymbol{\nabla}\pi)} \cdot \delta(\boldsymbol{\nabla}\pi) + \frac{\partial \mathscr{H}}{\partial t}\,dt\right]$$

$$= \int_\Omega d\mathbf{r}\left\{\left[\frac{\partial \mathscr{H}}{\partial \psi} - \boldsymbol{\nabla}\cdot\frac{\partial \mathscr{H}}{\partial(\boldsymbol{\nabla}\psi)}\right]\delta \psi\right.$$

$$\left. + \left[\frac{\partial \mathscr{H}}{\partial \pi} - \boldsymbol{\nabla}\cdot\frac{\partial \mathscr{H}}{\partial(\boldsymbol{\nabla}\pi)}\right]\delta \pi + \frac{\partial \mathscr{H}}{\partial t}\,dt\right\}, \tag{18}$$

where we have again interchanged derivatives and variations of the field, and integrated by parts, dropping surface terms. Comparing Eqs. 17 and 18, we find the equations of motion in the Hamiltonian formulation,

$$\dot{\psi}(\mathbf{r},\,t) = \frac{\partial \mathscr{H}}{\partial \pi} - \boldsymbol{\nabla}\cdot\frac{\partial \mathscr{H}}{\partial(\boldsymbol{\nabla}\pi)} \equiv \frac{d\mathscr{H}}{d\pi} \tag{19a}$$

and

$$-\dot{\pi}(\mathbf{r},\,t) = \frac{\partial \mathscr{H}}{\partial \psi} - \boldsymbol{\nabla}\cdot\frac{\partial \mathscr{H}}{\partial(\boldsymbol{\nabla}\psi)} = \frac{d\mathscr{H}}{d\psi}, \tag{19b}$$

and, in the presence of explicit time dependence,

$$\frac{\partial \mathscr{H}}{\partial t} = -\frac{\partial \mathscr{L}}{\partial t}, \tag{19c}$$

that is, any time dependence assumed in the Lagrangian density is carried over to the Hamiltonian density with a change of sign. In Eqs. 19a and b we have introduced a temporary notation for the particular combination of (functional) derivatives that appears there and which frequently arises in the context of functional dependence of fields and on their gradients.

We now define Poisson brackets for functionals F and G—say, of ψ, $\boldsymbol{\nabla}\psi$

and $\pi, \nabla\pi$. Using the generalized coordinates q_s and conjugate momenta p_s of Eq. 11, these brackets are

$$[F, G]_{PB} = \sum_s \left[\frac{\partial F}{\partial q_s} \frac{\partial G}{\partial p_s} - \frac{\partial G}{\partial q_s} \frac{\partial F}{\partial p_s} \right]$$

$$\Rightarrow \sum_s \left[\frac{df}{d\psi_s} \delta\mathbf{r}_s \cdot \frac{dg}{d\pi_s} - \frac{dg}{d\psi_s} \delta\mathbf{r}_s \cdot \frac{df}{d\pi_s} \right]$$

$$= \int d\mathbf{r} \left\{ \left[\frac{\partial f}{\partial \psi} - \nabla \cdot \frac{\partial f}{\partial(\nabla\psi)} \right] \left[\frac{\partial g}{\partial \pi} - \nabla \cdot \frac{\partial g}{\partial(\nabla\pi)} \right] \right.$$

$$\left. - \left[\frac{\partial g}{\partial \psi} - \nabla \cdot \frac{\partial g}{\partial(\nabla\psi)} \right] \left[\frac{\partial f}{\partial \pi} - \nabla \cdot \frac{\partial f}{\partial(\nabla\pi)} \right] \right\}, \qquad (20)$$

where the arrow indicates the definition of Poisson brackets for use in the continuum case, as motivated by the combination for the functional derivatives that has appeared above. Here f and g are the densities for F and G. For example

$$F(\psi, \nabla\psi; \pi, \nabla\pi) = \int f(\psi, \nabla\psi; \pi, \nabla\pi) \, d\mathbf{r}$$

$$\rightarrow \sum_s f(q_s, p_s) \, \delta\mathbf{r}_s , \qquad (21)$$

where we have shown the cellular decomposition, from which one sees that

$$\frac{\partial F}{\partial q_s} = \frac{\partial f}{\partial q_s} \delta\mathbf{r}_s \Rightarrow \frac{df}{d\psi_s} \delta\mathbf{r}_s , \qquad (22a)$$

and, recalling Eq. 12,

$$\frac{\partial F}{\partial p_s} = \frac{\partial f}{\partial p_s} \delta\mathbf{r}_s \Rightarrow \frac{df}{d\pi_s} , \qquad (22b)$$

both of which results were incorporated in Eq. 20. The equation of motion in Poisson-bracket formulation is thus

$$\frac{dF[\psi, \nabla\psi; \pi, \nabla\pi; t]}{dt} = [F, H]_{PB} + \frac{\partial F}{\partial t} . \qquad (23)$$

The Poisson brackets for the generalized coordinates and conjugate momenta at equal times are

$$[q_s(t), q_{s'}(t)]_{PB} = [p_s(t), p_{s'}(t)]_{PB} = 0 \qquad (24a)$$

and

$$[q_s(t), p_{s'}(t)]_{PB} = \delta_{ss'} , \qquad (24b)$$

which become for the field variables

$$[\psi(\mathbf{r}, t), \psi(\mathbf{r}', t)]_{\text{PB}} = [\pi(\mathbf{r},t), \pi(\mathbf{r}', t)]_{\text{PB}} = 0 \tag{25a}$$

and

$$[\psi(\mathbf{r}, t), \pi(\mathbf{r}', t)]_{\text{PB}} = \delta(\mathbf{r} - \mathbf{r}') . \tag{25b}$$

If the fields involved more than one component, this last would take the form

$$[\psi_\alpha(\mathbf{r}, t), \pi_\beta(\mathbf{r}', t)]_{\text{PB}} = \delta_{\alpha\beta}\delta(\mathbf{r} - \mathbf{r}') \tag{26}$$

on the assumption that the different components represent independent degrees of freedom. The Hamiltonian density would then involve a sum over the various components,

$$\mathcal{H} = \sum_\alpha \pi_\alpha \dot\psi_\alpha - \mathcal{L} , \tag{27}$$

and other generalizations of our above expressions are obvious.

The transition to quantized form for all these quantities takes place when we recall that the quantum commutator for a pair of quantum operators is to be obtained by multiplying the classical Poisson bracket result for the corresponding dynamical quantities by $i\hbar$ ($=i$ here, since $\hbar = 1$),

$$[A^q, B^q] = i\hbar[A^{\text{cl}}, B^{\text{cl}}]_{\text{PB}} = i[A^{\text{cl}}, B^{\text{cl}}]_{\text{PB}} . \tag{28}$$

Thus here we have, again at equal times,

$$[\psi(\mathbf{r}, t), \pi(\mathbf{r}', t)] = i\delta(\mathbf{r} - \mathbf{r}') , \tag{29}$$

so that ψ and π do not commute everywhere, a property that will be incorporated through the creation and annihilation operators of Chapter 1 in expansions such as that of Eq. 1. Of course some aspects of the method implied by Eq. 28 fail for quantities that do not possess a well-defined classical limit, and must be repaired accordingly on the basis of the general features we have seen. For example, fermion fields cannot have large occupation numbers and thus do not have classical limits, but parallel methods are developed for them using anticommutators, to express the Pauli principle, in place of commutators. It is on this last basis that one applies this second-quantization approach to the anticommuting fermion fields for the Schrödinger–Pauli equation in nonrelativistic situations or for the Dirac equation in relativistic ones. In fact, in this book most of the fields we encounter will be for fermions, simply because most of our considerations will pertain to systems of many electrons or many nucleons.

2.2 SCHRÖDINGER FIELDS

The fact that the time derivative appears linearly in the Schrödinger equation (as it does also, for example, in the Dirac equation) leads to a somewhat special feature in handling the field quantization for this equation. Normally one might expect the Lagrangian density $\mathscr{L}$ to involve the time derivative of the field bilinearly through a term like $\frac{1}{2}\dot{\psi}^2$ (thus preserving invariance under $t \rightarrow -t$), so that from Eq. 13 one would expect the conjugate field variable to be $\pi = \dot{\psi}$. But since we know that the dynamics of the Schrödinger equation will then relate $\dot{\psi}$ to ψ and its spatial derivatives, we cannot expect π to be independent of ψ. Thus to allow for an independent option we must consider complex fields for the Schrödinger case. (This would not be necessary for the Klein–Gordon equation, where the second derivative with respect to time appears and there is no problem in considering a real, scalar field. Recall, also, that even in first-quantized forms the Schrödinger case requires complex wave functions for this same reason.)

The complex fields are an obvious superposition of two real fields, ψ_1 and ψ_2, which are viewed as being subject to the treatment for many-component fields given in Eqs. 26 and 27—our basic rules of operation for the present situation. We define

$$\psi = \frac{1}{\sqrt{2}}\,(\psi_1 + i\psi_2)\,, \qquad \psi^\dagger = \frac{1}{\sqrt{2}}\,(\psi_1 - i\psi_2)\,. \tag{30a}$$

The conjugate fields are

$$\pi = \frac{1}{\sqrt{2}}\,(\pi_1 - i\pi_2)\,, \qquad \pi^T = \frac{1}{\sqrt{2}}\,(\pi_1 + i\pi_2)\,. \tag{30b}$$

Note that π^T is defined as the field conjugate to $\psi^\dagger$ and we do not necessarily have $\pi^T = \pi^\dagger$, since we do not know yet if π_1 and π_2 are real. This will be the case of course if the Lagrangian density is real, but that also is still unknown. From Eqs. 30, with Eq. 26 for $\alpha = 1, 2$ and $\beta = 1, 2$,

$$[\psi(\mathbf{r}, t), \pi(\mathbf{r}', t)] = i\delta(\mathbf{r} - \mathbf{r}') \tag{31a}$$

and

$$[\psi^\dagger(\mathbf{r}, t), \pi^T(\mathbf{r}, t)] = i\delta(\mathbf{r} - \mathbf{r}')\,; \tag{31b}$$

all the other combinations commute. The Hamiltonian density of Eq. 27 is

$$\mathscr{H} = \pi_1\dot{\psi}_1 + \pi_2\dot{\psi}_2 - \mathscr{L} = \pi\dot{\psi} + \pi^T\dot{\psi}^\dagger - \mathscr{L}\,. \tag{32}$$

The conjecture for the Lagrangian density which we must now make is again a bit lopsided because of the anticipated need for a result that is linear in $\partial/\partial t$. We thus assume

$$\mathcal{L} = i\psi^\dagger \dot\psi - \frac{1}{2m} \, \boldsymbol{\nabla}\psi^\dagger \cdot \boldsymbol{\nabla}\psi - \psi^\dagger(\mathbf{r}, t)V(\mathbf{r}, t)\psi(\mathbf{r}, t) \, , \qquad (33)$$

whence the Euler–Lagrange equation 8, resulting from varying ψ (and not $\psi^\dagger$, which is to be taken as independent because of the definitions in Eq. 30), gives

$$-\frac{1}{2m} \, \nabla^2\psi^\dagger + V\psi^\dagger = -i\dot\psi^\dagger \, , \qquad (34a)$$

while (independent) variations in $\psi^\dagger$ lead to

$$-\frac{1}{2m} \, \nabla^2\psi + V\psi = i\dot\psi \, . \qquad (34b)$$

These are indeed the Schrödinger equation and its Hermitian conjugate, as we had hoped, thus supporting our guess for $\mathcal{L}$.

The conjugate fields are now

$$\pi = \frac{\partial \mathcal{L}}{\partial \dot\psi} = i\psi^\dagger \quad \text{and} \quad \pi^T = \frac{\partial \mathcal{L}}{\partial \dot\psi^\dagger} = 0 \, . \qquad (35a, b)$$

This last result implies that π^T is not a valid, independent degree of freedom for the field and that Eq. 31b, involving $\psi^\dagger$ and π^T, cannot be fulfilled. This causes us no special difficulty, however, since the Hamiltonian density of Eq. 32 becomes

$$\mathcal{H} = \pi\dot\psi - \mathcal{L} \qquad (36a)$$

and the problematic $\psi^\dagger$ dependence can be eliminated from $\mathcal{L}$ through Eq. 35a to write $\mathcal{H}$ as a functional of ψ and π, as we require,

$$\mathcal{H} = -\frac{i}{2m} \, \boldsymbol{\nabla}\pi \cdot \boldsymbol{\nabla}\psi - i\pi V\psi \, . \qquad (36b)$$

Thus the Hamiltonian is

$$H = \int_\Omega \mathcal{H} \, d\mathbf{r} = \int_\Omega \psi^\dagger \left[-\frac{1}{2m} \, \nabla^2 + V(\mathbf{r}) \right] \psi \, d\mathbf{r} \, , \qquad (36c)$$

where we have integrated by parts and artificially reintroduced $\psi^\dagger$ in order to show the usual structure well known from first quantization.

In order to construct a number operator for the Schrödinger case we consider

$$N \equiv \int_\Omega \psi^\dagger \psi \, d\mathbf{r} = -i \int_\Omega \pi\psi \, d\mathbf{r} \, , \qquad (37)$$

which, as the volume integral of what promises to represent particle density, should serve to count the particles in the system. We note that it is a Hermitian operator. If N is to be a number operator then, for the dynamics of Eq. 34b, in which no particles are created or destroyed, we expect it to be a constant operator. To check this we calculate from Eqs. 23 and 28 (and with the help of Eqs. 1.22 and 31a)

$$\dot{N} = -i[N, H]$$

$$= i \left[\int_{\Omega} \pi\psi \, d\mathbf{r}, \int_{\Omega} \left(\frac{1}{2m} \, \mathbf{\nabla}'\pi \cdot \mathbf{\nabla}'\psi + \pi V\psi \right) d\mathbf{r}' \right]$$

$$= - \int_{\Omega} d\mathbf{r} \int_{\Omega} d\mathbf{r}' \left\{ \frac{1}{2m} \left[\pi(\mathbf{r}, t)(\mathbf{\nabla}'\delta(\mathbf{r}' - \mathbf{r})) \cdot \mathbf{\nabla}'\psi(\mathbf{r}', t) \right. \right.$$

$$\left. - (\mathbf{\nabla}'\pi(\mathbf{r}', t)) \cdot (\mathbf{\nabla}'\delta(\mathbf{r}' - \mathbf{r}))\psi(\mathbf{r}, t) \right]$$

$$+ [\pi(\mathbf{r}, t)V(\mathbf{r}', t)\delta(\mathbf{r}' - \mathbf{r})\psi(\mathbf{r}', t)$$

$$\left. - \pi(\mathbf{r}', t)V(\mathbf{r}', t)\delta(\mathbf{r}' - \mathbf{r})\psi(\mathbf{r}, t)] \right\} = 0 , \qquad (38)$$

so indeed N is a constant operator.

To make contact with the creation and annihilation operators of Chapter 1, we expand the field operators over a complete, orthonormal set of single-particle states. The label α for these states again refers to a complete set of commuting quantum numbers for the single-particle modes. We denote by $u_\alpha(\mathbf{r})$ the wave functions for these modes. These of course contain spin functions if the particles possess spin, and in general incorporate state vectors for whatever spaces are required by the internal degrees of freedom of the single-particle states. Then the field decompositions are

$$\psi(\mathbf{r}, t) = \sum_\alpha a_\alpha(t)u_\alpha(\mathbf{r}) \qquad (39a)$$

and

$$\psi^\dagger(\mathbf{r}, t) = \sum_\alpha a_\alpha^\dagger(t)u_\alpha^\dagger(\mathbf{r}) , \qquad (39b)$$

where the creation and annihilation operators are time dependent, and thus do not refer to the Schrödinger picture but, say, to the Heisenberg picture or to the interaction picture (see, also, the comment below Eq. 1 and the actual usage of these pictures in Chapter 7). They may be projected from the fields using the orthonormality of the set $\{u_\alpha(\mathbf{r})\}$ according to

$$a_\alpha(t) = \int_\Omega u_\alpha^\dagger(\mathbf{r})\psi(\mathbf{r}, t) \, d\mathbf{r} \qquad (40a)$$

and

$$a_\alpha^\dagger(t) = \int_\Omega u_\alpha(\mathbf{r})\psi^\dagger(\mathbf{r}, t)\, d\mathbf{r} \,. \tag{40b}$$

We can easily verify that at any given moment in time they satisfy the commutation relations of Eqs. 1.19, since, for example,

$$[a_\alpha(t), a_\beta^\dagger(t)] = \int_\Omega d\mathbf{r} \int_\Omega d\mathbf{r}'\, u_\alpha^\dagger(\mathbf{r})u_\beta(\mathbf{r}')[\psi(\mathbf{r}, t), \psi^\dagger(\mathbf{r}', t)]$$

$$= \int_\Omega d\mathbf{r} \int_\Omega d\mathbf{r}'\, u_\alpha^\dagger(\mathbf{r})u_\beta(\mathbf{r}')\delta(\mathbf{r} - \mathbf{r}')$$

$$= \int_\Omega d\mathbf{r}\, u_\alpha^\dagger(\mathbf{r})u_\beta(\mathbf{r}) = \delta_{\alpha\beta} \,, \tag{41}$$

where we have used Eq. 35a and the field commutators, Eq. 31a. Similarly, but more easily, $[a_\alpha, a_\beta] = [a_\beta^\dagger, a_\beta^\dagger] = 0$. Last, expanded over creation and annihilation operators, the number operator is

$$N = \int_\Omega \psi^\dagger\psi\, d\mathbf{r} = \sum_{\alpha\beta} a_\alpha^\dagger a_\beta \int_\Omega u_\alpha^\dagger(\mathbf{r})u_\beta(\mathbf{r})\, d\mathbf{r} = \sum_\alpha a_\alpha^\dagger a_\alpha \,, \tag{42}$$

which is just the counting operator of Eq. 1.21.

We have here developed the field results for the case of bosons, and thus the discussion has everywhere involved commutation relations and not anticommutators. For fermions exactly parallel consequences ensue, but the theory replaces all the commutators of the creation and annihilation operators $a^\dagger$ and a or the fields $\psi^\dagger$ and ψ with the anticommutators in order to embody the exclusion principle. The fermions have half-integer spin, and so their functions $u_\alpha(\mathbf{r})$ must of necessity contain spinors (and again whatever other state vectors are required by internal degrees of freedom). For a relativistic situation, described say by the Dirac equation for the spin-$\frac{1}{2}$ case, these spinors could be the four-component Dirac spinors. We are by and large concerned with nonrelativistic situations, however, for which two-component Pauli–Schrödinger spinors suffice.

2.3 PARTICLE PRODUCTION AND ABSORPTION AND THE ROLE OF FIELDS IN MEDIATING FORCES

In the coming chapters we explore in detail the description of physical phenomena in the language of second quantization, both as relates to the handling of correct statistics for a fixed number of identical particles and for purposes of changing the number of particles in the system. We here note briefly, in order to see the power of the method that is emerging, the consequences of the theory in the context of particle production and absorption. Suppose, for example, that we wish to consider an interaction involving a source or current, possibly provided by fermions, that changes the number of bosons present by increments or decrements of one boson at a time. Physical cases of this sort might be electrons in an atom that can emit or absorb a photon in changing their state, or a nucleon that produces or annihilates a pion, or an electron in a solid that absorbs or emits a phonon. Although in this example we are primarily concerned with changing the number of bosons, we shall treat all the fields as quantized, so that for the fermions we shall have automatic treatment of antisymmetrization.

An interaction having the characteristics we seek, called generically a Yukawa interaction, has the structure

$$V = g \int_\Omega d\mathbf{r} \, [\psi^\dagger(\mathbf{r}, t)\psi(\mathbf{r}, t) \cdot \phi(\mathbf{r}, t)$$

$$+ \psi^\dagger(\mathbf{r}, t)\psi(\mathbf{r}, t) \cdot \phi^\dagger(\mathbf{r}, t)] \,, \qquad (43)$$

where $\psi, \psi^\dagger$ are fermion fields and $\phi, \phi^\dagger$ are boson fields, g being the coupling constant for the interaction which describes the strength of the boson production or annihilation and is supposed to be known from experiment. [In Eq. 43 we have taken the simplest possible form of an interaction that changes bosons one by one through the linear appearance of the field ϕ. The interaction is a scalar in the fermion space and contains a scalar boson field. We could imagine substantially more complicated interaction structures. For instance $(\psi^\dagger \boldsymbol{\sigma}\psi) \cdot \nabla\phi$ is a (pseudo)vector in the fermion spinor space coupled to a (pseudo)scalar boson in order to produce a scalar interaction Hamiltonian. Or, for a scalar case in configuration space but involving a boson of isovector character through its internal degrees of freedom, we could have $(\psi^\dagger \vec{\tau}\psi) \cdot \vec{\phi}$, where the arrow refers to isospin space in which the fermion isospinor combination $(\psi^\dagger \vec{\tau}\psi)$ is an isovector and $\vec{\phi} = \{\phi_1, \phi_2, \phi_3\}$ is a three-component field in isospin space; the resulting interaction is an isoscalar which will guarantee charge independence for this system. One can also encounter both these situations together in the form of an interaction containing $(\psi^\dagger \vec{\tau}\boldsymbol{\sigma}\psi) \cdot \nabla\vec{\phi}$, where the fermion fields carry both spinors and isospinors and the boson field is a pseudoscalar isovector. Such an interaction is appropriate for nonrelativistic Yukawa coupling of the three-isocomponent π^+, π^0, π^- field of pseudoscalar mesons to nucleons.]

Our main purpose at the moment is not to explore in depth the possibilities for incorporating internal degrees of freedom, but rather to consider the production and absorption features in Eq. 43. These may be made even more explicit by expanding all the fields there in terms of operators $a, a^\dagger$. Then, suppressing possible time dependence,

$$V = g \sum_{\alpha\beta\sigma} \int_\Omega d\mathbf{r} \, u_\beta^\dagger(\mathbf{r})u_\alpha(\mathbf{r})[w_\sigma(\mathbf{r})b_\sigma + w_\sigma^\dagger(\mathbf{r})b_\sigma^\dagger]a_\beta^\dagger a_\alpha \,, \qquad (44)$$

where $u, u^\dagger$ are fermion spinor functions, $w, w^\dagger$ are state functions for the bosons, $a, a^\dagger$ annihilate and create the fermions, and $b, b^\dagger$ do the same for the bosons. Since in V the combination $a_\beta^\dagger a_\alpha$ appears, the number of fermions is not changed. The fermions are, however, transferred from state to state by the interaction. The action of V does raise or lower the number of bosons by one through the linear appearance of $b^\dagger$ and b. This situation is depicted in Figure 2.1. (At this stage this figure and its companions are intended only as pictures to aid comprehension. Later, we shall develop systematic techniques for calculating such processes; see Section 11.5.)

If the Yukawa interaction of Eq. 43 or 44 appears as part of a calculation in second-order perturbation theory, that is, if it acts twice in the course of the process in question, it can describe boson–fermion scattering, as in Figure 2.2. The Yukawa interaction can also appear as a mechanism for producing forces between the fermions in the situation of Figure 2.3, where one fermion emits a boson which is then absorbed by the other. This repeated action of V changes the energy of the two-fermion system through the exchange of the virtual boson, and that energy change can be interpreted as a potential acting between the two fermions. Indeed, it was for this purpose—in the particular context of the nucleon–nucleon force—that Yukawa introduced the general interaction form of V. The diagrams of Figure 2.3 relate closely to the two-particle interactions discussed in Section 1.3 and shown in Figure 1.1. Although such interactions need not arise from boson exchange, they may do so. They are treated as time-independent potentials if the boson emission from one fermion and absorption on the other occur simultaneously. For the potential of Section 1.3, acting between two fermions, the boson degree of freedom—if such there was—has been eliminated after V of Eqs. 43 and 44 was allowed to act twice. But if the force indeed has its origins in boson exchange, then this feature may appear explicitly when the two-fermion center-of-mass energy is greater than the boson rest-mass energy. Then the (third-order in V) production process drawn in Figure 2.4 may occur, leading to the presence of a physical boson in the final state.

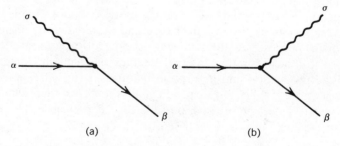

FIGURE 2.1 (a) Absorption and (b) production of bosons as given in the first and second terms of Eq. 44.

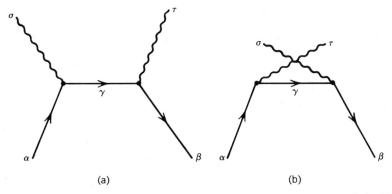

FIGURE 2.2 Boson–fermion scattering as described by two appearances of the Yukawa interaction of Figure 2.1; (b) is referred to as the "crossed" case.

The case of electromagnetism provided the original manifestation of an interaction with a structure like that of V in Eqs. 43 and 44; it was on that basis that the Yukawa interaction was generalized for other dynamics as well. For electromagnetism, to which we shall now be turning in greater detail, the graphs of Figure 2.1 represent photon absorption or emission, as in the atomic transition of an electron say, while Figure 2.2 corresponds to Compton scattering. These two processes were very central in the early exploration of quantum phenomena, and the language of second quantization allows for systematic treatment of them. Figure 2.3 gives the Coulomb force between two charged particles, and Figure 2.4 represents bremsstrahlung following upon a Coulomb interaction.

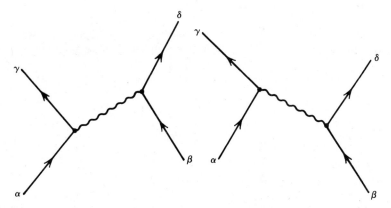

FIGURE 2.3 Fermion–fermion interaction, in two different time orderings, as mediated by boson exchange through the Yukawa interaction.

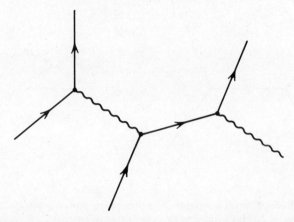

FIGURE 2.4 **Production of a nonvirtual boson in the final state through the action of V after a fermion–fermion scattering has occurred, mediated by two appearances of V.**

EXERCISES

2.1 Consider a Lagrangian density

$$\mathcal{L} = \tfrac{1}{2}\dot{\phi}^2 - \tfrac{1}{2}(\boldsymbol{\nabla}\phi)^2 - \tfrac{1}{2}m^2\phi^2$$

defined in terms of a real field $\phi(\mathbf{r}, t)$.
(a) Derive and identify the Euler–Lagrange equation for this field.
(b) Derive the conjugate field $\pi(\mathbf{r}, t)$ and the Hamiltonian H.
(c) Find the field commutation relations $[H, \phi]$ and $[H, \pi]$.
(d) If the field is expanded in the complete set of plane waves,

$$\phi(\mathbf{r}, t) = \sum_{\alpha} \left[e^{i\mathbf{k}\cdot\mathbf{r}} f_{\alpha}(t) a_{\alpha} + e^{-i\mathbf{k}\cdot\mathbf{r}} f_{\alpha}^*(t) a_{\alpha}^\dagger \right],$$

explain the nature of the a_{α}'s and $a_{\alpha}^\dagger$'s, find the functional dependence of f_{α} on t, and find the normalization factor for f_{α}.
(e) Show that

$$H = \sum_{\alpha} \omega_{\alpha}\left(a_{\alpha}^\dagger a_{\alpha} + \tfrac{1}{2}\right).$$

2.2 For a particular complex, classical field

$$\phi(\mathbf{r}, t) = \frac{1}{\sqrt{2}}\left[\phi_1(\mathbf{r}, t) + i\phi_2(\mathbf{r}, t)\right]$$

the Lagrangian density is

$$\mathcal{L} = \tfrac{1}{2}\dot{\phi}^*\dot{\phi} - \tfrac{1}{2}\boldsymbol{\nabla}\phi^*\cdot\boldsymbol{\nabla}\phi - \tfrac{1}{2}m^2\phi^*\phi - \frac{\lambda}{4}(\phi^*\phi)^2, \qquad \lambda > 0.$$

(a) Derive the Euler–Lagrange equations for the field.

(b) Derive the equations of motion of the field by constructing the conjugate field $\pi(\mathbf{r}, t)$, the Hamiltonian, and Poisson-bracket relations.

(c) Show from the equations of motion that the (charge) operator

$$Q = i \int (\pi^* \phi^* - \pi\phi) \, d\mathbf{r}$$

is conserved, $dQ/dt = 0$. (How is this modified if the fields are quantized?)

(d) Note that if $m^2 > 0$ then the energy minimum for constant field ϕ is at $\phi = 0$. If $m^2 = -\mu^2 < 0$ (imaginary "mass"), then show that the minimum is at $|\phi|^2 = |v|^2 = -m^2/\lambda > 0$, where $v = (v_1 + iv_2)/\sqrt{2}$. Expanding about this minimum and defining

$$\phi_M = \frac{1}{\sqrt{2}|v|} [v_1(\phi_1 - v_1) + v_2(\phi_2 - v_2)]$$

and

$$\phi_G = \frac{1}{\sqrt{2}|v|} [-v_2(\phi_1 - v_1) + v_1(\phi_2 - v_2)],$$

show that ϕ_M corresponds to a field of mass $\sqrt{2}\mu$, while ϕ_G is massless (a Goldstone boson).

(e) Note that the original Lagrangian density is invariant under the replacement

$$\phi \to \phi' = e^{i\Lambda}\phi, \qquad \Lambda = \text{constant}.$$

What is the situation near the minima for the two cases $m^2 > 0$ and $m^2 < 0$?

2.3 (a) Show that if a Lagrangian density $\mathcal{L}[\phi]$, ϕ complex, is invariant under the phase transformation

$$\phi \to \phi' = e^{i\Lambda}\phi, \qquad \Lambda = \text{constant},$$

then the current and density defined by

$$\mathbf{j} = -i\left[\frac{\partial\mathcal{L}}{\partial(\nabla\phi)} \phi - \frac{\partial\mathcal{L}}{\partial(\nabla\phi^*)} \phi^* \right]$$

and

$$\rho = -i\left[\frac{\partial\mathcal{L}}{\partial\dot\phi} \phi - \frac{\partial\mathcal{L}}{\partial\dot\phi^*} \phi^* \right]$$

are conserved, that is,

$$\mathbf{\nabla} \cdot \mathbf{j} + \frac{\partial \rho}{\partial t} = 0 \,.$$

(This is known as Noether's theorem for the case of current conservation.)

(b) How does this relate to parts (c) and (e) of Exercise 2.2?

(c) Show generally that in the presence of current conservation

$$\frac{dQ}{dt} = 0 \qquad \text{for} \quad Q = \int \rho(\mathbf{r}, t) \, d\mathbf{r} \,.$$

2.4 For the Schrödinger field, Eq. 33, find the conserved current associated with the invariance property in Exercise 2.3, and find the quantity which is conserved in time. To what do these quantities correspond in the language of first quantization?

CHAPTER 3

Electromagnetic Fields

We now apply the methods of the previous two chapters—the number representation and field quantization—to the electromagnetic case. The electromagnetic field provided the historical context in which the general properties of bosons were first understood; it still serves well the pedagogical purpose of allowing a thorough exploration of the broad features of boson behavior. The phenomena in question include blackbody radiation, photon absorption and spontaneous and induced emission, fluctuations in boson populations, the phenomenon of Hanbury Brown and Twiss, laser behavior, and so forth. Electromagnetic theory further gives the major prototype for other theories of interactions between bosons and fermions, namely weak interactions (with which it has been unified) and hadronic interactions (with which it may ultimately be unified). Above all, electromagnetic theory gives an understanding of the vast range of physical phenomena that arise from the interaction of charged particles with the electromagnetic field and from the properties of the field itself.

3.1 QUANTIZATION OF THE TRANSVERSE RADIATION FIELD

The equations of motion of the electromagnetic field are Maxwell's equations. Fortunately for our purpose and for physics at large, these have precisely the same structure in the quantum context as they do classically, requiring only an extension of interpretation to encompass quantization. For the moment, all fields are treated classically; only after Eq. 33 will a quantum treatment start. In vacuum Maxwell's four equations read, first

$$\nabla \cdot \mathbf{E}(\mathbf{r}, t) = \rho(\mathbf{r}, t) \,, \tag{1}$$

which is Gauss's law or the differential form of Coulomb's law relating the electric field $\mathbf{E}$ to its source, the charge density ρ; second,

$$\nabla \times \mathbf{B}(\mathbf{r}, t) = \mathbf{j}(\mathbf{r}, t) + \frac{\partial \mathbf{E}(\mathbf{r}, t)}{\partial t} \,, \tag{2}$$

Ampère's law connecting the magnetic field $\mathbf{B}$ to its source, the current $\mathbf{j}$, and containing Maxwell's displacement current $\partial \mathbf{E}/\partial t$; third,

$$\nabla \times \mathbf{E}(\mathbf{r}, t) + \frac{\partial \mathbf{B}(\mathbf{r}, t)}{\partial t} = 0 \,, \tag{3}$$

Faraday's law pertaining to the dynamics of the free electromagnetic field; and, fourth,

$$\nabla \cdot \mathbf{B}(\mathbf{r}, t) = 0 \,, \tag{4}$$

another law for the free field which states that there are no magnetic monopoles. (Although the possible existence of magnetic monopoles has been raised repeatedly, an affirmative answer to this question should not change the description of the phenomena within our scope here.) Note that in Eqs. 1–4 we have again used units such that the velocity of light equals unity, $c = 1$. Further, we have used rationalized units, that is, there are no explicit factors of 4π in Eqs. 1 and 2 that relate the fields to sources. This is done in order to be consistent with the conventional usage for fields other than the electromagnetic one. With these choices of units, the electromagnetic coupling constant, or fine-structure constant, satisfies $\alpha \equiv e^2/4\pi \cong 1/137.04$, where e is the fundamental charge.

As relations auxiliary to Maxwell's equations, we have the continuity equation for charges and currents,

$$\nabla \cdot \mathbf{j}(\mathbf{r}, t) + \frac{\partial \rho(\mathbf{r}, t)}{\partial t} = 0 \,, \tag{5}$$

and the equation for the force $\mathbf{F}$ exerted by the field on a distribution of charge and current,

$$\mathbf{F} = \int \rho(\mathbf{r}', t)\mathbf{E}(\mathbf{r}', t) \, d\mathbf{r}' + \int \mathbf{j}(\mathbf{r}', t) \times \mathbf{B}(\mathbf{r}', t) \, d\mathbf{r}' \,. \tag{6}$$

In order to carry out the procedure of second quantization we must identify appropriate field variables, construct a Lagrangian leading to the correct dynamics (i.e. producing Maxwell's equations for the fields), and generate conjugate variables, the Hamiltonian, and Poisson-bracket relations which are replaced by commutators to achieve second quantization. At this stage we shall find, as is to be expected on the basis of our experience in Chapters 1 and 2, that an oscillator-like boson-counting device is attached to each mode of the electromagnetic field (wave vector **k** and spin projection or polarization state λ).

The formal treatment of the electromagnetic field is much simplified if potentials are introduced. From Eq. 4 we know that the magnetic field **B** can be derived from a vector potential **A** through

$$\mathbf{B}(\mathbf{r}, t) = \nabla \times \mathbf{A}(\mathbf{r}, t) . \tag{7}$$

Then, from Eq. 3,

$$\nabla \times \left(\mathbf{E} + \frac{\partial \mathbf{A}}{\partial t} \right) = 0 , \tag{8}$$

and we can conclude that the combination in parentheses is derivable from a scalar potential $\phi(\mathbf{r}, t)$ according to

$$\nabla \phi(\mathbf{r}, t) = -\mathbf{E}(\mathbf{r}, t) - \frac{\partial \mathbf{A}(\mathbf{r}, t)}{\partial t} . \tag{9}$$

Thus Eq. 1 becomes

$$\nabla^2 \phi + \frac{\partial}{\partial t} (\nabla \cdot \mathbf{A}) = -\rho , \tag{10}$$

and Eq. 2 is

$$\nabla \times (\nabla \times A) = -\nabla^2 \mathbf{A} + \nabla(\nabla \cdot \mathbf{A}) = \mathbf{j} - \frac{\partial}{\partial t} \left(\nabla \phi + \frac{\partial \mathbf{A}}{\partial t} \right) . \tag{11}$$

We note that Eqs. 10 and 11 are coupled in the sense that the scalar and the vector potentials ϕ and **A** appear in both of them. On the other hand there is a remaining freedom with regard to the definition of ϕ and **A** which we can exploit to decouple these equations. This is the freedom of performing a gauge transformation. If, in the definitions of the potentials in Eqs. 7 and 9, we make the replacements, for arbitrary $\Lambda(\mathbf{r}, t)$,

$$\mathbf{A}' = \mathbf{A} + \nabla \Lambda(\mathbf{r}, t) , \tag{12}$$

and

$$\phi' = \phi - \frac{\partial \Lambda(\mathbf{r}, t)}{\partial t} ; \tag{13}$$

the **B** and **E** fields are unchanged, since the curl of Eq. 7 annihilates the gradient of a scalar, and the two changes of Eqs. 12 and 13 cancel in producing **E** through Eq. 9. We now have a number of options in selecting the gauge function Λ. For example, we may choose Λ so that the scalar and vector potentials satisfy the *Lorentz condition*

$$\nabla \cdot \mathbf{A} + \frac{\partial \phi}{\partial t} = 0 \, . \tag{14}$$

This condition when exploited in Eqs. 10 and 11 yields for the potentials equations of the form

$$\Box \phi = -\rho \, , \qquad \Box \mathbf{A} = -\mathbf{j} \, , \tag{15a, b}$$

where we have introduced the d'Alembertian operator $\Box \equiv \nabla^2 - \partial^2/\partial t^2$. In contexts where covariance is important the Lorentz condition is very natural and highly useful since, identifying $(\mathbf{A}, \phi)$ as the components of a four-vector, as are also $(\mathbf{j}, \rho)$, it is easily seen that Eqs. 14 and 15 are all covariant.

For us, at this point, the relativistic properties of our equations are not central. Instead we are more interested in studying the radiation dynamics of real photons. This leads us to choose the *transverse* or *Coulomb* gauge defined by requiring

$$\nabla \cdot \mathbf{A} = 0 \, . \tag{16}$$

This condition has the effect of reducing the three independent fields implied by the three components of the vector **A** to two independent components which correspond to the two states of polarization (left or right circular polarization or plane polarization in two orthogonal planes) that we know to occur in electromagnetic radiation. We shall also see below that the condition of Eq. 16 implies—as it should—that the polarization is transverse; the third, longitudinal component has been eliminated by Eq. 16. This situation is related to the fact that the photon has zero mass. The connection can be seen through a simple physical argument which we make in rough form here: A vector field with its three components generally represents a particle of spin one, the three independent components mapping over to the three spin projections of a spin-one situation, $-1, 0, 1$. However if a particle has zero mass in the context of special relatively, it cannot possess a spin projection that is not either parallel or antiparallel to its direction of motion, since any other angle of orientation would be different in different Lorentz frames whereas a massless particle moves with the speed of light and should appear the same in all frames. Thus one possible spin projection or one independent field component or direction is eliminated, in our present case through the condition 16, which gives one relationship between the three components of **A**.

Using the relationship in 16, Eq. 10 becomes

$$\nabla^2 \phi = -\rho \, , \tag{17}$$

the static Poisson equation, with the immediate Coulomb solution

$$\phi(\mathbf{r}, t) = \frac{1}{4\pi} \int \frac{\rho(\mathbf{r}', t)}{|\mathbf{r} - \mathbf{r}'|} \, d\mathbf{r}' \, . \tag{18}$$

Thus the Coulomb part of the problem has been separated out for immediate treatment and is not handled as a full dynamical variable. The full dynamics with second quantization will apply to the vector potential $\mathbf{A}$; it is this quantity that contains the radiation degrees of freedom. From Eqs. 11 and 16, it satisfies

$$\Box \mathbf{A} = -\mathbf{j} + \frac{1}{4\pi} \frac{\partial}{\partial t} \nabla \int \frac{\rho(\mathbf{r}', t)}{|\mathbf{r} - \mathbf{r}'|} \, d\mathbf{r}'$$

$$= -\mathbf{j} - \frac{1}{4\pi} \nabla \int \frac{\nabla' \cdot \mathbf{j}(\mathbf{r}', t)}{|\mathbf{r} - \mathbf{r}'|} \, d\mathbf{r}' = \mathbf{j}^{\mathrm{tr}}(\mathbf{r}, t) \, , \tag{19}$$

where we have used Eq. 18 and the continuity equation 5 to cast the right-hand side into the form of a source $\mathbf{j}^{\mathrm{tr}}(\mathbf{r}, t)$ involving the transverse current only. Note that we can apply the divergence operation to Eq. 19, whence from the transversality condition imposed on $\mathbf{A}$ in Eq. 16 we must expect that the right-hand side has vanishing divergence,

$$\nabla \cdot \mathbf{j}^{\mathrm{tr}}(\mathbf{r}, t) = 0 \, . \tag{20}$$

That this is indeed the case is immediately seen from

$$\nabla^2 \frac{1}{|\mathbf{r} - \mathbf{r}'|} = -4\pi\delta(\mathbf{r} - \mathbf{r}') \tag{21}$$

(the mathematical expression of Coulomb's law for a point source). Thus the right-hand side of Eq. 19 is also transverse, as incorporated in the notation.

The extra feature of gauge freedom is a peculiarity of the electromagnetic field that—as we shall note briefly later on—relates to the masslessness of the photon. Maxwell's equations have automatically produced a theory in which the photon will turn out to have zero mass. This aspect of electromagnetism requires some nonroutine treatment, but all the other features of second quantization, to which we now turn, pertain generally to boson systems.

We first study the properties of plane-wave solutions for $\mathbf{A}$ in the absence of transverse currents, that is, for $\mathbf{j}^{\mathrm{tr}}(\mathbf{r}, t) = 0$. This corresponds to the case of a free photon field. The equation of motion is

$$\Box \mathbf{A} = \left(\nabla^2 - \frac{\partial^2}{\partial t^2}\right)\mathbf{A}(\mathbf{r}, t) = 0 \, , \tag{22}$$

with the side condition of transversality, Eq. 16. The plane-wave solution is of the form

$$\mathbf{A}(\mathbf{r}, t) = N_k \boldsymbol{\epsilon}_k e^{i\mathbf{k}\cdot\mathbf{r} - i\omega_k t} \, , \qquad k = \omega_k \, , \tag{23}$$

for photon wave vector or momentum $\mathbf{k}$, and angular velocity or energy* ω_k, recalling that $\hbar = 1$ in our units. Equation 23 contains an as yet arbitrary normalization factor N_k and a polarization unit vector $\boldsymbol{\epsilon}_k$ which, from the transversality condition 16, must indeed be transverse to $\mathbf{k}$:

$$\mathbf{k} \cdot \boldsymbol{\epsilon}_k = 0 \, . \tag{24}$$

To take into account the two independent directions perpendicular to a given wave vector $\mathbf{k}$ we introduce a polarization label $\lambda = 1, 2$ for the unit vector $\boldsymbol{\epsilon}_{k\lambda}$. We assume that periodic boundary conditions are imposed in a large cube of sides L so that the wave vector $\mathbf{k}$ is of the form

$$\mathbf{k} = \frac{2\pi}{L}\,\mathbf{n} \, , \qquad \mathbf{n} = \{n_x, n_y, n_z\} \, , \tag{25}$$

for a triad of integers n_x, n_y, n_z. The density of states is then

$$d\mathbf{n} = L^3 \frac{d\mathbf{k}}{(2\pi)^3} \tag{26a}$$

or in terms of a spherical form with solid angle $d\hat{\mathbf{k}}$,

$$n^2\, dn\, d\Omega = L^3 \frac{k^2\, dk}{(2\pi)^3}\, d\hat{\mathbf{k}} \, . \tag{26b}$$

The electric and magnetic fields are easily derived from Eqs. 7, 9, and 23, yielding

$$\mathbf{E}_{k\lambda} = -\frac{\partial \mathbf{A}}{\partial t} = iN_k k \boldsymbol{\epsilon}_{k\lambda} e^{i\mathbf{k}\cdot\mathbf{r} - i\omega_k t} \tag{27}$$

and

$$\mathbf{B}_{k\lambda} = \nabla \times \mathbf{A} = iN_k (\mathbf{k} \times \boldsymbol{\epsilon}_{k\lambda}) e^{i\mathbf{k}\cdot\mathbf{r} - i\omega_k t} \tag{28}$$

---

*Note that if the photon had a nonzero mass, Eq. 22 would be replaced by $(\Box - m_\gamma^2)\mathbf{A} = 0$ and the relation between energy and momentum would be $\omega_k = (k^2 + m_\gamma^2)^{1/2}$ instead of $\omega_k = k$.

for each photon mode $\mathbf{k}$, λ. The appearance of complex fields is a consequence of having used periodic boundary conditions with complex exponentials.

For the quantization of the electromagnetic field (that is, actually for the transverse potential $\mathbf{A}$, whence the quantization of $\mathbf{E}$ and $\mathbf{B}$ follows from Eqs. 27 and 28) we again consider a source-free region and take as our guess for the Lagrangian density the form $(k, l = 1, 2, 3)$

$$\mathcal{L} = \frac{1}{2}\left(\frac{\partial \mathbf{A}}{\partial t}\right)^2 - \frac{1}{4}\sum_{kl}\left(\frac{\partial A_l}{\partial r_k} - \frac{\partial A_k}{\partial r_l}\right)^2. \qquad (29)$$

We test this guess by verifying that it leads to the correct equations of motion—Maxwell's equations—for the field. Had we been developing the theory in covariant form, our choice at this point would have been aided by the requirement that the Lagrangian density $\mathcal{L}$ be a Lorentz scalar. We would further have wanted it to be quadratic in $\mathbf{A}$ so that the resulting equations of motion would be linear. These conditions would have strongly limited our options. (One could of course take powers of the form for $\mathcal{L}$ in Eq. 29 or add terms that are total four-derivatives of functions whose variations vanish at the boundary surfaces, since these would not change the results of the variational procedure.)

We now apply the Euler–Lagrange equation 2.8 to Eq. 29 for each successive component of $\mathbf{A}$, temporarily ignoring the transversality constraint. (Alternatively, we could use the method of Lagrange multipliers to deal with it.) We then have

$$-\frac{\partial^2}{\partial t^2}A_l + \sum_k \frac{\partial}{\partial r_k}\left(\frac{\partial A_l}{\partial r_k} - \frac{\partial A_k}{\partial r_l}\right) = 0, \qquad (30)$$

and if we now impose Eq. 16, the last term vanishes since $\sum_k \partial A_k/\partial r_k = 0$, and we obtain $\Box \mathbf{A} = 0$ as in Eq. 22. This justifies the claim that $\mathcal{L}$ contains the correct dynamics. In other words, this Lagrangian density is equivalent to Maxwell's equations for the free electromagnetic field in the vacuum, bearing in mind that two of those equations, namely Eqs. 3 and 4, are automatically incorporated in the use of potentials, while a third one, Eq. 1, is treated outside the dynamical framework of transverse radiation. [Note that $\mathcal{L}$ is numerically equal to $\frac{1}{2}(\mathbf{E}^2 - \mathbf{B}^2)$, which is a relativistic invariant, so that it is possible to develop a covariant theory along the lines we are using here if this were of importance in the treatment.]

The conjugate field variable is easily derived from Eq. 29 as

$$\boldsymbol{\pi}(\mathbf{r}, t) \equiv \frac{\partial \mathcal{L}}{\partial \dot{\mathbf{A}}} = \frac{\partial \mathbf{A}}{\partial t} = -\mathbf{E}, \qquad (31)$$

and the Hamiltonian is then

$$H = \int_{L^3} (\boldsymbol{\pi} \cdot \dot{\mathbf{A}} - \mathscr{L}) \, d\mathbf{r} = \frac{1}{2} \int_{L^3} [\boldsymbol{\pi}^2 + (\boldsymbol{\nabla} \times \mathbf{A})^2] \, d\mathbf{r}$$

$$= \frac{1}{2} \int_{L^3} (\mathbf{E}^2 + \mathbf{B}^2) \, d\mathbf{r} . \tag{32}$$

In this equation the variables of interest are $\boldsymbol{\pi}$ and $\mathbf{A}$, and the calculational result in terms of electric and magnetic fields is introduced partly for notational convenience and partly because it shows the immediate equivalence of the Hamiltonian to the well-known energy content of the electromagnetic field.

We now expand the field variables and their conjugates over plane-wave states and creation and annihilation operators, choosing a particular normalization N_k in Eq. 23 for future convenience; at this point the second-quantized treatment of the dynamic quantities begins:

$$\mathbf{A}(\mathbf{r}, t) = \sum_{\mathbf{k}} \sum_{\lambda=1,2} \sqrt{\frac{1}{2\omega_k}} \, [a_{\mathbf{k}\lambda} \mathbf{u}_{\mathbf{k}\lambda}(\mathbf{r}) e^{-i\omega_k t} + a^{\dagger}_{\mathbf{k}\lambda} \mathbf{u}^*_{\mathbf{k}\lambda}(\mathbf{r}) e^{i\omega_k t}] \tag{33}$$

and

$$\boldsymbol{\pi}(\mathbf{r}, t) = \dot{\mathbf{A}} = i \sum_{\mathbf{k}} \sum_{\lambda=1,2} \sqrt{\frac{\omega_k}{2}} \, [-a_{\mathbf{k}\lambda} \mathbf{u}_{\mathbf{k}\lambda}(\mathbf{r}) e^{-i\omega_k t} + a^{\dagger}_{\mathbf{k}\lambda} \mathbf{u}^*_{\mathbf{k}\lambda}(\mathbf{r}) e^{i\omega_k t}] , \tag{34}$$

where $a_{\mathbf{k}\lambda}$ and $a^{\dagger}_{\mathbf{k}\lambda}$ are creation and annihilation operators for the modes shown,

$$\mathbf{u}_{\mathbf{k}\lambda}(\mathbf{r}) = L^{-3/2} \boldsymbol{\epsilon}_{\mathbf{k}\lambda} e^{i\mathbf{k}\cdot\mathbf{r}} , \tag{35}$$

normalized so that

$$\int_{L^3} \mathbf{u}^*_{\mathbf{k}'\lambda'} \cdot \mathbf{u}_{\mathbf{k}\lambda} \, d\mathbf{r} = \delta_{\mathbf{k}\mathbf{k}'} \delta_{\lambda\lambda'} . \tag{36}$$

Note that the fields in Eqs. 33 and 34 are Hermitian. These fields must satisfy appropriate commutation relations, as in Eqs. 2.26 and 2.28. It is easily verified that

$$[A_j(\mathbf{r}, t), A_l(\mathbf{r}', t)] = [\pi_j(\mathbf{r}, t), \pi_l(\mathbf{r}', t)] = 0 . \tag{37}$$

Somewhat more challenging is the nonvanishing combination,

$$[A_j(\mathbf{r}, t), \pi_l(\mathbf{r}', t)] = \frac{i}{2} \sum_{\mathbf{kk}'} \sum_{\lambda\lambda'} \sqrt{\frac{\omega_{k'}}{\omega_k}} \{ -(\mathbf{u}^*_{\mathbf{k}\lambda}(\mathbf{r}))_j (\mathbf{u}_{\mathbf{k}'\lambda'}(\mathbf{r}'))_l$$

$$\times e^{i(\omega_k - \omega_{k'})t}[a^\dagger_{\mathbf{k}\lambda}, a_{\mathbf{k}'\lambda'}]$$

$$+ (\mathbf{u}_{\mathbf{k}\lambda}(\mathbf{r}))_j (\mathbf{u}^*_{\mathbf{k}'\lambda'}(\mathbf{r}'))_l e^{i(\omega_{k'} - \omega_k)t}[a_{\mathbf{k}\lambda}, a^\dagger_{\mathbf{k}'\lambda'}]\}$$

$$= i \sum_{\mathbf{k}} \sum_{\lambda=1,2} (\boldsymbol{\epsilon}_{\mathbf{k}\lambda})_j (\boldsymbol{\epsilon}_{\mathbf{k}\lambda})_l \frac{1}{2L^3} [e^{i\mathbf{k}\cdot(\mathbf{r}'-\mathbf{r})} + e^{i\mathbf{k}\cdot(\mathbf{r}-\mathbf{r}')}] . \quad (38)$$

Now the combination involved in the sum over polarization states spans two of the three available spatial dimensions, the third direction being that of $\mathbf{k}$ itself (see Eq. 24),

$$\sum_{\lambda=1,2} (\boldsymbol{\epsilon}_{\mathbf{k}\lambda})_j (\boldsymbol{\epsilon}_{\mathbf{k}\lambda})_l = \sum_{\lambda=1,2,3} (\boldsymbol{\epsilon}_{\mathbf{k}\lambda})_j (\boldsymbol{\epsilon}_{\mathbf{k}\lambda})_l - (\boldsymbol{\epsilon}_{\mathbf{k}3})_j (\boldsymbol{\epsilon}_{\mathbf{k}3})_l$$

$$= \delta_{jl} - \frac{k_j k_l}{k^2} . \quad (39)$$

Thus in Eq. 38 we require for the first of these terms

$$i\delta_{jl} \frac{1}{L^3} \sum_{\mathbf{k}} e^{i\mathbf{k}\cdot(\mathbf{r}-\mathbf{r}')} = i\delta_{jl} \int \frac{d\mathbf{k}}{(2\pi)^3} e^{i\mathbf{k}\cdot(\mathbf{r}-\mathbf{r}')} = i\delta_{jl}\delta(\mathbf{r} - \mathbf{r}') , \quad (40)$$

while for the second

$$-i \frac{1}{L^3} \sum_{\mathbf{k}} \frac{k_j k_l}{k^2} e^{i\mathbf{k}\cdot(\mathbf{r}-\mathbf{r}')} = -i \frac{1}{L^3} \nabla_j \nabla'_l \sum_{\mathbf{k}} \frac{1}{k^2} e^{i\mathbf{k}\cdot(\mathbf{r}-\mathbf{r}')}$$

$$= -i\nabla_j \nabla'_l \int \frac{d\mathbf{k}}{(2\pi)^3} \frac{e^{i\mathbf{k}\cdot(\mathbf{r}-\mathbf{r}')}}{k^2}$$

$$= -i\nabla_j \nabla'_l \frac{1}{4\pi} \frac{1}{|\mathbf{r} - \mathbf{r}'|} ; \quad (41)$$

this last form can be shown directly from the calculus of residues or verified by means of Eq. 21. Thus

$$[A_j(\mathbf{r}, t), \pi_l(\mathbf{r}', t)] = i\delta^{\text{tr}}_{jl}(\mathbf{r} - \mathbf{r}') , \quad (42a)$$

where the transverse δ-function is

$$\delta^{\text{tr}}_{jl}(\mathbf{r} - \mathbf{r}') \equiv \delta_{jl}\delta(\mathbf{r} - \mathbf{r}') - \frac{1}{4\pi} \nabla_j \nabla'_l \frac{1}{|\mathbf{r} - \mathbf{r}'|} . \quad (42b)$$

It is the appearance of this transverse construct that saves us from paradoxes that would result if we took over Eqs. 2.26 with 2.28 directly for A_j and π_l of the electromagnetic field without making allowances for the special features resulting from the need to limit the electromagnetic field to two transverse components. For example, if we supposed that the commutator were

$$[A_j(\mathbf{r}, t), \pi_l(\mathbf{r}', t)] = -[A_j(\mathbf{r}, t), E_l(\mathbf{r}', t)] = i\delta_{jl}(\mathbf{r} - \mathbf{r}') \quad \text{(wrong!)}, \quad (43)$$

then the application of $\Sigma_j \nabla_j$ or of $\Sigma_l \nabla_l'$ to this relation would give zero on the left-hand side because of Eq. 16 or 1 (with vanishing source here), and a nonzero result on the right. The transverse δ-function of eq. 42b, however, is such that

$$\sum_j \nabla_j \delta_{jl}^{\text{tr}}(\mathbf{r} - \mathbf{r}') = \nabla_l \delta(\mathbf{r} - \mathbf{r}') - \frac{1}{4\pi} \nabla_l' \nabla^2 \frac{1}{|\mathbf{r} - \mathbf{r}'|}$$

$$= (\nabla_l + \nabla_l')\delta(\mathbf{r} - \mathbf{r}') = 0 , \qquad (44)$$

again using Eq. 21.

The Hamiltonian, or the energy content of the field, is given by Eq. 32 as

$$H = \frac{1}{2} \int_{L^3} (\mathbf{E}^2 + \mathbf{B}^2) \, dr = \frac{1}{2} \sum_{\mathbf{k}\lambda} \omega_k (a_{\mathbf{k}\lambda}^\dagger a_{\mathbf{k}\lambda} + a_{\mathbf{k}\lambda} a_{\mathbf{k}\lambda}^\dagger)$$

$$= \sum_{\mathbf{k}\lambda} \omega_k (a_{\mathbf{k}\lambda}^\dagger a_{\mathbf{k}\lambda} + \tfrac{1}{2}) , \qquad (45)$$

where the first step uses vector identities for

$$(\mathbf{k} \times \boldsymbol{\epsilon}_{\mathbf{k}\lambda}) \cdot (\mathbf{k}' \times \boldsymbol{\epsilon}_{\mathbf{k}'\lambda'})|_{\mathbf{k}'=-\mathbf{k}} = -\boldsymbol{\epsilon}_{\mathbf{k}\lambda} \cdot \boldsymbol{\epsilon}_{\mathbf{k}'\lambda'} k^2 ,$$

to annihilate the terms in $a_{\mathbf{k}\lambda} a_{\mathbf{k}'\lambda'} e^{-i(\omega_k + \omega_{k'})t}$ and $a_{\mathbf{k}\lambda}^\dagger a_{\mathbf{k}'\lambda'}^\dagger e^{i(\omega_k + \omega_{k'})t}$, while the last step uses the commutation relations of the a, $a^\dagger$. This expression then contains within it the quantization condition of Planck that each photon mode contributes energy ω_k for each photon populating that mode. This explains the choice of normalization introduced in Eq. 33. The Hamiltonian thus has the form of Eq. 1.38 with the exception of the fact that it contains a summation over the zero-point energies of the mode oscillators. These energies are the precise equivalent here of the zero-point motion found for the ground state in the usual mechanical oscillator. The sum over them gives an infinite, constant term in the energy, which we shall ignore on the grounds that we are only interested in the energy *differences* that result from adding or subtracting photons. Put another way, we insist that the vacuum state be a state of zero energy by definition and achieve this by dropping the additive constant $\frac{1}{2}\Sigma\omega_k$ in Eq. 45.

Since the photon field $\mathbf{A}$ will produce or annihilate photons one at a time, we can use Eqs. 1.27 to summarize the action of this field on each mode, given the initial population of the mode, namely

$$\langle n_{k\lambda} + 1 | \mathbf{A}(\mathbf{r}, t) | n_{k\lambda} \rangle = \sqrt{n_{k\lambda} + 1} \, \sqrt{\frac{1}{2\omega_k L^3}} \, \boldsymbol{\epsilon}_{k\lambda}^* \, e^{-i\mathbf{k}\cdot\mathbf{r} + i\omega_k t} \qquad (46a)$$

and

$$\langle n_{k\lambda} - 1 | \mathbf{A}(\mathbf{r}, t) | n_{k\lambda} \rangle = \sqrt{n_{k\lambda}} \, \sqrt{\frac{1}{2\omega_k L^3}} \, \boldsymbol{\epsilon}_{k\lambda} \, e^{i\mathbf{k}\cdot\mathbf{r} - i\omega_k t} . \qquad (46b)$$

Here we have allowed for the possibility that in some applications (e.g., for circular polarization) it may be desirable to use a complex unit vector to describe the polarization state, whence $\boldsymbol{\epsilon}_{k\lambda}^*$ in Eq. 46a. For many cases of interest the photon population change is—in one direction or the other—between the vacuum state for the mode in question and a state of unit occupancy of the mode; then the number-counting radicals in Eqs. 46 are unity. If more photons are present, then the transition is enhanced by these square-root factors. This feature was first treated—even to the quantitive level—by Einstein in 1916, long before second quantization (or even *first* quantization) was developed. We shall return to these points below after discussing the interaction of the electromagnetic field with charged matter in the next subsection.

Before leaving the quantization of the electromagnetic field we wish to draw a clear line between those features that arose from the restriction of the radiative electromagnetic field to two components and those that are general to boson systems. In the former category are the gauge freedom of Eqs. 12 and 13 and the transversality that we found in our particular gauge in Eqs. 16 and 24. This reduced the degrees of freedom for the photon polarization vector from three to two ($\lambda = 1,2$), since the possibility $\boldsymbol{\epsilon}_{k3} \| \mathbf{k}$ was ruled out by Eq. 24. Had we chosen to expand $\mathbf{A}$ in *angular*-momentum eigenstates, rather than in *linear*-momentum eigenstates as in Eq. 33, we would have determined that the photon has intrinsic spin $S = 1$ (in addition to the units of orbital angular momentum it may carry). One would expect to find three polarization states corresponding to the three projections of $S = 1$, namely $S_z = 1, 0, -1$, but one of these is in fact eliminated by $m_\gamma = 0$. On the other hand, the quantization that we found for the field is general, applying for all bosons, as are the statistical properties contained in the commutation relations for $a, a^\dagger$ and expressions such as Eqs. 46. These features will be the same (after suitable modifications are made for internal degrees of freedom) for any particle of boson character, including, for example, mesons with spin zero or one.

3.2 THE INTERACTION OF THE ELECTROMAGNETIC FIELD WITH CHARGED MATTER

We consider the behavior of a spinless point charged particle of mass m and charge q in the presence of an electromagnetic field. It is well known in nonrelativistic quantum theory that the Hamiltonian in this case is

$$H = \frac{(\mathbf{p} - q\mathbf{A})^2}{2m} + q\phi = \frac{\mathbf{p}^2}{2m} - q\,\frac{\mathbf{p}}{m}\cdot\mathbf{A} + \frac{q^2\mathbf{A}^2}{2m} + q\phi \,, \qquad (47)$$

where $\mathbf{p}$ is the particle momentum operator conjugate to the position variable ($\mathbf{p} = -i\boldsymbol{\nabla}$). The quantities ϕ and $\mathbf{A}$ are the electromagnetic scalar and vector potentials. These are taken as classical, external fields for problems such as the hydrogen atom (where $\phi = e/r$, $q = -e$, and $\mathbf{A} = 0$, e being the absolute magnitude of the electron charge) or a charged particle in an external magnetic field (where $\mathbf{A} \neq 0$ and $\phi = 0$). By "external" we mean here that the fields are provided by an outside agency and do not figure as part of the physical system whose dynamics we are considering. Now when we wish to take into account photon creation or annihilation as part of the dynamical treatment, the vector potential $\mathbf{A}$ will also include the radiation field whose second quantization we have just discussed (though ϕ in the transverse or radiation gauge is still just given as the Coulomb field arising from the charges).

The Hamiltonian with the interaction between charges and electromagnetic field—now including radiation or photons—is still just as in Eq. 47, essentially as implied by the classical theory. The first term in Eq. 47 is just the usual nonrelativistic kinetic energy, while the second term is the interaction energy containing the convection piece $q\mathbf{p}/m$ and the transverse vector potential $\mathbf{A}$. The part of H quadratic in $\mathbf{A}$, namely $q^2\mathbf{A}^2/2m$, is important in certain situations where an external field of sufficient strength to require its inclusion is present. It may also enter in radiation problems where two photons are involved, since after second quantization that is the effect of a term of the form $\mathbf{A}^2$. It is then treated by a straightforward extension of the techniques developed here; since by and large we shall not deal with such situations, we do not consider it further at this point. Last, H contains the Coulomb energy $q\phi$ for a charged particle in the presence of a scalar potential.

Thus of the terms in H those pertaining to the single-photon interaction Hamiltonian are

$$H' = q\phi - q\,\frac{\mathbf{p}}{m}\cdot\mathbf{A} \,. \qquad (48)$$

If we think in terms of charge and current distributions ρ and $\mathbf{j}$, this generalizes to

$$H' = \int (\rho\phi - \mathbf{j} \cdot \mathbf{A}) \, d\mathbf{r} . \tag{49}$$

This may be inferred from the classical case, where $\rho = q\delta(\mathbf{r} - \mathbf{r}_p)$ and $\mathbf{j} = q(\mathbf{p}/m)\delta(\mathbf{r} - \mathbf{r}_p)$ for particle position $\mathbf{r}_p$, or from the matrix element in quantum mechanics for the single-particle case described by a wave function $\psi(\mathbf{r})$, where

$$\rho(\mathbf{r}) = q\psi^*(\mathbf{r})\psi(\mathbf{r}) \tag{50}$$

and

$$\mathbf{j}(\mathbf{r}) = \frac{q}{2im} \{ \psi^*(\mathbf{r})\nabla\psi(\mathbf{r}) - [\nabla\psi^*(\mathbf{r})]\psi(\mathbf{r}) \} . \tag{51}$$

Equation 49 is more general than these origins in that for other more complex systems, such as a many-particle case for example, it will still apply but with a charge density and current appropriate to the system in question. These may even be left as operator forms, with H' to be taken between ket and bra for the full system. The density and current are of course expected to satisfy a continuity equation—local charge conservation

$$\frac{\partial\rho}{\partial t} + \nabla \cdot \mathbf{j} = 0 \tag{52}$$

—if they have been appropriately defined for whatever the system under consideration.

As we have implied, the Coulomb part of the interaction in Eq. 49, namely

$$H'_{\text{Coulomb}} = \int \rho\phi \, d\mathbf{r} , \tag{53}$$

contains the standard static interaction that one treats for example in the quantum study of the hydrogen atom by studying the motion of a charged particle in a Coulomb potential. The interaction of the charged particle with the (transverse) or vector potential $\mathbf{A}$ allows for the description of the motion of a particle in a static magnetic field, for instance. But what is much more to the point for our present purposes is that this interaction term

$$H'_{\text{tr}} = - \int \mathbf{j} \cdot \mathbf{A} \, d\mathbf{r} \tag{54}$$

permits the handling of the absorption or production of photons through coupling to the current $\mathbf{j}$. The current in Eq. 54 leads to the same result upon integration as would the transverse current

$$j_k^{tr} \equiv \sum_{l=1}^{3} \int \delta_{kl}^{tr}(\mathbf{r} - \mathbf{r}')j_l(\mathbf{r}', t) \, d\mathbf{r}' , \tag{55}$$

in terms of the transverse δ-function of Eq. 42b. This current satisfies

$$\mathbf{\nabla} \cdot \mathbf{j}^{tr} = 0 , \tag{56}$$

as is immediately seen from Eq. 42b with 21. In terms of it, the interaction in Eq. 54 can also be written

$$H_{tr}' = -\int \sum_{lk} j_l(\mathbf{r}', t) \left[\delta_{lk}\delta(\mathbf{r} - \mathbf{r}') - \frac{1}{4\pi} \mathbf{\nabla}_l' \mathbf{\nabla}_k \frac{1}{|\mathbf{r} - \mathbf{r}'|} \right]$$

$$\times A_k(\mathbf{r}, t) \, d\mathbf{r} \, d\mathbf{r}' ; \tag{57}$$

this follows because integrating by parts on the last term to transfer the action of $\mathbf{\nabla}_k$ to the potential and using Eq. 16 causes the last term to vanish and reduces Eq. 57 to 54.

The Lagrangian form that takes this transverse interaction into account is

$$L = L_{\text{free field}} + \int \mathbf{j}^{tr}(\mathbf{r}, t) \cdot \mathbf{A}(\mathbf{r}, t) \, d\mathbf{r} , \tag{58}$$

for which by the usual variational procedure we immediately obtain the equation of motion for the field

$$\Box \mathbf{A} = -\mathbf{j}^{tr} , \tag{59}$$

which is consistent with the transversality condition of Eq. 16, that is, the divergences of both sides of Eq. 59 manifestly vanish. The new conjugate field variable for L is

$$\boldsymbol{\pi} = \frac{\partial \mathcal{L}}{\partial \dot{\mathbf{A}}} = \frac{\partial \mathcal{L}_{\text{free field}}}{\partial \dot{\mathbf{A}}} = \boldsymbol{\pi}_{\text{free field}} , \tag{60}$$

a conclusion that greatly simplifies our considerations here, but of course depends on the specific Lagrangian we are treating. As a consequence of this simplification all our previous results for quantization of the free electromagnetic field apply here in the presence of the transverse interaction as well.

Last, we note—though we shall make no use of it here—that the structure of the interaction in Eq. 49 is the scalar contraction of the two four-vectors $(\mathbf{j}, \rho)$ for the four-current and $(\mathbf{A}, \phi)$ for the four-potential, so that this theory has suitable elements of covariance for use in a relativistic situation. The transverse gauge condition is not a covariant form, however, and to improve on this situation requires a more elaborate treatment of the gauge problem.

Once we know the form of the interaction Hamiltonian we are immediately in a position to calculate the transition probability w_{n0} for a system of charges in interaction with the electromagnetic field on the basis of Fermi's "golden rule".

$$w_{n0} = 2\pi |\langle n|H'|0\rangle|^2 \rho_n , \qquad (61)$$

where the transition is between an initial state 0 and a final state n for the combined system of charges and photons, and ρ_n is the density of final states for the total system which involves a continuum of final states. The transition matrix element is for the interaction Hamiltonian of Eq. 54, but we drop the "transverse" subscript on H'. For photon emission into the continuum the initial state is $|0\rangle = |\alpha;0\rangle$, where α is the initial state of the charged particles and we have indicated that no photons are present to begin with, while the final state is $|n\rangle = |\beta;1_{k\lambda}\rangle$, where β is the final state of the charges and where one photon, of momentum $\mathbf{k}$ and polarization state λ, has been emitted. Separating the spaces of the charges and the photons, we have

$$\langle n|H'|0\rangle = -\int \langle \beta|\mathbf{j}(\mathbf{r},t)|\alpha\rangle \cdot \langle 1_{k\lambda}|\mathbf{A}(\mathbf{r},t)|0\rangle \, d\mathbf{r} , \qquad (62)$$

where, quite generally, the time dependence in the current is

$$\langle \beta|\mathbf{j}(\mathbf{r},t)|\alpha\rangle = \langle \beta|\mathbf{j}(\mathbf{r})|\alpha\rangle \, e^{i(E_\beta - E_\alpha)t} \qquad (63)$$

for the usual time phase of a matrix element in quantum mechanics (or exploiting invariance under translations in time). Here E_α and E_β are the energy eigenvalues of the initial and final states of the system of charges. The photon matrix element is given by Eq. 46a, in this case for $n_{k\lambda} = 0$. Note that energy conservation causes the oscillatory time factors to cancel, a feature which is anticipated in the "golden rule" of Eq. 61 and incorporated there.

Equations 61 and 62 provide the rigorous justification for what is sometimes called "semiclassical radiation theory." There the electromagnetic radiation field $\mathbf{A}$ is not quantized, but it is tacitly assumed that it can change the number of photons from zero to one or one to zero, for creation and annihilation respectively, with the counting factors of Eq. 46 consequently equal to unity. Further, the time dependence of the radiation field is there put in as a driving term, and the theory generates photon absorption or induced emission while spontaneous emission of photons (whereby the system ejects a photon without a time-dependent driving term) remains artificial. Once $\mathbf{A}$ is second-quantized, all these ingredients of the theory emerge automatically, as we see here.

Of course, if $n_{k\lambda}$ photons had initially been present in the mode $k\lambda$, the transition probability w_{nj} between any two states n and j would be enhanced by a factor $n_{k\lambda} + 1$ from Eq. 46a, that is $(n \leftarrow j)$

$$w_{nj}(n_{k\lambda} + 1 \leftarrow n_{k\lambda}) = (n_{k\lambda} + 1) \cdot w_{nj}(1_{k\lambda} \leftarrow 0), \tag{64}$$

for the emission of a photon when $n_{k\lambda}$ were initially present. The absorption of a photon is naturally proportional to the number of photons initially present (see Eq. 46b), so $(j \leftarrow n)$

$$w_{jn}(n_{k\lambda} - 1 \leftarrow n_{k\lambda}) = n_{k\lambda} w_{jn}(0 \leftarrow 1_{k\lambda}). \tag{65}$$

Last, the two transition rates for $0 \leftrightarrow 1_{k\lambda}$ photon occupation between arbitrary states j and n of the system of charges are equal,

$$w_{jn}(0 \leftarrow 1_{k\lambda}) = w_{nj}(1_{k\lambda} \leftarrow 0). \tag{66}$$

This comes about because the matrix elements of the form of Eq. 62 that enter in the transition rates will differ only by phase factors on grounds of time-reversal invariance for the two directions of photon $+ S(\alpha) \leftrightarrow S(\beta)$, where $S(\eta)$ denotes the system of charges in the state η. Further, the factor ρ_n for phase space reverses its role between emission and absorption: For emission ρ_n is the density of final states spread over the photon continuum. For absorption one envisages a distribution of photon frequencies within an incident beam, and in deriving the "golden rule" one sums over this *initial* continuum of states in order to describe a transition which excites the charge system from one level to another. The density of states ρ_n thus enters, where n is the initial state for absorption $n \rightarrow j$ or photon $+ S(\alpha) \rightarrow S(\beta)$, and is the final state for emission $j \rightarrow n$ or $S(\beta) \rightarrow S(\alpha) + $ photon, where ρ_n enters as a final-state phase-space factor. As a result of this detailed-balancing symmetry, the ratio of emission to absorption is given by the ratio of occupancy factors only,

$$\frac{w_{nj}(n_{k\lambda} + 1 \leftarrow n_{k\lambda})}{w_{jn}(n_{k\lambda} - 1 \leftarrow n_{k\lambda})} = \frac{n_{k\lambda} + 1}{n_{k\lambda}}. \tag{67}$$

3.3 ASSEMBLAGES OF BOSONS

3.3.1 Blackbody Radiation

Taking photons as an example, we now explore features arising from Bose–Einstein statistics for ensembles of bosons. Even though these bosons are noninteracting, we shall find that their statistical behavior generates new properties. The first of these exploits the ratio of emission to absorption between a pair of levels n and j within the atoms of a gas, say, in order to understand the necessary conditions for the atoms and the radiation field to coexist in equilibrium. The energies of the levels are E_α and E_β, with $E_\alpha < E_\beta$, as shown in Figure 3.1, so that photon emission or absorption can

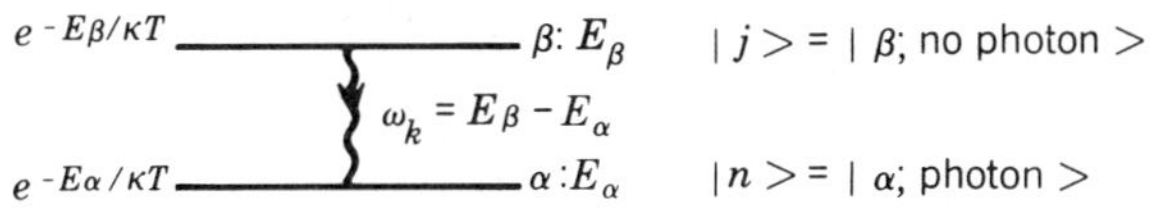

$e^{-E_\beta/\kappa T}$ _____ $\beta: E_\beta$ $|j> = |\beta;$ no photon $>$

$\omega_k = E_\beta - E_\alpha$

$e^{-E_\alpha/\kappa T}$ _____ $\alpha: E_\alpha$ $|n> = |\alpha;$ photon $>$

FIGURE 3.1 Two levels α and β in an atomic system, with energies E_α, E_β and Boltzmann occupation factors $e^{-E_\alpha/\kappa T}$, $e^{-E_\beta/\kappa T}$, where κ is the Boltzmann constant and T is the temperature.

proceed through $S(\beta) \leftrightarrow S(\alpha) +$ photon with energy conservation, the photon frequency or energy being

$$\omega_k = E_\beta - E_\alpha . \tag{68}$$

Now the relative population of the atomic levels when they are in equilibrium at temperature T is given by the Boltzmann factors $e^{-E_\alpha/\kappa T}$ and $e^{-E_\beta/\kappa T}$, where κ is Boltzmann's constant. When emission and absorption are possible, general equilibrium requires that there be a balance between them, so

$$w_{nj}(\bar{n}_{k\lambda} + 1 \leftarrow \bar{n}_{k\lambda}) e^{-E_\beta/\kappa T} = w_{jn}(\bar{n}_{k\lambda} - 1 \leftarrow \bar{n}_{k\lambda}) e^{-E_\alpha/\kappa T} , \tag{69}$$

where $\bar{n}_{k\lambda}$ refers to the average occupancy of the photon mode $k\lambda$ at a given instant. Then, from Eqs. 64 to 68,

$$\frac{\bar{n}_{k\lambda} + 1}{\bar{n}_{k\lambda}} = e^{(E_\beta - E_\alpha)/\kappa T} = e^{\omega_k/\kappa T} , \tag{70}$$

$$\bar{n}_{k\lambda} = \frac{1}{e^{\omega_k/\kappa T} - 1} , \tag{71}$$

which is Planck's law for blackbody radiation, precisely the equilibrium situation we imagined. Einstein in 1916 established the necessity of the occupancy factors of Eqs. 64 and 65—without any of the considerations of formal second quantization—by reversing the argument as it was used here and showing that Planck's law, together with the Boltzmann distribution, requires that photon emission be proportional to $n_{k\lambda} + 1$ and absorption to $n_{k\lambda}$ when $n_{k\lambda}$ photons are present initially. This in turn requires the notion of *induced emission*, that is, in addition to the spontaneous photon decay represented by "1" in $n_{k\lambda} + 1$, the very presence of $n_{k\lambda}$ photons in the mode $k\lambda$ will speed up the photon *emission* by $n_{k\lambda}$, just as the presence of $n_{k\lambda}$ photons hastens absorption of one of them by such a factor.

From Planck's law, Eq. 71, we can derive the energy density (per unit volume) for photons of a given energy ω at temperature T,

$$\rho(\omega_k, T) = \sum_{\lambda=1,2} \int \frac{d\hat{\mathbf{k}}}{(2\pi)^3} k^2 \frac{dk}{d\omega_k} \frac{\omega_k}{e^{\omega_k/\kappa T} - 1}$$

$$= 2 \times 4\pi \left(\frac{\omega_k}{2\pi}\right)^3 \frac{1}{e^{\omega_k/\kappa T} - 1} = \frac{1}{\pi^2} \frac{\omega_k^3}{e^{\omega_k/\kappa T} - 1} , \qquad \omega_k = k , \tag{72}$$

where we have included a factor of the density of states, with solid-angle integration over the direction of **k** leading to a factor of 4π, and a summation over the two polarization states λ, which contributes a factor of two to the result. The total radiated intensity at temperature T is

$$I(T) = \int_0^\infty \rho(\omega_k, T)\, d\omega_k = \frac{\pi^2(\kappa T)^4}{15} = \sigma T^4 , \qquad (73)$$

where $\sigma = \pi^2 \kappa^4/15$ is Stefan's constant. This result for the intensity agrees with measurement in magnitude; that is, the coefficient of the T^4 dependence for the intensity is indeed given by Stefan's constant. Among other things, this verifies that photons have two spin projection values rather than the three that would otherwise be expected for a spin-one particle. In our terms we thus corroborate that photons are transverse. As we have noted briefly in Section 3.1, this is a consequence of the vanishing mass of the photon. The vanishing of the mass rules out any spin projection that is not parallel or antiparallel to the photon momentum **k**, since intermediate cases, involving as it were an arbitrary angle between photon spin and **k**, would not appear the same in all frames, as they must for a particle traveling with speed c.

3.3.2 Lasers

The situation described by Eqs. 69 and 70, and leading to Planck's law in Eq. 71, pertains to a condition of equilibrium, whence the relevance of the Boltzmann factors in Eq. 70 to describe the population of the atomic states. A quite different case results if one can bring about a significant deviation from the Boltzmann factors by using optical pumping to create a population inversion whereby an excited level is populated far in excess of what is reached in the Boltzmann distribution. This is the scheme that allows for the action of a laser, sketched schematically in Figure 3.2 for a system with

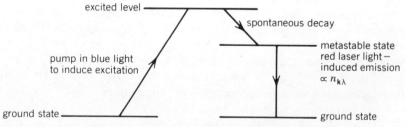

FIGURE 3.2 A schematic picture of a three-level laser. Energetic, blue light impinges on atoms in their ground state and elevates them to an excited state. This then decays spontaneously to a metastable state (whose presence depends of course on a correct choice of material). With time more and more atoms find themselves in this metastable state, whereupon the presence of a photon (from a spontaneous decay of a metastable state, say) in the appropriate mode to give decay to the ground state will trigger an avalanche of induced emission in this mode.

three active levels. The general strategy of the laser is to find a system with, say, three levels, the middle one of which is metastable. One then shines sufficiently energetic photons on the system (blue light, for example) to excite the upper level, which in turn decays spontaneously to the metastable state, so that a population of that state builds up with time: Then a photon matched to the energy difference between the ground state and the meta-stable state, for instance a photon from spontaneous decay of the metastable state, will induce the emission of another photon in this mode so that an exponential avalanche of photons, all in the same mode, builds up. The process is further aided by enclosing the active atoms between two parallel mirrors to avoid wasting photons that otherwise would escape from the system. The creation of a population inversion, moreover, need not involve optical absorption, but can instead be achieved through other mechanisms for the initial pumping, such as by passing the relevant molecules through an inhomogeneous electric field, or by using spin-resonance methods.

Once the all-important population inversion is achieved, the change in energy density as the monochromatic photons propagate through the medium is, as a function of time,

$$\frac{dD(\omega_k, t)}{dt} \propto N_\beta w(n_{k\lambda} + 1 \leftarrow n_{k\lambda}) - N_\alpha w(n_{k\lambda} - 1 \leftarrow n_{k\lambda})$$

$$\propto (N_\beta - N_\alpha)D(\omega_k, t) , \tag{74}$$

where N_α, N_β are the populations of the lower and upper level and we have noted that the rates for emission or for absorption are proportional, essentially, to the number of photons initially present in the relevant mode and thus are proportional to the energy density $D(\omega_k, t)$ itself; see Eqs. (64), (65), and (72). Thus from Eq. 74

$$D(\omega_k, t) = D(\omega_k, t = 0)e^{\gamma t} , \tag{75}$$

where

$$\gamma \propto N_\beta - N_\alpha > 0 , \tag{76}$$

yielding an exponentially growing cascade if population inversion is ach-ieved.

3.3.3 Photon Fluctuations

The special features of boson statistics, as opposed to statistical features based on random events, can be made more evident by considering fluctua-tions in distribution numbers. We might at first suppose that photons were "randomly" distributed in their occupation of various modes, that is, that they do not obey boson statistics but instead satisfy a Poisson statistical distribution. (The Poisson case is chosen arbitrarily to show what happens for conventional statistically random situations.) The probability of finding n photons in a given mode (we suppress the mode label $\mathbf{k}\lambda$ everywhere) is then not related to Eq. 71 but rather given by

$$p(n, \mu) = \frac{\mu^n}{n!} e^{-\mu} , \tag{77}$$

where μ is the average number of photons in the mode, as can be verified from

$$\bar{n} \equiv \sum_{n=0}^{\infty} np(n, \mu) = \sum_{n=0}^{\infty} n \frac{\mu^n}{n!} e^{-\mu} = e^{-\mu} \mu \frac{d}{d\mu} e^{\mu} = \mu . \tag{78}$$

Now the average of the occupation number squared is

$$\overline{n^2} = \sum_{n=0}^{\infty} n^2 p(n, \mu) = e^{-\mu} \left(\mu^2 \frac{d^2}{d\mu^2} + \mu \frac{d}{d\mu} \right) e^{\mu} = \mu^2 + \mu , \tag{79}$$

so that the fluctuations in occupation number are equal to the average occupation number,

$$(\Delta n)^2 = \overline{n^2} - \bar{n}^2 = \mu = \bar{n} . \tag{80}$$

However, we know that photons do obey Bose–Einstein statistics. How will this change the fluctuations? The relevant distribution for boson occupation numbers is dependent on the energy of the mode occupied and is given by

$$b(n, \omega) = (1 - e^{-\omega/\kappa T}) e^{-n\omega/\kappa T} . \tag{81}$$

[This quantity is simply the canonical ensemble partition function for n bosons as derived in statistical mechanics, here normalized so that $\sum_{n=0}^{\infty} b(n, \omega) = 1$; see Eqs. 7.58 and 7.59 below. Note that the Poisson distribution in Eq. 77 was taken to represent a situation of random statistics, but could also be viewed as representing the partition function for classical particles, similarly normalized.] In the Poisson case, we expressed the distribution in terms of the average number of photons in the mode, $p(n, \mu) = p(n, \bar{n})$; that could be done for this distribution as well, $b(n, \omega) = \bar{n}^n(\bar{n} + 1)^{-n-1}$, but it is more straightforward for us here to incorporate directly the fact that the photon energy determines this average number through Planck's law, Eq. 71. We must now verify that Eq. 81 indeed yields this average occupancy. To simplify notation we introduce

$$x = e^{-\omega/\kappa T} , \tag{82}$$

in terms of which

$$\bar{n} = \sum_{n=0}^{\infty} nb(n, \omega) = \sum_{n=0}^{\infty} n(1-x)x^n = (1-x)x \frac{d}{dx} \frac{1}{1-x}$$

$$= \frac{x}{1-x} = \frac{1}{e^{\omega/\kappa T} - 1} , \tag{83}$$

just as in Eq. 71. Then

$$\overline{n^2} = \sum_{n=0}^{\infty} n^2 b(n, \omega) = (1-x)\left(x^2 \frac{d^2}{dx^2} + x \frac{d}{dx}\right)\frac{1}{1-x}$$

$$= \frac{x(1+x)}{(1-x)^2} \tag{84}$$

and

$$(\Delta n)^2 = \overline{n^2} - \bar{n}^2 = \frac{x}{(1-x)^2} = \bar{n}(1+\bar{n}) \tag{85}$$

instead of Eq. 80 for the Poisson distribution. (This result for photon fluctuations was obtained by Einstein, using the known Planck distribution of Eq. 71, before the evolution of modern quantum theory.) Thus instead of having the usual result for random systems that

$$\frac{\Delta n}{\bar{n}} \xrightarrow[\bar{n} \to \infty]{} \frac{1}{\sqrt{\bar{n}}} , \tag{86}$$

we find for bosons

$$\frac{\Delta n}{\bar{n}} \xrightarrow[\bar{n} \to \infty]{} 1 . \tag{87}$$

The bosons are not at all randomly distributed, and indeed in the equilibrium situation exhibit a correlational tendency: The presence of a photon in equilibrium in a given mode makes it more likely that another photon will be found in that mode even though there is no interaction between the photons. This tendency is sometimes referred to as photon "bunching", although that term is more precisely reserved for the tendency of photons from equilibrium sources to correlate in *time*. In certain nonequilibrium situations, there is also "antibunching" in time. [See, for example, H. Paul, Rev. Mod. Phys. **54**, 1061 (1982).]

3.3.4 The Effect of Hanbury Brown and Twiss

Another manifestation of this tendency of photons to correlate is found in the effect of Hanbury Brown and Twiss, in which correlations are measured in the distribution of photons emitted by a body into the various available modes, and these correlations are related to the size of the body. To see the central point in rough outline, consider the boson fluctuations of Eq. 85, where we now require a mode label k,

$$(\Delta n_k)^2 = \bar{n}_k(1 + \bar{n}_k) , \tag{88}$$

and suppose that in a particular measurement we sum, inevitably, over a certain number of adjacent states within the pass band α of our photon detector, the frequencies in α being between ω_α and $\omega_\alpha + \delta\omega_\alpha$. It is this consideration of a bin of finite width within which photons are measured that introduces the density of states for emitted photons. This in turn is what causes the size of the emitting system to enter the problem. Note that we tacitly assume that emission is in a broad, continuous band.

The number of photons measured in the bin α is

$$N_\alpha \equiv \sum_{k \in \alpha} n_k . \tag{89}$$

The different modes for photons (i.e. different cases of k) are statistically independent, so that the *band* fluctuations are

$$(\Delta N_\alpha)^2 = \rho_\alpha \, \delta\omega_\alpha \, \bar{n}_\alpha(1 + \bar{n}_\alpha) , \tag{90}$$

where $\bar{n}_\alpha$ is the common, average value for n_k with k in α, and ρ_α is the density of states for the bin α,

$$\rho_\alpha = 2 \times 4\pi \, \frac{L^3 \omega_\alpha^2}{(2\pi)^3} , \tag{91}$$

as we recall for example from Eq. 72. The average occupation value for the bin α is of course

$$\bar{N}_\alpha = \rho_\alpha \, \delta\omega_\alpha \, \bar{n}_\alpha , \tag{92}$$

and the average energy value for the bin is

$$\bar{E}_\alpha = \bar{N}_\alpha \omega_\alpha , \tag{93}$$

so that the number fluctuations are

$$(\Delta N_\alpha)^2 = \bar{N}_\alpha\left(1 + \frac{\bar{N}_\alpha}{\rho_\alpha \, \delta\omega_\alpha}\right) , \tag{94}$$

and the energy fluctuations are

$$(\Delta E_\alpha)^2 = \omega_\alpha \bar{E}_\alpha + \frac{\pi^2 \bar{E}_\alpha^2}{L^3 \omega_\alpha^2 \, \delta\omega_\alpha} . \tag{95}$$

Thus once one has selected a band frequency and width, ω_α and $\delta\omega_\alpha$, measurement of the average energy and fluctuations within the band yields information on the system size L for the system emitting the photons.

The same issue can be treated from a somewhat different point of view if we consider a system of volume V that ejects two identical particles (we now extend our considerations to fermions as well), say of momentum $\mathbf{k}$ and $\mathbf{k}'$, by some incoherent mechanism. This might come about, for example, as the result of some reaction that deposits energy in a certain volume of the system, raising the temperature locally there. For the incoherent ejection the pertinent construct describing the joint emission of the particles will have the general form

$$f(\mathbf{k}, \mathbf{k}') = \int_V d\mathbf{r} \int_V d\mathbf{r}'\, D(\mathbf{r}, \mathbf{r}')$$

$$\times \tfrac{1}{2}\left| e^{-i\mathbf{k}\cdot\mathbf{r}} e^{-i\mathbf{k}'\cdot\mathbf{r}'} \pm e^{-i\mathbf{k}'\cdot\mathbf{r}} e^{-i\mathbf{k}\cdot\mathbf{r}'} \right|^2 . \qquad (96)$$

The function $D(\mathbf{r}, \mathbf{r}')$, with support for $\mathbf{r}, \mathbf{r}'$ in V, embodies the incoherent statistical nature of the ejection mechanism envisioned here, and thus is the implicit link to the treatment in Eqs. 88–95 based on equilibrium of a system at a given temperature.* The rest of the integrand contains a suitably symmetrized or antisymmetrized plane-wave state for two ejected bosons or fermions (see Figure 3.3). Here we broaden our considerations to fermions in order to emphasize the similarity between bosons and fermions with regard to the general aspects of the correlations present, while showing the

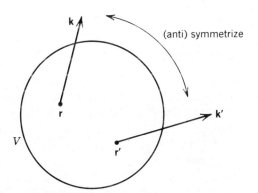

FIGURE 3.3 The ejection of two identical particles in plane-wave states of momentum k and k' from points r and r' within a system of volume V; the wave function must be symmetrized or antisymmetrized depending on whether the particles are bosons or fermions.

*On the phase averaging implied by incoherence in the source, see G. N. Fowler and R. M. Weiner, Phys. Lett. **70B**, 201 (1977).

distinctions between them arising from the difference in statistics. To focus on this point we have treated the fermions as if in a spin-parallel condition and so possessing a fully antisymmetric spatial wave function.

The (anti)symmetrization in Eq. 96 leads to an interference effect,

$$f(\mathbf{k}, \mathbf{k}') = \int_V d\mathbf{r} \int_V d\mathbf{r}' \, D(\mathbf{r}, \mathbf{r}')$$

$$\times [1 \pm \text{Re}(e^{i\mathbf{k}'\cdot\mathbf{r}} e^{i\mathbf{k}\cdot\mathbf{r}'} e^{-i\mathbf{k}\cdot\mathbf{r}} e^{-i\mathbf{k}'\cdot\mathbf{r}'})]$$

$$= \int_V d\mathbf{r} \int_V d\mathbf{r}' \, D(\mathbf{r}, \mathbf{r}')\{1 \pm \cos[(\mathbf{k} - \mathbf{k}') \cdot (\mathbf{r} - \mathbf{r}')]\} \,, \quad (97)$$

where the second term indicates a correlation between $\mathbf{k}, \mathbf{k}'$ that reflects information on the system size $|\mathbf{r} - \mathbf{r}'|$ as averaged with a weighting of the incoherent particle ejection mechanism D. Note that the quantity in braces in Eq. 97 is maximal for $\mathbf{k} = \mathbf{k}'$ for bosons, but zero there for fermions, as one would expect from considerations of "bunching" or the exclusion principle, respectively.

The result of Eq. 97 shows that the connection between the bin ocupancy of ejected identical particles and the system size pertains to fermions as well as to bosons. For bosons, in fact, the phenomenon can also be derived on the basis of classical properties of the electromagnetic field [see, for example, R. Hanbury Brown and R. Q. Twiss, Nature **178**, 1447 (1956), and E. Purcell, Nature **178**, 1449 (1956)]. The method is so general that it can be applied to photon emission by stars to measure star size [for example, R. Hanbury Brown and R. Q. Twiss, Nature **177**, 27 (1956)] or to pion emission by highly excited nuclei to measure the size of the nuclear hot spot [for example, S. Y. Fung *et al.*, Phys. Rev. Lett. **41**, 1592 (1978)].

3.3.5 The Coherent State

The term *the coherent state*—with the definite article—refers to a particular form of quantum state,* with somewhat unusual properties, that allows us to discuss in a straightforward manner the classical limit of the second-quantized electromagnetic field. Note that the usual intuition concerning the classical limit, namely that it should obtain when the quantum numbers are large $(n \rightarrow \infty)$, does not readily apply here, since the diagonal elements of the vector potential $\langle n|\mathbf{A}|n \rangle$, from which the $\mathbf{E}$ and $\mathbf{B}$ fields are derived, vanish for any value of n, as will be recalled from Eqs. 46. Since none of the second-quantized operators $\mathbf{A}$, $\mathbf{E}$, or $\mathbf{B}$ of Eqs. 33 with 27 and 28 commutes with the photon-number operators $N_{\mathbf{k}\lambda} = a_{\mathbf{k}\lambda}^{\dagger} a_{\mathbf{k}\lambda}$, it is clear that one cannot expect to have well-defined field strengths, as in the classical theory,

---

*See, for example, R. J. Glauber, Phys. Rev. **131**, 2766 (1963).

together with sharp photon number. Thus one must either contemplate extensions to off-diagonal elements or relax the sharpness of the number of bosons in the state under consideration. The use of the coherent state follows the second route. (This, incidentally, prepares the way somewhat for the exploitation of states without well-defined particle number which will appear in Chapter 5 in the treatment of pairing phenomena. We do not here explore possible uses of the coherent state for treating classical statistical features of ensembles of photons.) The coherent state is a pure, unmixed quantum state—but without sharp number eigenvalue—that yields the classical limit.

We consider a single-mode oscillator or the boson occupation of a single quantum state. For any complex number α, we define *the coherent state*

$$|\alpha\rangle \equiv \sum_{n=0}^{\infty} \frac{\alpha^n}{\sqrt{n!}} \, e^{-\frac{1}{2}|\alpha|^2} |n\rangle \,, \tag{98}$$

where $|n\rangle$ is a Fock-space ket for occupation number n. (If many single-particle modes enter, then one defines a coherent state, Eq. 98, for each mode; we here suppress any mode label.) Since Eq. 98 sums over all possible values of the occupation number, it does not of course have well-defined boson number. For $\alpha = 0$, the coherent state is indeed the ground state $|\alpha = 0\rangle = |n = 0\rangle = |0\rangle$; among other things this means that there is no ambiguity in notation brought about by the definition of Eq. 98. For $\alpha \neq 0$ the coherent state is a superposition of states of different energies (as well as boson occupation numbers, of course). The probability for finding a given energy E_n is

$$P(E_n) = |\langle n|\alpha\rangle|^2 = \frac{|\alpha|^{2n}}{n!} \, e^{-|\alpha|^2} \,, \tag{99}$$

which is a Poisson distribution (whence the potential usefulness of the coherent state for describing certain statistical situations involving many photons).

The overlap of two states of the form of Eq. 98 for different complex-number labels α and β is

$$|\langle \alpha|\beta\rangle|^2 = \left| \sum_{n=0}^{\infty} \frac{(\alpha^*\beta)^n}{n!} \right|^2 e^{-|\alpha|^2} e^{-|\beta|^2} = \left| e^{\alpha^*\beta} \right|^2 e^{-|\alpha|^2} e^{-|\beta|^2}$$

$$= e^{-|\alpha - \beta|^2} \neq 0 \,, \tag{100}$$

so such states are not orthogonal, though their overlap decreases as the complex-number labels are farther removed from each other on the complex plane. If we annihilate a quantum in the mode of the coherent state,

$$a|\alpha\rangle = \sum_{n=1}^{\infty} \sqrt{n}|n-1\rangle \frac{\alpha^n}{\sqrt{n!}} e^{-\frac{1}{2}|\alpha|^2}$$

$$= \sum_{n=0}^{\infty} |n\rangle \frac{\alpha^{n+1}}{\sqrt{n!}} e^{-\frac{1}{2}|\alpha|^2} = \alpha|\alpha\rangle . \qquad (101)$$

Thus the coherent state is an eigenstate of a. Since a is not Hermitian, it can have complex eigenvalues—in this case α—and nonorthogonal eigenfunctions as in Eq. 100.

We note further that

$$\langle\alpha|a^\dagger = \langle\alpha|\alpha^* \qquad (102a)$$

and

$$\langle\alpha|a^{\dagger m}a^n|\alpha\rangle = \alpha^{*m}\alpha^n , \qquad (102b)$$

and for a function F of the creation and annihilation operators that can be expanded in a Taylor series

$$\langle\alpha|F(a^\dagger, a)|\alpha\rangle = \langle 0|F(a^\dagger + \alpha^*, a + \alpha)|0\rangle . \qquad (103)$$

In particular, for oscillator position and momentum operators of a mechanical oscillator with mass m and frequency ω (see Eqs. 1.15)

$$x \equiv (2m\omega)^{-1/2}(a + a^\dagger) , \qquad p \equiv -i(\tfrac{1}{2}m\omega)^{1/2}(a - a^\dagger) , \qquad (104a, b)$$

we have

$$\langle\alpha|x|\alpha\rangle = \left(\frac{2}{m\omega}\right)^{1/2} \text{Re } \alpha , \qquad \langle\alpha|p|\alpha\rangle = (2m\omega)^{1/2} \text{Im } \alpha , \qquad (105a, b)$$

and

$$(\Delta x)^2 = (2m\omega)^{-1} , \qquad (\Delta p)^2 = \tfrac{1}{2} m\omega , \qquad (106a, b)$$

so that the uncertainty combination for the coherent state,

$$\Delta x \, \Delta p = \tfrac{1}{2} , \qquad (107)$$

is the minimal possible value. This suggests that the coherent state is as close as possible to a classical state for the oscillator. (Note also that the number fluctuations for the coherent state behave as for the Poisson case; see Section 3.3.3.)

The time evolution dictated by the oscillator Hamiltonian of the coherent state, assuming that at time $t = 0$ the state has the value α,

$$|\psi(t = 0)\rangle = |\alpha\rangle , \qquad (108)$$

is

$$\begin{aligned}
|\psi(t)\rangle &= e^{-iHt}|\psi(t = 0)\rangle \\
&= \sum_n |n\rangle e^{-in\omega t} \frac{\alpha^n}{\sqrt{n!}} e^{-\frac{1}{2}|\alpha|^2} \\
&= \sum_n |n\rangle \frac{(\alpha e^{-i\omega t})^n}{\sqrt{n!}} e^{-\frac{1}{2}|\alpha|^2} = |\alpha e^{-i\omega t}\rangle .
\end{aligned} \qquad (109)$$

Thus the coherent state remains the coherent state with time, and its complex-number label α follows a circular path in the complex plane with angular velocity ω. This is of course the same as the behavior of the classical oscillator when Eqs. 105 are used to translate this parameter into position and momentum. Moreover, the uncertainties in x and p remain fixed while their expectation values oscillate with the classical angular velocity ω. The energy expectation value is

$$\langle \alpha|H|\alpha\rangle = \omega|\alpha|^2 , \qquad (110)$$

which is the energy of the classical oscillator with fixed amplitude α, the phase oscillating in time as in Eq. 109, just as for the classical case. Thus it is reasonable to conclude that the coherent state represents the quantum state that is closest to the description of the classical situation.

In order to use the coherent state to treat the classical limit of the electromagnetic field, we define

$$\mathbf{A}^{cl}(\mathbf{r}, t) \equiv \langle \alpha_{\mathbf{k}\lambda}|\mathbf{A}(\mathbf{r}, t)|\alpha_{\mathbf{k}\lambda}\rangle , \qquad (111)$$

where $\mathbf{A}$ is the second-quantized vector-potential operator and we shall assume that $\mathbf{A}^{cl}$ represents the classical field. For each mode the complex-number label for the corresponding coherent state is

$$\alpha_{\mathbf{k}\lambda} = |\alpha_{\mathbf{k}\lambda}|e^{i\phi_{\mathbf{k}\lambda}} = \sqrt{\bar{n}_{\mathbf{k}\lambda}}\,e^{i\phi_{\mathbf{k}\lambda}} , \qquad (112a)$$

where we have introduced the average occupation number of the cell, $\bar{n}_{\mathbf{k}\lambda}$, this last form arising from the use of Eq. 103b in

$$\bar{n}_{\mathbf{k}\lambda} = \langle \alpha_{\mathbf{k}\lambda}|a^\dagger_{\mathbf{k}\lambda}a_{\mathbf{k}\lambda}|\alpha_{\mathbf{k}\lambda}\rangle = |\alpha_{\mathbf{k}\lambda}|^2 . \qquad (112b)$$

The system energy is

$$\langle\{\alpha\}|H|\{\alpha\}\rangle = \sum_{k\lambda} \bar{n}_{k\lambda}\omega_k \,, \tag{113}$$

that is, we again find that the total energy is given by the sum of the number of quanta in each bin times the angular velocity of the bin.

The field quantity of Eq. 111 is

$$\mathbf{A}^{cl}(\mathbf{r}, t) = \sum_{k\lambda} \sqrt{\frac{1}{2\omega_k L^3}} [\sqrt{\bar{n}_{k\lambda}}e^{i\phi_{k\lambda}}\boldsymbol{\epsilon}_{k\lambda}e^{i\mathbf{k}\cdot\mathbf{r}-i\omega_k t}$$

$$+ \sqrt{\bar{n}_{k\lambda}}e^{-i\phi_{k\lambda}}\boldsymbol{\epsilon}_{k\lambda}^*e^{-i\mathbf{k}\cdot\mathbf{r}+i\omega_k t}] \tag{114}$$

and involves average occupation numbers $\bar{n}_{k\lambda}$ for the photon modes $\mathbf{k}\lambda$. From the quantity of Eq. 114 one can proceed to construct all the well-known physics of classical electromagnetic theory, thus showing how classical electromagnetism can be viewed as a limiting case of the quantized theory.

EXERCISES

3.1 For a massive vector field $A_\mu = (\mathbf{A}, \phi)$ having the Lagrangian density

$$\mathcal{L} = \tfrac{1}{2}(\boldsymbol{\nabla}\phi + \dot{\mathbf{A}})^2 - \tfrac{1}{2}(\boldsymbol{\nabla}\times\mathbf{A})^2 - \tfrac{1}{2}m^2(\mathbf{A}^2 - \phi^2) \,,$$

    (a) derive equations of motion and show that the Lorentz condition is automatically fulfilled;

    (b) carry out the quantization procedure in parallel to the electromagnetic case.

3.2 Explain in detail the origin of Eq. 66, including the role of the density-of-states factors there.

3.3 Beginning with the statement of general equilibrium, the Boltzmann distribution (Eq. 69), and Planck's law for blackbody radiation (Eq. 71), derive the photon occupancy factors of Eqs. 64 and 65.

3.4 [For the mathematically inclined: Do the integral in Eq. 73. (Hints for one possible method: Use the expansion for $(1 - e^{-x})^{-1}$; consider the Fourier series $\Sigma (1/n^4) \cos nt$.)]

3.5 **(a)** Calculate fluctuations for assemblages of particles satisfying (i) Fermi statistics, and (ii) Boltzmann statistics.

    (b) Calculate energy fluctuations for both.

3.6 Let a function $f(a, a^\dagger)$ be defined by its power series in these operators, and let z be a complex number. Show the following relations:

(a) $a^m|n\rangle = [\sqrt{n!}/(n-m)!](a^\dagger)^{n-m}|0\rangle$,

(b) $[a, f(a, a^\dagger)] = \partial f/\partial a^\dagger$ and $[a^\dagger, f(a, a^\dagger)] = -\partial f/\partial a$,

(c) $e^{za}|0\rangle = |0\rangle$ and $e^{za^\dagger}|0\rangle = \sum_{n=0}^{\infty}(z^n/\sqrt{n!})|n\rangle$,

(d) $e^{za}f(a, a^\dagger)|0\rangle = f(a, a^\dagger + z)|0\rangle$,

(e) $e^{za^\dagger a}f(a, a^\dagger)e^{-za^\dagger a} = f(ae^{-z}, a^\dagger e^z)$.

3.7 Prove the completeness of the set of coherent states,

$$\int |\alpha\rangle\langle\alpha| \frac{d^2\alpha}{\pi} = 1,$$

where integration is over the entire complex plane.

3.8 (a) Show for the electromagnetic field that

$$[N_{k\lambda}, A] \neq 0, \qquad [N_{k\lambda}, E] \neq 0, \qquad [N_{k\lambda}, B] \neq 0,$$

where $N_{k\lambda}$ is the number operator for photons of momentum **k** and polarization state λ. What is the meaning of this result?

(b) Show that the classical angular-momentum content **J** of the electromagnetic field can be divided into two terms,

$$\mathbf{J} = \int d\mathbf{r}\, \mathbf{r} \times (\mathbf{E} \times \mathbf{B}) = \mathbf{L} + \mathbf{S} = \int d\mathbf{r} \sum_{i=1}^{3} E_i(\mathbf{r} \times \nabla A_i) + \int d\mathbf{r}\, \mathbf{E} \times \mathbf{A}.$$

(c) Identifying this last term, in which there appears no explicit **r**-dependence, as the helicity content of the field, show that its quantized analog is

$$\mathbf{S} = \sum_{k,\lambda} \lambda k a_{k\lambda}^\dagger a_{k\lambda}.$$

CHAPTER **4**

Hartree–Fock Methods

We now turn to the consideration of systems of fermions with a fixed number of particles. The use of the number representation here is primarily to allow for easy antisymmetrization amongst the fermions. The central theoretical device that forms our point of departure for handling many-fermion systems is the Hartree–Fock self-consistent field. The idea here is that each fermion is viewed as traveling in an average potential. The potential is produced by the interactions between the fermion of interest and all the other fermions, and the averaging is over the motion of these other fermions. Self-consistency enters in that each fermion upon which we concentrate successively is itself helping to produce the analogous average field for each other fermion, so that orbits must be found that are compatible ultimately for all the fermions simultaneously.

If one succeeds in constructing a self-consistent field, and to the degree that it represents a satisfactory approximation to the whole physical behavior of the system, one has replaced the many-particle problem with the single-particle motions of the fermions in the field. Of course many such single-particle excitations may be possible, so that the dynamics of the system may still be intricate.

For many purposes the Hartree–Fock field is not an adequate approximation to the dynamical situation and at best represents a point of departure for the application of further theoretical methods. These methods may be of the nature of a perturbation-theory expansion, or, if that is inadequate, may use Tamm–Dancoff techniques in which one treats states nearby in energy through nonperturbative methods and more remote excitations through perturbation theory. They may exploit variational methods or specially designed transformations, and so forth. For many systems there exist collective, many-particle effects that make it imperative to go beyond the Hartree–Fock result in one of these nonperturbative fashions. (We shall discuss an important example of this—pairing effects—in the next chapter.) However, even in such situations the Hartree–Fock method provides a line of approach that is capable of generalization to deal with more complex phenomena.

The Hartree–Fock methods that we develop here have been applied to all manner of many-fermion systems, including the electrons in atoms, the nucleons in nuclei, and the electrons in solids. After treating the general formalism, we explore these applications briefly. The atomic case is made conceptually easier by the presence of a preexisting central field—the Coulomb field of the atomic nucleus—which dominates the character of the self-consistent field into which electron–electron interactions are incorporated. For nuclei the entirety of the self-consistent field is built up from itself, and the great difference between nuclear forces and Coulomb forces (the former being vastly stronger and of short range) changes the qualitative nature of the result considerably. For solids the atomic nuclei provide an important component of the field felt by individual electrons, just as for atoms, but the electrons then determine the binding between the ions and thus the nature of the solid itself.

In order to provide a general picture of how Hartree–Fock methods work, we first develop an intuitive view of them. We work in configuration space, and, for the moment, ignore antisymmetrization; the case without antisymmetrization is referred to as the Hartree method. The Hamiltonian for the N-particle system is made up of a one-particle piece and a two-particle interaction term,

$$H = \sum_{i=1}^{N} t_i + \sum_{i<j} v_{ij} , \tag{1}$$

where the one-particle terms t_i might refer to kinetic-energy operators, $t_i = -\nabla_i^2/2m_i$, or might incorporate some other one-particle interaction such as the central Coulomb field provided by the nucleus in the atomic case. $t_i = -\nabla_i^2/2m_i + V_i(\mathbf{r}_i)$. The two-particle interactions is v_{ij}. (One could imagine extending the method to three- or more-particle interaction terms, say for the nuclear case if there are three-particle forces; there is no great formal problem in doing so, but in practice this has not often been considered.) We wish to solve the Schrödinger equation for the Hamiltonian of Eq. 1 in terms of a many-particle wave function Ψ,

$$H\Psi(\mathbf{r}_1, \mathbf{r}_2, \ldots, \mathbf{r}_N) = E\Psi(\mathbf{r}_1, \mathbf{r}_2, \ldots, \mathbf{r}_N) . \tag{2}$$

The Hartree method insists on a radically simplified structure for the solutions of Eq. 2, namely, it requires that Ψ be taken as a simple product of single-particle wave functions,

$$\Psi(\mathbf{r}_1, \mathbf{r}_2, \ldots, \mathbf{r}_N) = \phi_1(\mathbf{r}_1)\phi_2(\mathbf{r}_2) \cdots \phi_N(\mathbf{r}_N). \tag{3}$$

This product wave function is completely devoid of any correlational information: The only influence that a particle j has on the motion of a particle k is through its participation, along with the other particles, in producing an average potential $\mathcal{U}_k(\mathbf{r}_k)$ for the second particle. The many-particle problem is reduced to a single-particle situation by assuming that the functions $\phi_k(\mathbf{r}_k)$ satisfy a single-particle Schrödinger equation containing this average potential in it,

$$[t_k + \mathcal{U}_k(\mathbf{r}_k)]\phi_k(\mathbf{r}_k) = \epsilon_k \phi_k(\mathbf{r}_k), \qquad k = 1, 2, \ldots, N, \tag{4a}$$

where t_k is a single-particle operator, for example the kinetic energy. The self-consistent aspect enters here in that, for each fermion k, the potential $\mathcal{U}_k$ is produced by averaging the two-particle interaction v_{jk} over the motion of the other fermions $j = 1, 2, \ldots, k-1, k+1, \ldots, N$

$$\mathcal{U}_k(\mathbf{r}_k) \equiv \sum_{j \neq k} \int \phi_j^*(\mathbf{r}_j) v_{jk}(\mathbf{r}_j, \mathbf{r}_k) \phi_j(\mathbf{r}_j) \, d\mathbf{r}_j; \tag{4b}$$

note that the average potential is different for each fermion in the Hartree approximation. More explicitly, the self-consistent equations for the Hartree case are

$$\left[-\frac{1}{2m_k} \nabla_k^2 + \sum_{j \neq k} \int \phi_j^*(\mathbf{r}_j) v_{jk}(\mathbf{r}_j, \mathbf{r}_k) \phi_j(\mathbf{r}_j) \, d\mathbf{r}_j \right] \phi_k(\mathbf{r}_k)$$

$$= \epsilon_k \phi_k(\mathbf{r}_k), \qquad k = 1, 2, \ldots, N, \tag{4c}$$

where the single-particle operator t_k has been taken to be the kinetic energy for a particle of mass m_k. In Eq. 4c it is especially clear that each ϕ_k, $k = 1, 2, \ldots, N$, is selected consistently with the $N-1$ other functions ϕ_j, while these in turn reflect, of course, the behavior of the particular ϕ_k we are seeking. It is clear from the introduction of the self-consistent potential that our original problem of Eqs. 1 and 2 can always be written exactly as

$$H = \sum_{i=1}^{N} (t_i + \mathcal{U}_i) + \left[\sum_{i<j} v_{ij} - \sum_{i=1}^{N} \mathcal{U}_i \right] = H_{\text{s.c.}} + H_{\text{res}}, \tag{5}$$

where we have added and subtracted the self-consistent potential to obtain a one-particle form, which is the Hartree self-consistent term, and a two-particle piece, called the *residual interaction*, which is small on the average, by virtue of Eq. 4b, and may, perhaps, be successfully treated by perturbative methods.

If, for identical fermions, we now introduce antisymmetrization into Eq. 4a—still at the intuitive level—we obtain the Hartree–Fock version of the self-consistent field,

$$-\frac{1}{2m}\nabla_k^2\phi_k(\mathbf{r}_k) + \sum_{j=1}^{N}\left[\int d\mathbf{r}_j\,\phi_j^*(\mathbf{r}_j)v(\mathbf{r}_j,\mathbf{r}_k)\phi_j(\mathbf{r}_j)\cdot\phi_k(\mathbf{r}_k)\right.$$

$$\left.-\int d\mathbf{r}_j\,\phi_j^*(\mathbf{r}_j)v(\mathbf{r}_j,\mathbf{r}_k)\phi_k(\mathbf{r}_j)\cdot\phi_j(\mathbf{r}_k)\right]$$

$$=\epsilon_k\phi_k(\mathbf{r}_k),\quad(6a)$$

where, because the particles are identical, we have taken the same mass m and interaction v everywhere. The antisymmetrization has been put into Eq. 6a by taking the one place in the Hartree equation 4c where there is a bilinear combination $\phi_j(\mathbf{r}_j)\phi_k(\mathbf{r}_k)$ and there replacing that combination by an antisymmetrized form $\phi_j(\mathbf{r}_j)\phi_k(\mathbf{r}_k) - \phi_k(\mathbf{r}_j)\phi_j(\mathbf{r}_k)$; this step will be justified formally (see Eqs. 11–15) after we develop an approach to this problem in the language of second quantization which allows us to treat antisymmetrization almost automatically. This form includes the previous self-consistent potential $\mathcal{U}$ as well as a potential $\mathcal{W}$, resulting from antisymmetrization, which is nonlocal. Thus the one-particle Schrödinger equation in the Hartree–Fock case is

$$-\frac{1}{2m}\nabla^2\phi_k(\mathbf{r}) + \mathcal{U}(\mathbf{r})\phi_k(\mathbf{r}) - \int d\mathbf{r}'\,\mathcal{W}(\mathbf{r}',\mathbf{r})\phi_k(\mathbf{r}') = \epsilon_k\phi_k(\mathbf{r}),\quad(6b)$$

where

$$\mathcal{U}(\mathbf{r}) \equiv \sum_{j=1}^{N}\int \phi_j^*(\mathbf{r}')v(\mathbf{r}',\mathbf{r})\phi_j(\mathbf{r}')\,d\mathbf{r}'\quad(6c)$$

and

$$\mathcal{W}(\mathbf{r}',\mathbf{r}) = \sum_{j=1}^{N}\phi_j^*(\mathbf{r}')v(\mathbf{r}',\mathbf{r})\phi_j(\mathbf{r}).\quad(6d)$$

Note that the term in Eq. 6a with $j = k$ cancels when exchange is introduced, so that in the definitions of Eqs. 6c and 6d the summation is over all particles and in Eq. 6b there is no reference in the self-consistent potentials to the fermion of interest, k: The self-consistent field is the same for all the fermions.

These intuitive results help to obtain a preliminary understanding of the workings of the self-consistent-field methods, but to have a more precise formulation we must derive the Hartree–Fock equations from more basic principles. Furthermore it is clear that a convenient incorporation of antisymmetrization will help with this formulation. For these reasons we now turn to the treatment of this problem in Fock space.

4.1 THE VARIATIONAL TREATMENT OF SELF-CONSISTENT EQUATIONS IN FOCK SPACE

In Fock space the one- and two-particle pieces of the Hamiltonian become, as in Eqs. 1.40 and 1.47,

$$H = \sum_{\alpha\beta} \langle \alpha|t|\beta \rangle a_\alpha^\dagger a_\beta + \frac{1}{2} \sum_{\alpha\beta\gamma\delta} \langle \alpha\beta|v|\gamma\delta \rangle a_\alpha^\dagger a_\beta^\dagger a_\delta a_\gamma , \qquad (7)$$

where we have expanded on some as yet unspecified single-fermion basis $\alpha, \beta, \ldots$. We now assume that the ground state of the N-particle system can be well approximated by the analog of a product wave function as in Eq. 3. The analog will take into account antisymmetrization so that it becomes a Slater determinant, or in the language of second quantization a product of creation operators for the filled levels,

$$|\Phi\rangle = \prod_{\mu \leq F} a_\mu^\dagger |0\rangle , \qquad (8)$$

where F denotes the Fermi surface (the last filled single-particle level). Having limited rather severely the structure of the approximate ground-state solution $|\Phi\rangle$, there remain to be chosen the single-particle states, labeled μ in Eq. 8, for which $|\Phi\rangle$ is to be a good approximation; this will also fix the single-particle basis over which the Hamiltonian H in Eq. 7 is being expanded. The guiding criterion we apply towards this end is that of the variational principle where the quantities to be adjusted are the single-particle states themselves.

We consider variations in $|\Phi\rangle$ such that

$$\delta \langle \Phi|H|\Phi\rangle = 0 , \quad \text{or} \quad \langle \delta\Phi|H|\phi\rangle = 0 . \qquad (9a, b)$$

[Note that, in deriving the second version (Eq. 9b) from the first (Eq. 9a), we are exploiting the fact that $|\Phi\rangle$ is to be thought of generally as complex and thus contains two separate functions—the real and the imaginary parts—to be varied. There is therefore a companion equation to 9b, namely $\langle \Phi|H|\delta\Phi\rangle = 0$, but since this merely leads to the Hermitian conjugate equation of 9b, we do not consider it separately here.] In order for a variation of $|\Phi\rangle$ to be nontrivial, yet stay within the constraint of the Hartree–Fock assumption, it must have the structure

$$|\delta\Phi\rangle = \eta a_\sigma^\dagger a_\lambda |\Phi\rangle , \qquad (10)$$

where η is a small parameter $|\eta| \ll 1$ governing the smallness of the variation. This is equivalent to performing the variation on each occupied orbit λ separately. Since the change of a single-particle wave function can be achieved by acting on it by a one-body operator, we may choose $\eta a_\sigma^\dagger a_\lambda$ as such an operator, with an independent variational parameter η for each σ, λ. The modification implied by Eq. 10 is to promote one fermion from an occupied to an unoccupied level. This is basically the only thing we can change in $|\Phi\rangle$, since if we were to act on it with $a_\lambda^\dagger |\Phi\rangle$, where $\lambda \le F$ refers to a filled level, or *hole* state, the variation would vanish by reason of the exclusion principle $a_\lambda^\dagger a_\lambda^\dagger = 0$. If we act to annihilate a particle in an unfilled level, or *particle* state, $\sigma > F$, we again get zero. Thus only changes like Eq. 10 will not vanish, and they must involve one annihilation and one creation if we are working with a system having a fixed number of fermions. The variations we have in mind to perform will eventually fix the nature of the as yet undermined single-particle states $\alpha, \beta, \ldots, \lambda, \sigma$, but the formal change made in Fock space is through the rearranged occupation implied by Eq. 10. Of course, one could use a variation that is a sum over terms like that of Eq. 10 which still corresponds to a simple product wave function in the sense of Eq. 8 (see Eq. 36a). However, as long as we treat each term of the form of Eq. 10 as an independent variation, there is no difference in the result.

In terms of changes of the form of Eq. 10, the variational principle, Eq. 9b, requires

$$\langle \Phi | a_\lambda^\dagger a_\sigma H | \Phi \rangle = 0 \,, \qquad (11a)$$

or

$$\sum_{\alpha\beta} \langle \alpha | t | \beta \rangle \langle \Phi | a_\lambda^\dagger a_\sigma a_\alpha^\dagger a_\beta | \Phi \rangle$$

$$+ \frac{1}{2} \sum_{\alpha\beta\gamma\delta} \langle \alpha\beta | v | \gamma\delta \rangle \langle \Phi | a_\lambda^\dagger a_\sigma a_\alpha^\dagger a_\beta^\dagger a_\delta a_\gamma | \Phi \rangle$$

$$= \sum_{\alpha\beta} \langle \alpha | t | \beta \rangle \, \delta_{\alpha\sigma} \delta_{\beta\lambda}$$

$$+ \frac{1}{2} \sum_{\alpha\beta\gamma\delta} \langle \alpha\beta | v | \gamma\delta \rangle [\bar\delta_{\alpha\gamma} \delta_{\beta\sigma} \delta_{\delta\lambda} - \bar\delta_{\alpha\delta} \delta_{\beta\sigma} \delta_{\gamma\lambda} - \bar\delta_{\beta\gamma} \delta_{\alpha\sigma} \delta_{\delta\lambda} + \bar\delta_{\beta\delta} \delta_{\alpha\sigma} \delta_{\gamma\lambda}]$$

$$= \langle \sigma | t | \lambda \rangle + \frac{1}{2} \sum_{\mu \le F} [\langle \mu\sigma | v | \mu\lambda \rangle - \langle \mu\sigma | v | \lambda\mu \rangle$$

$$- \langle \sigma\mu | v | \mu\lambda \rangle + \langle \sigma\mu | v | \lambda\mu \rangle]$$

$$= \langle \sigma | t | \lambda \rangle + \sum_{\mu \le F} \langle \mu\sigma | v | \mu\lambda \rangle_A \,, \qquad (11b)$$

where the barred Kronecker delta symbol refers to filled levels only,

$$\bar{\delta}_{\alpha\beta} = \delta_{\alpha\beta \leq F} \, . \tag{12a}$$

In Eq. 11b, the antisymmetrization notation of Eq. 1.50 has been used,

$$|\alpha\beta\rangle_A = |\alpha\beta\rangle - |\beta\alpha\rangle \, , \tag{12b}$$

and we have exploited the symmetry noted in Eq. 1.49, namely

$$\langle \alpha\beta | v | \gamma\delta \rangle = \langle \beta\alpha | v | \delta\gamma \rangle \, . \tag{12c}$$

The final result of our variational calculation in Eq. 11 is indeed the very same Hartree–Fock self-consistent equations that we encountered in the intuitive development of Eqs. 6, but now written in terms of matrix elements defined on a set of single-particle basis states. If we define a single-particle Hamiltonian as

$$\langle \alpha | h | \beta \rangle \equiv \langle \alpha | (t + \mathcal{U}) | \beta \rangle \, , \tag{13}$$

where

$$\langle \alpha | \mathcal{U} | \beta \rangle \equiv \sum_{\mu \leq F} \langle \alpha\mu | v | \beta\mu \rangle_A \tag{14}$$

is the single-particle potential (and here includes the exchange term), then the Hartree–Fock single-particle basis is to be the set of states that diagonalize h, that is, we require

$$\langle \alpha | h | \beta \rangle = \langle \alpha | (t + \mathcal{U}) | \beta \rangle = \epsilon_\alpha \delta_{\alpha\beta} \tag{15}$$

for all the states α, β, ... Note that the variational-principle result of Eq. 11 strictly requires only that h vanish between single-particle states when one such state is above the Fermi surface and the other below; the stronger restriction of Eq. 15 will of course ensure that that is the case as well. Naturally, if one is only interested in ground-state properties, it suffices to solve the self-consistent equations implied by Eq. 15 only for the filled orbitals, which are the only ones required in Eq. 8. If excitations are considered, then one must proceed to solve Eq. 15 for the space of excitations under consideration as well.

It is convenient to incorporate distinctions concerning the positions of single-particle levels relative to the Fermi surface into our notation. We use for this a system that has gained some acceptance in the area, namely we reserve for filled states or hole states the Greek letters

$$\lambda, \mu, \nu \leq F \tag{16a}$$

and for empty single-particle levels, or particle states,

$$\rho, \sigma, \tau > F \, , \tag{16b}$$

while $\alpha, \beta, \gamma, \delta, \ldots$ are generic labels referring to states above or below the Fermi surface. For some purposes it is also convenient to redefine the zero of energy relative to the Fermi surface and to take the filled Fermi sea as the new vacuum state. This leads to the introduction of a transformed set of creation and annihilation operators such that for levels in the Fermi sea, that is for hole states,

$$b_\lambda^\dagger = -a_\lambda \,, \quad b_\lambda = -a_\lambda^\dagger \,, \qquad \lambda \le F \,, \tag{17}$$

while levels above the Fermi sea we leave unchanged. (The minus sign in Eq. 17 is arbitrary and is introduced here to conform to later usage in Chapter 5; in Chapter 7—see Eqs. 7.20 and 7.21—we in fact generalize this phase consideration somewhat.) The commutation relations for the hole states are then

$$\{b_\lambda, b_\mu^\dagger\} = \{a_\lambda^\dagger, a_\mu\} = \delta_{\lambda\mu} \,, \tag{18a}$$

and

$$\{b_\lambda^\dagger, b_\mu^\dagger\} = \{b_\lambda, b_\mu\} = 0 \,, \tag{18b, c}$$

so that fermion anticommutation relations are preserved by this transformation. The state $|\Phi\rangle$ is indeed a vacuum state for these new operators, since

$$b_\lambda|\Phi\rangle = -a_\lambda^\dagger \prod_{\mu \le F} a_\mu^\dagger |0\rangle = 0 \,, \qquad \lambda \le F \,; \tag{19}$$

of course for states above the Fermi sea it continues to be true that

$$a_\rho|\Phi\rangle = a_\rho \prod_{\mu \le F} a_\mu^\dagger |0\rangle = 0 \,, \qquad \rho > F \,. \tag{20}$$

Below we shall make use of these operators when symmetry with respect to the Fermi surface introduces a simplification. They also represent a particularly simple version of a more general transformation of the single-particle basis to what are called *quasiparticles*, which will be very useful later (see Eqs. 5.42 and 5.44).

Having found the best possible Slater-determinant ground state for our system—in the sense of the variational principle—we can also calculate the total system energy for it,

$$\begin{aligned}
E_0 &\equiv \langle\Phi|H|\Phi\rangle = \sum_{\mu \le F} \langle\mu|t|\mu\rangle + \frac{1}{2} \sum_{\mu,\lambda \le F} \langle\mu\lambda|v|\mu\lambda\rangle_A \\
&= \sum_{\mu \le F} \epsilon_\mu - \frac{1}{2} \sum_{\lambda,\mu \le F} \langle\mu\lambda|v|\mu\lambda\rangle_A \\
&= \sum_{\mu \le F} \left[\epsilon_\mu - \frac{1}{2} \langle\mu|\mathcal{U}|\mu\rangle\right] \,,
\end{aligned} \tag{21}$$

where we have used Eqs. 14 and 15. Note that the total energy in the Hartree–Fock ground state is not the sum of the single-particle energies ϵ_λ of Eq. 15, since the Hartree–Fock energy is defined in terms of an artificial one-particle potential, whereas the true energy derives from the original two-particle interactions. This leaves open for the moment the question of the interpretation of the ϵ_λs, and we return to this momentarily. It also makes clear that the single-particle operator

$$\mathcal{H} \equiv \sum_{\alpha\beta} \langle \alpha|h|\beta \rangle a_\alpha^\dagger a_\beta = \sum_\alpha \epsilon_\alpha a_\alpha^\dagger a_\alpha \qquad (22a)$$

is not to be interpreted as yielding the total energy when it acts on $|\Phi\rangle$, although $|\Phi\rangle$ is of course an eigenstate of this operator,

$$\mathcal{H}|\Phi\rangle = \left(\sum_{\mu \leq F} \epsilon_\mu \right)|\Phi\rangle . \qquad (22b)$$

Note that if we introduce the operators of Eq. 17, this operator becomes

$$\mathcal{H} = \sum_{\rho > F} \epsilon_\rho a_\rho^\dagger a_\rho + \sum_{\lambda \leq F} \epsilon_\lambda b_\lambda b_\lambda^\dagger$$

$$= \sum_{\rho > F} \epsilon_\rho a_\rho^\dagger a_\rho - \sum_{\lambda \leq F} \epsilon_\lambda b_\lambda^\dagger b_\lambda + \sum_{\lambda \leq F} \epsilon_\lambda , \qquad (22c)$$

so that, counting relative to the Fermi surface, the particle states $\rho > F$ refer to additive ϵ_ρ while the hole states $\lambda \leq F$ refer to subtractive ϵ_λ, and there is a constant energy $\Sigma \, \epsilon_\lambda$ attached to $\mathcal{H}$. In the form of Eq. 22c each of the two number-operator terms annihilates $|\Phi\rangle$, which acts as a vacuum for a_ρ and b_λ, and only the constant term survives, as in Eq. 22b. The construct $\mathcal{H}$ may therefore be useful for counting energy contributions and changes for single-particle levels even though its action on $|\Phi\rangle$ differs from the total Hartree–Fock energy by the term $-\frac{1}{2}\Sigma \, \langle \lambda\mu|v|\lambda\mu \rangle_A$. It offers a point of departure that may be used, for example, in a perturbation-theory approach when one goes beyond Hartree–Fock (e.g., see Section 5.3 on the ground state of systems involving pairing).

4.2 SYSTEM EXCITATIONS AND THE MEANING OF SINGLE-PARTICLE ENERGIES

The Hartree-Fock ground state is $|\Phi\rangle$ of Eq. 8 with the single-particle basis of Eq. 15. To describe excited states of the many-fermion systems we could perform the variational calculation once again, this time demanding that the new state that minimizes the expectation value of H be orthogonal to the ground state, and thus proceed iteratively to generate excitations. This would be a tedious procedure to carry out and would also lose the simple insight based on retaining the Hartree–Fock single-particle levels generated for the ground state and describing excitations in terms of them. (It would, however, produce a valid calculation of an upper bound on the excited-state energy.) This latter approach implies that we discuss excitations in terms of the promotion of a fermion from a filled to an unfilled level as in the construct

$$|\Phi_\mu^\sigma\rangle = a_\sigma^\dagger a_\mu |\Phi\rangle = -a_\sigma^\dagger b_\mu^\dagger |\Phi\rangle ,\qquad(23)$$

a *particle–hole* (1p–1h) state; higher degrees of excitation then involve the promotion of two particles,

$$|\Phi_{\mu\nu}^{\sigma\tau}\rangle = a_\sigma^\dagger a_\tau^\dagger a_\mu a_\nu |\Phi\rangle = a_\sigma^\dagger a_\tau^\dagger b_\mu^\dagger b_\nu^\dagger |\Phi\rangle ,\qquad(24)$$

a *two-particle–two-hole* (2p–2h) state, and so forth. These states are naturally not eigenstates of the Hamiltonian (although they are eigenstates of $\mathcal{H}$ of Eq. 22). In general one often finds a large number of states of the 1p–1h type of Eq. 23—and an even larger number for 2p–2h states, and so on—because of the many ways to select hole and particle orbitals to produce the class of excitations in question. Moreover, many of these states tend to have similar energies since the hole orbitals in question may be nearly degenerate, as may be the particle orbitals. The standard way to deal with this situation in quantum mechanics is by the method of nearly degenerate perturbation theory, in which one truncates the problem in question to a finite space of state vectors and diagonalizes the secular Hamiltonian matrix within that space. This is the origin of the shell model ("with configuration interaction" as it is sometimes called) in atomic and nuclear physics, where the fermions in question are electrons and nucleons, respectively. The simplest level of truncation is to restrict consideration to the space of 1p–1h states, and indeed if particle orbitals lie well above hole orbitals the 1p–1h states may be much lower than 2p–2h configurations, which will come at an energy sufficiently high to achieve the raising of two particles from filled to unfilled levels. In treating a space of one-particle–one-hole excitations it is necessary to construct the secular Hamilton matrix for that space,

$$\langle\Phi_\mu^\sigma|H|\Phi_\nu^\tau\rangle = \delta_{\sigma\tau}\delta_{\mu\nu}(E_0 + \epsilon_\sigma - \epsilon_\mu) - \langle\sigma\nu|v|\tau\mu\rangle_A ,\qquad(25)$$

(see Exercise 4.4) and to diagonalize it to obtain approximate eigenvectors for the excitation.

The first term in Eq. 25 contains the total Hartree–Fock energy E_0 of Eq. 21 and an excitation energy relative to it of $\epsilon_\sigma - \epsilon_\mu$, the particle energy less the hole energy (compare also Eq. 22c). The last term of Eq. 25 is the particle–particle interaction matrix element. It appears with a sign reversal to give the correct result for the particle–hole matrix element. This sign is related to the fact that the hole state represents the absence of a particle and hence its contribution to the energy balance is to be subtracted from the total of the summands of the various orbitals. Eq. 25 also illustrates the nature of the hole in switching sides of a matrix element when one goes from Fock space to conventional space, that is, ν in the ket and μ in the bra in the Fock-space matrix element of Eq. 25 become ν in the bra and μ in the ket in $-\langle \sigma\nu|v|\tau\mu \rangle_A$. A pictorial representation of this matrix element is given in Figure 4.1 and may be compared with the particle–particle interaction in Figure 1.1. Figure 4.1 contains two parts: a direct graph and an exchange graph, representing the two pieces of the antisymmetrized wave function $|\tau\mu\rangle_A$ in Eq. 25, as defined in Eq. 12b.

The expression in Eq. 25 provides the beginnings of an understanding concerning the precise role of the Hartree–Fock single-particle energies ϵ_α of Eq. 15: Consider a diagonal matrix element, $\sigma = \tau$ and $\mu = \nu$, in which a particle is taken from a hitherto occupied state μ to a particle state σ having the property that it lies just at the beginning of the continuum for the system with $N-1$ particles bound together and one particle removed to infinity. That is, we have $\epsilon_\sigma = 0$ and the fermion in this state is just barely moving off from the rest of the $N-1$ residual system. The energy of this state is represented by Eq. 25. The interaction matrix element $\langle \sigma\nu|v|\tau\mu \rangle_A$ in that equation will vanish because there will be only an infinitesimal overlap of the continuum wave function, which spreads all over space, with the finite extent of the bound system, i.e., the $1/\sqrt{\Omega}$ in the normalization of the continuum wave functions together with the finite range of the interaction v and the limited spatial support of the bound-state wave functions will yield a vanishingly small value for $\langle \sigma\nu|v|\tau\mu \rangle_A$. The excitation energy of this state is then $-\epsilon_\mu$, which may be thought of as the removal energy for taking the fermion from the bound orbit μ into the continuum.

The point can be approached a little differently if we consider a state $|\Phi_\mu\rangle$ that differs from our Hartree–Fock ground state $|\Phi\rangle$ by having one less fermion namely, the hole state μ is unoccupied,

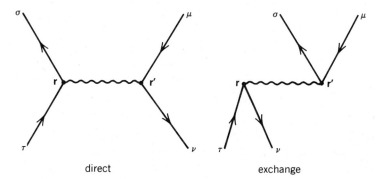

direct   exchange

FIGURE 4.1   Direct and exchange parts of the particle–hole interaction matrix element of Eq. 25.

$$|\Phi_\mu\rangle = a_\mu|\Phi\rangle = -b_\mu^\dagger|\Phi\rangle = \prod_{\substack{\lambda \leq F \\ \lambda \neq \mu}} a_\lambda^\dagger|0\rangle \ . \tag{26}$$

The total energy for this state can be immediately read off from Eq. 21 and is

$$E_\mu = \langle\Phi_\mu|H|\Phi_\mu\rangle = \sum_{\substack{\lambda \leq F \\ \lambda \neq \mu}} \langle\lambda|t|\lambda\rangle + \frac{1}{2}\sum_{\substack{\lambda,\nu \leq F \\ \lambda \neq \mu \\ \nu \neq \mu}} \langle\lambda\nu|v|\lambda\nu\rangle_A$$

$$= E_0 - \left[\langle\mu|t|\mu\rangle + \sum_\nu \langle\mu\nu|v|\mu\nu\rangle_A\right] = E_0 - \epsilon_\mu \ . \tag{27}$$

Thus

$$E_\mu - E_0 = -\epsilon_\mu \tag{28}$$

is in this sense the single-particle binding energy or separation energy for a fermion in the state μ. Note that the removal procedure of Eqs. 26 and 27 is not iterative. If one continues to remove particles in this way, the underlying relevance of the original Hartree–Fock solution is destroyed as soon as more than a few particles are removed relative to the starting number, and one can ascribe little meaning to the result. In particular the ultimate implication of such an iteration, namely that the total system energy is $\Sigma \, \epsilon_\mu$, is wrong, as we saw in Eq. 21.

In fact the correct relation between the removal energies and the total energy E_0 is expressed by the sum rule derived in Exercise 4.3(b),

$$E_0 = \tfrac{1}{2}\langle\Phi|T|\Phi\rangle + \frac{1}{2}\sum_\alpha \mathcal{E}_\alpha \ , \tag{29a}$$

with

$$\mathcal{E}_\alpha = \langle\Phi|a_\alpha^\dagger[a_\alpha, H]|\Phi\rangle \ . \tag{29b}$$

For the Hartree–Fock case, it is an easy exercise to show that $\mathcal{E}_\mu = \epsilon_\mu$ for $\mu \leq F$, $\mathcal{E}_\alpha = 0$ otherwise, and that (29a) becomes

$$E_0 = \frac{1}{2}\sum_{\mu \leq F}[\langle\mu|t|\mu\rangle + \epsilon_\mu] \ . \tag{30}$$

So the interpretation of $-\epsilon_\mu$ as the removal energy for the orbit μ is justified. For a bound orbit, this energy is positive ($\epsilon_\mu < 0$).

Note that the one-hole state $|\Phi_\mu\rangle$ is not a Hartree–Fock state for the system of $N - 1$ particles. What is usually called Koopmans's theorem states that for the least-bound orbit ($\mu = F$), the removal energy ϵ_F gives a good approximation to the difference of ground-state energies for N and $N-1$ particles (often called the *separation* energy):

$$\epsilon_F \cong E_0(N) - E_0(N-1) , \qquad (31)$$

where the corrections are of order $N^{-1} \times$ (orbital energies).

Before leaving the description of excitations by means of the particle-hole states of Eqs. 23 and 24, we illustrate the overall scheme for the Hamiltonian matrix that is implied by our results. This will have the structure

$$
\begin{array}{c}
 \\
\text{Ground state} \left\{ \rule{0pt}{12pt}\right. \\
 \\
\text{1p-1h} \left\{ \rule{0pt}{30pt}\right. \\
 \\
 \\
\text{2p-2h} \left\{ \rule{0pt}{30pt}\right. \\

\end{array}
\left[
\begin{array}{cc|cc|cc}
\overbrace{E_0}^{\text{Ground state}} & 0 & 0 & V_{2\text{p-}2\text{h,}0} & V'_{2\text{p=}2\text{h,}0} \\
\hline
0 & E_0 + \epsilon_\rho - \epsilon_\lambda & -\langle \sigma\lambda|v|\rho\lambda\rangle_A & V_{1\text{p-}1\text{h, }2\text{p-}2\text{h}} & \cdots \\
 & -\langle \rho\lambda|v|\rho\lambda\rangle_A & & & \\
0 & -\langle \rho\lambda|v|\sigma\lambda\rangle_A & E_0 + \epsilon_\sigma - \epsilon_\lambda & \cdots & \cdots \\
 & & -\langle \sigma\lambda|v|\sigma\lambda\rangle_A & & \\
\hline
V_{0,\,2\text{p-}2\text{h}} & V_{2\text{p-}2\text{h,}1\text{p-}1\text{h}} & V'_{2\text{p-}2\text{h,}1\text{p-}1\text{h}} & E_0 + \epsilon_\rho + \epsilon_\sigma & \cdots \\
 & & & -\epsilon_\lambda - \epsilon_\mu & \\
 & & & +V_{2\text{p-}2\text{h,}2\text{p-}2\text{h}} & \\
V'_{0,2\text{p-}2\text{h}} & V''_{2\text{p-}2\text{h,}1\text{p-}1\text{h}} & \cdots & \cdots & \cdots
\end{array}
\right]
$$

$$ (32)$$

Just outside the borders of the matrix the pertinent space (ground state, one-particle–one-hole, two-particle–two-hole, . . .) is indicated. The bordering zeros for matrix elements between the ground state and the one-particle–one-hole excitations occur because of our basic variational requirement, Eq. 11, which states that just such elements must vanish. This situation is sometimes referred to as Brillouin's theorem. It prevents, as it were, the "immediate" mixture of one-particle–one-hole states with the ground state, which would tend to detract from the purity of that state. As it is, the two-particle–two-hole states do have matrix elements to the ground state and so will be admixed with it. Of course the 2p–2h states have larger diagonal energies because of the appearance of two particle separation energies and two hole separation energies, $\epsilon_\rho + \epsilon_\sigma - \epsilon_\lambda - \epsilon_\mu$ say, with consequently less admixture of such configurations into the ground state. Indeed, the whole approach to excitations here is based on an underlying approximation known as the *Tamm–Dancoff approximation*, which supposes that as one goes to higher and higher degrees of excitation (3p–3h, 4p–4h, and so on), the separation of energies along the diagonal gets larger while matrix elements of v stay roughly the same in magnitude (or even decrease if overlap between single-particle functions worsens suitably), and the net result is to yield smaller admixtures of configurations lying at higher energies. This Tamm–Dancoff approximation is the basis of the use of the shell model for the description of excitations in atomic or nuclear physics and is successful to the degree that the single-particle orbitals are well separated in energies and the configuration interaction falls off with increasing orbitals. Even if these conditions are met there may still be intruder states driven down in energy by the large interactions or large number of configurations in some higher-energy states which consequently appear much below their expected position. In more promising situations, however, one may truncate the matrix so that it becomes finite and diagonalize it to produce excitation energies and state vectors for the system. Transition probabilities between the ground and excited states can then be calculated from these eigenvectors. For example, for a one-particle transition operator

$$F = \sum_{\alpha\beta} \langle \alpha | f | \beta \rangle a_\alpha^\dagger a_\beta , \tag{33}$$

the amplitude to excite a particular particle–hole state from the ground state is

$$\sum_{\substack{\sigma > F \\ \mu \leq F}} x_{\sigma\mu} \langle \Phi_\mu^\sigma | F | \Phi \rangle = \sum_{\substack{\sigma > F \\ \mu \leq F}} x_{\sigma\mu} \langle \sigma | f | \mu \rangle , \tag{34}$$

where $x_{\sigma\mu}$ are the components of the state vector in question.

4.3 STABILITY OF THE HARTREE–FOCK GROUND STATE

Consideration of the structure of the secular Hamiltonian matrix in Eq. 32 inevitably raises the issue of the degree to which the excitations inherent in the Hartree–Fock method may detract from the quality of the ground-state solution $|\Phi\rangle$ itself. We now explore this matter further, since it points to some of the significant limitations of the method. First note that the most modest requirement for the Hartree–Fock solution is that the ground state be a stable minimum so that the varied states of Eq. 10 yield energies E_0' that are above those of the Hartree–Fock ground-state energy E_0. To lowest nonvanishing order in η this requires, from Eqs. 11, 21, and 25,

$$E_0' \equiv \frac{\langle \Phi + \delta\Phi | H | \Phi + \delta\Phi \rangle}{\langle \Phi + \delta\Phi | \Phi + \delta\Phi \rangle}$$

$$= E_0 + |\eta|^2 [\epsilon_\sigma - \epsilon_\lambda - \langle \sigma\lambda | v | \sigma\lambda \rangle_A]$$

$$> E_0 , \tag{35a}$$

for each particle–hole configuration choice σ, λ. Thus the diagonal particle–hole energy of Eq. 25 must be positive,

$$\epsilon_\sigma - \epsilon_\lambda - \langle \sigma\lambda | v | \sigma\lambda \rangle_A > 0 ; \tag{35b}$$

if this were not the case, then the 1p–1h subspace of the secular matrix in Eq. 32 would contain diagonal elements below the ground-state energy, implying that the relevant configurations may take over from the ground state as state of lowest energy. The inequality 35b is likely to be fulfilled for systems with a large number of fermions, $N \gg 1$, since we expect that $\epsilon \propto N$ for a system with fixed volume in which we increase the number of particles, while the interaction matrix element in Eq. 35b is essentially independent of N, so that the positive single-particle energy difference $\epsilon_\sigma - \epsilon_\lambda$ tends to prevail strongly over the matrix element.* Of course, for a large number of particles there will also be a large number of possible excitation configurations, which may tend to erode the position of the ground state through the admixture of many excitation components with it.

A widespread feature of the failure of stability in the Hartree–Fock ground state is a change in some basic characteristic of that state. For example, in nuclear physics one may work with an underlying assumption of a spherical ground state only to find that the system actually prefers an average potential that is deformed. If one insists on a spherical Hartree–Fock basis, this is manifested by large admixtures of excited configurations in the ground state, that is, the sort of instability we have noted. A better procedure is likely to be the use of a deformed Hartree–Fock basis which will embody this feature from the start. It may be possible to study the onset of such a situation by varying some parameter: For example, nuclei tend to be spherical near closed shells but may undergo a "phase transition"—not a true phase transition, of course, since we are discussing finite systems, not a thermodynamic limiting case—as particles are added outside the closed shell. (Such a change may take place even within a given system as higher excitations are considered. Thus the ground state may be spherical but excited states may not be easily obtainable through a Tamm–Dancoff method based on it, instead requiring the generation of a deformed Hartree–Fock basis.)

The inequalities of Eqs. 35a and 35b are based on pure particle–hole variations of the form of Eq. 10 (i.e., on the choice of a particular σ and λ). More generally a nonvanishing variation of the original Slater determinant, or "product" state vector in the sense of Eq. 8, is

$$|\Phi'\rangle = \prod_{\mu \leq F} \prod_{\rho > F} (1 + c_{\rho\mu} a_\rho^\dagger a_\mu)|\Phi\rangle$$

$$= \exp\left\{ \sum_{\mu \leq F} \sum_{\rho > F} c_{\rho\mu} a_\rho^\dagger a_\mu \right\} |\Phi\rangle , \tag{36a}$$

where the set of coefficients $c_{\rho\mu}$ replace the previous single variational parameter η ($c_{\rho\mu} = \eta \delta_{\rho\sigma} \delta_{\mu\lambda}$ there) and are to be considered small for small

---

*Of course the volume may also increase with N, but the ratio of energies discussed is unchanged.

variations; the exponential form in Eq. 36a results after recalling $a_\alpha^\dagger a_\alpha^\dagger = a_\alpha a_\alpha = 0$. Up to and including the order $|c|^2$—the lowest nonvanishing order in c—the normalization of this state is

$$\langle \Phi' | \Phi' \rangle = 1 + \sum_{\rho\mu} |c_{\rho\mu}|^2 , \tag{36b}$$

while the energy matrix element is

$$\langle \Phi' | H | \Phi' \rangle = \langle \Phi | \left[1 + \sum_{\rho\mu} c_{\rho\mu}^* a_\mu^\dagger a_\rho + \frac{1}{2} \sum_{\rho\mu\sigma\nu} c_{\rho\mu}^* c_{\sigma\nu}^* a_\mu^\dagger a_\rho a_\nu^\dagger a_\sigma \right]$$

$$\times \left[\sum_{\alpha\beta} \langle \alpha | t | \beta \rangle a_\alpha^\dagger a_\beta + \frac{1}{2} \sum_{\alpha\beta\gamma\delta} \langle \alpha\beta | v | \gamma\delta \rangle a_\alpha^\dagger a_\beta^\dagger a_\delta a_\gamma \right]$$

$$\times \left[1 + \sum_{\tau\lambda} c_{\tau\lambda} a_\tau^\dagger a_\lambda + \frac{1}{2} \sum_{\tau\lambda\phi\kappa} c_{\tau\lambda} c_{\phi\kappa} a_\tau^\dagger a_\lambda a_\phi^\dagger a_\kappa \right] | \Phi \rangle$$

$$= E_0 \langle \Phi' | \Phi' \rangle + \sum_{\rho\mu} (\epsilon_\rho - \epsilon_\mu) |c_{\rho\mu}|^2$$

$$+ \sum_{\rho\mu\tau\lambda} [- \langle \rho\lambda | \epsilon | \tau\mu \rangle_A c^*{}_{\rho\mu} c_{\tau\lambda}$$

$$+ \tfrac{1}{2} \langle \mu\lambda | v | \rho\tau \rangle_A c_{\rho\mu} c_{\tau\lambda}$$

$$+ \tfrac{1}{2} \langle \rho\tau | v | \mu\lambda \rangle_A c_{\rho\mu}^* c_{\tau\lambda}^*] , \tag{37a}$$

where we have assumed the use of the Hartree–Fock single-particle basis and dropped all those terms that then vanish because of Eq. 11 or 15. Thus, with the Hartree–Fock choice, we have

$$E_0' \equiv \frac{\langle \Phi' | H | \Phi' \rangle}{\langle \Phi' | \Phi' \rangle}$$

$$= E_0 + \sum_{\rho\mu} (\epsilon_\rho - \epsilon_\mu) |c_{\rho\mu}|^2$$

$$+ \sum_{\rho\mu\tau\lambda} [- \langle \rho\lambda | v | \tau\mu \rangle_A c_{\rho\mu}^* c_{\tau\lambda}$$

$$+ \tfrac{1}{2} \langle \mu\lambda | v | \rho\tau \rangle_A c_{\rho\mu} c_{\tau\lambda}$$

$$+ \tfrac{1}{2} \langle \rho\tau | v | \mu\lambda \rangle_A c_{\rho\mu}^* c_{\tau\lambda}^*] , \tag{37b}$$

as the generalization of Eq. 35a. For a stable Hartree–Fock ground state, $E_0' \geq E_0$, the quadratic form here must be nonnegative. Introducing the square matrix A with indices $(\rho\mu)$ and $(\tau\lambda)$ according to

$$A_{\rho\mu ; \tau\lambda} \equiv (\epsilon_\rho - \epsilon_\mu) \delta_{\rho\tau} \delta_{\mu\lambda} - \langle \rho\lambda | v | \tau\mu \rangle_A , \tag{38a}$$

the previous matrix of particle–hole excitations in Eq. 25, and

$$B_{\rho\mu\,;\tau\lambda} \equiv \langle \rho\tau|v|\mu\lambda \rangle_A \,, \tag{38b}$$

we rewrite Eq. 37b in matrix notation with column matrices c^* and c as

$$E_0' = E_0 + \tfrac{1}{2}(c^*, c)\begin{pmatrix} A & B \\ B^* & A^* \end{pmatrix}\begin{pmatrix} c \\ c^* \end{pmatrix}. \tag{39}$$

If the square matrix here is brought to diagonal form by a unitary transformation, then the resulting eigenequations are

$$Ac + Bc^* = \omega c \tag{40a}$$

and

$$B^*c + A^*c^* = \omega c^* \,, \tag{40b}$$

or

$$(\epsilon_\rho - \epsilon_\mu)c_{\rho\mu} + \sum_{\tau\lambda}[-\langle \rho\lambda|v|\tau\mu \rangle_A c_{\tau\lambda} + \langle \rho\tau|v|\mu\lambda \rangle_A c_{\tau\lambda}^*] = \omega c_{\rho\mu} \tag{40c}$$

and

$$(\epsilon_\rho - \epsilon_\mu)c_{\rho\mu}^* + \sum_{\tau\lambda}[\langle \mu\lambda|v|\rho\tau \rangle_A c_{\tau\lambda} - \langle \tau\mu|v|\rho\lambda \rangle_A c_{\tau\lambda}^*] = \omega c_{\rho\mu}^* \,, \tag{40d}$$

and for stability the eigenvalues ω in these equations must be positive or zero. (Actually $\omega = 0$ is an unsatisfactory, degenerate situation, but may be unavoidable because of symmetry considerations to which we shortly turn.) This stability condition may be seen as a side condition on the Hartree–Fock solution for it to be of value. It will enter as a central ingredient in the discussion of symmetry considerations in the Hartree–Fock method and, later, in the context of the random-phase approximation.

4.4 SYMMETRIES, BROKEN SYMMETRIES, AND SUM RULES WITHIN THE HARTREE–FOCK METHOD

The Hamiltonian of Eq. 7 usually possesses certain general symmetries. For example, if the forces involve in coordinate space only the relative locations of the particles, then the Hamiltonian is invariant under linear displacements of the entire system as generated by the total linear-momentum operator

$$\mathbf{P} = \sum_{\alpha\beta} \langle \alpha | \mathbf{p} | \beta \rangle a_\alpha^\dagger a_\beta , \tag{41}$$

and if the forces are rotationally invariant, then the system will be unmodified by the rotations generated through the total angular-momentum operator

$$\mathbf{J} = \sum_{\alpha\beta} \langle \alpha | \mathbf{j} | \beta \rangle a_\alpha^\dagger a_\beta , \qquad \mathbf{j} \equiv \mathbf{r} \times \mathbf{p} + \tfrac{1}{2}\boldsymbol{\sigma} , \tag{42}$$

where $\mathbf{p}$, $\mathbf{r}$, $\boldsymbol{\sigma}$, and $\mathbf{j}$ are the single-fermion momentum, position, spin, and total-angular-momentum operators. The system invariance is expressed by the vanishing of the commutator for the relevant operator with the Hamiltonian,

$$[H, \mathbf{P}] = 0 \tag{43}$$

or

$$[H, \mathbf{J}] = 0. \tag{44}$$

For the Hartree–Fock method, however, these symmetries may be destroyed in the process of constructing the self-consistent field. For example, this field is fixed in space, centered around a particular point—that is the particles move in a potential that is located somewhere—and so translational invariance is lost; it is not easily regained, because in the course of the self-consistent procedure each particle coordinate $\mathbf{r}_i$, $i = 1, 2, \ldots, N$, has been dealt with on its own merits, so that one cannot readily express the state $|\Phi\rangle$ in terms of a center-of-mass coordinate and internal coordinates relative to it. Similarly, if the self-consistent field is nonspherical, then it will be constructed with a particular orientation in space, and it will be difficult to separate out three variables to describe this orientation and other variables to label particle positions relative to the aligned frame. A symmetry possessed by the original Hamiltonian—say translational or rotational invariance—has been broken by the choice of the (Hartree–Fock) ground state with which we work. (We encountered one variety of such situations in the context of scalar field theory in Exercise 2.2, where we noted that Goldstone's theorem requires the existence of an excitation degenerate with the ground state in such a case. Here that excitation will be the infinitely slow dragging or cranking of a nucleus for the respective cases of translational or rotational invariance.)

We now develop these ideas in somewhat greater detail. This material is a little peripheral to the main lines of development of the Hartree–Fock method, but it is very useful in completing the notions of Hartree–Fock stability discussed in the previous section and in motivating the random-phase approximation, to be discussed in Section 4.6. Consider the total system transformation generated by operators such as those of Eq. 41 or 42,

$$e^{iG} = \exp\left\{i \sum_{\alpha\beta} \langle \alpha|g|\beta\rangle a_\alpha^\dagger a_\beta\right\},\qquad(45a)$$

where

$$G = \mathbf{P}, \mathbf{J}, \ldots \quad \text{and} \quad g = \mathbf{p}, \mathbf{j}, \ldots,\qquad(45b, c)$$

G and g being Hermitian. The transformed ground state is

$$|\Phi'\rangle = e^{iG}|\Phi\rangle\qquad(46)$$

and has the structure of the variations in Eq. 36 if we make the identification

$$c_{\rho\mu} = i\langle \rho|g|\mu\rangle \quad \text{and} \quad c_{\rho\mu}^* = -i\langle \mu|g|\rho\rangle,\qquad(47)$$

and the energy in the transformed state is

$$E_0' = \langle\Phi'|H|\Phi'\rangle = \langle\Phi|H|\Phi\rangle + i\langle\Phi|[H, G]|\Phi\rangle$$

$$+ \frac{i^2}{2}\langle\Phi|[[H, G], G]|\Phi\rangle,\qquad(48)$$

to order G^2. For the one-particle operators G we are discussing, the linear terms vanish—just as in Eq. 37a—by Brillouin's theorem, since, from Eq. 11,

$$\langle\Phi|[H, G]|\Phi\rangle = \sum_{\rho\mu}[\langle\Phi|H|\Phi_\mu^\rho\rangle\langle\Phi_\mu^\rho|G|\Phi\rangle$$

$$-\langle\Phi|G|\Phi_\mu^\rho\rangle\langle\Phi_\mu^\rho|H|\Phi\rangle] = 0;\qquad(49)$$

note that the one-particle structure of G restricts us to 1p–1h intermediate states here. The quadratic form in Eq. 48 has the identical structure to that in Eq. 37a, with the identification of Eq. 47. Thus if we view G as generating a variation in the ground-state wave function, and if G is a constant of the motion so that

$$[H, G] = 0,\qquad(50)$$

as in Eqs. 43 or 44, then it follows from Eq. 48 that $E_0' = E_0$ and $|\Phi'\rangle$ is degenerate with the ground state; this is the unavoidable vanishing eigenvalue $\omega = 0$ that we noted in the context of Eqs. 40 (called a Goldstone boson in other contexts). This degeneracy merely reflects the fact that a translated or rotated self-consistent solution would work just as well as the particular one we happen to have hit upon for our Hartree–Fock solution. If $|\Phi\rangle$ were an eigenstate of G, this situation would not be problematic, because the transformed state in Eq. 46 would be trivially equivalent to the original state and G would not generate true variations; but that of course is not generally the case.

One might hope that the fact that G is a constant of the motion would lead to a self-consistent potential $\mathcal{U}$ possessing the symmetry in question (under translations for $G = \mathbf{P}$, or rotations for $G = \mathbf{J}$, and so on), but as we have already noted, this is not to be expected. In fact, from Eq. 50 it is not difficult to show (see Exercise 4.7) that the commutator of the self-consistent field $\mathcal{U}$ satisfies

$$\langle \rho | [g, \mathcal{U}] | \mu \rangle = \sum_{\alpha\lambda} [\langle \alpha\rho | v | \mu\lambda \rangle_A \langle \lambda | g | \alpha \rangle$$
$$- \langle \lambda\rho | v | \mu\alpha \rangle_A \langle \alpha | g | \lambda \rangle] \neq 0 , \tag{51}$$

so that $\mathcal{U}$ does not carry the symmetry in question, as we have already anticipated below Eq. 44.

A related problem arises for the time dependence of operators when treated in the Hartree–Fock approach. Suppose we have a one-particle operator F, as in Eq. 33, which satisfies the commutator equation of motion

$$-i\dot{F} = [H, F] . \tag{52}$$

For the ground-state expectation value one finds

$$\langle \Phi | - i\dot{F} | \Phi \rangle = \sum_{\rho\mu} [\langle \Phi | F | \Phi_\mu^\rho \rangle \langle \Phi_\mu^\rho | H | \Phi \rangle$$
$$- \langle \Phi | H | \Phi_\mu^\rho \rangle \langle \Phi_\mu^\rho | F | \Phi \rangle] = 0 , \tag{53}$$

where the one-particle structure of F restricts the intermediate states to 1p–1h states and Brillouin's theorem (or, equivalently, the variational principle) requires that the matrix elements of H that appear vanish. Thus the diagonal ground-state expectation value of the time derivative of F vanishes, as one would anticipate. However, matrix elements between the ground state and a particle–hole excitation yield an unexpectedly complicated result, namely

$$\langle \Phi_\mu^\rho | (-i\dot{F}) | \Phi \rangle = \langle \Phi_\mu^\rho | [H, F] | \Phi \rangle$$
$$= (\epsilon_\rho - \epsilon_\mu) \langle \Phi_\mu^\rho | F | \Phi \rangle$$
$$+ \sum_{\alpha\lambda} [\langle \lambda | f | \alpha \rangle \langle \alpha\rho | v | \mu\lambda \rangle_A$$
$$- \langle \rho\lambda | v | \alpha\mu \rangle_A \langle \alpha | f | \lambda \rangle] , \tag{54}$$

and one cannot cavalierly suppose that the time derivative merely brings down a factor of the difference of single-particle energies in matrix elements of 1p–1h states.

As a result of this, energy-weighted sum rules also do not hold directly in the Hartree–Fock method. The conventional derivation of such sum rules considers manipulations applied to a double commutator,

$$
\langle \Phi | [F, [H, F]] | \Phi \rangle = \sum_i [\langle \Phi | F | \Phi_i \rangle \langle \Phi_i | [H, F] | \Phi \rangle
$$

$$
- \langle \Phi | [H, F] | \Phi_i \rangle \langle \Phi_i | F | \Phi \rangle]
$$

$$
= \sum_i (E_i - E_0) |\langle \Phi_i | F | \Phi \rangle|^2 , \qquad (55)
$$

where $\{\Phi_i\}$ is a complete set of eigenstates of the Hamiltonian H with eigenvalues E_i. In the Hartree–Fock method one does not have available eigenstates of H. Instead, if F is a one-particle operator, the connecting intermediate states are, of necessity, $|\Phi_\mu^\sigma\rangle$, and the analog of $\langle \Phi_i | [H, F] | \Phi \rangle = (E_i - E_0) \langle \Phi_i | F | \Phi \rangle$ is Eq. 54 and does not at all permit the simple structure of Eq. 55.

Thus in Hartree–Fock problems where symmetry considerations are important, one must develop specialized techniques to deal with them. If time dependence or sum rules enter, one must treat them carefully within the framework set by the Hartree–Fock theory. A generalization of Hartree–Fock theory based on a particular approximative approach is available that reduces many of these ills. This method is called the *random-phase approximation* and is treated briefly at the end of this chapter. It cannot of course be regarded as a panacea for ills of the Hartree–Fock method, since it must be evaluated on its own merits as an approximation technique.

4.5 SOME EXAMPLES AND APPLICATIONS

4.5.1 Hartree–Fock Methods and Atomic Systems

It is well known that for atoms the Hartree–Fock method provides the conceptual underpinnings for the usual picture of atomic shells, the filling of which in accordance with the exclusion principle leads to an explanation of the periodic table of the chemical elements. For many purposes it is sufficient to anticipate that the resulting self-consistent field will lead to single-particle states nl of principal quantum number n and good orbital angular momentum l. The occupation of such a shell by spin $s = \frac{1}{2}$ fermions then admits $2(2l + 1)$ electrons, one each for the spin projections $m_s = \pm \frac{1}{2}$ and orbital angular momentum projections $m_l = -l, \ldots, -1, 0, 1, \ldots, l$. This leads to the occupation scheme shown in Table 4.1, with its obvious

TABLE 4.1 General Character of Shells and Their Occupancy for an Atomic Self-Consistent Potential

Shell number	Electron states	Number of states	Atoms generated
1	$1s$	2	H, He
2	$2s, 2p$	$2 \times (1+3) = 8$	Li, Be, B, C, N, O, F, Ne
3	$3s, 3p$	$2 \times (1+3) = 8$	Na, Mg, Al, Si, P, S, Cl, Ar
4	$4s, 3d, 4p$	$2 \times (1+5+3) = 18$	$\cdots$
5	$5s, 4d, 5p$	$2 \times (1+5+3) = 18$	$\cdots$
6	$6s, 4f, 5d, 6p$	$2 \times (1+7+5+3) = 32$	Rare earths: fill internal $4f$
7	$7s, 6d, 5f, \ldots$		

correspondence to properties of valency and inert elements that are known phenomenologically and embodied in the periodic table.

For more detailed or more subtle features one requires the actual properties of the Hartree–Fock solutions, for instance the single-particle wave functions. (These are often obtained in practice by performing the relatively straightforward Hartree calculation—without antisymmetriza-tion—and then incorporating antisymmetrization effects through an approx-imative approach inferred from Fermi-gas behavior known as the Hartree–Fock–Slater method.) For example, to understand the phenomenon of the rare earths one must take into account the radial behavior of the single-particle wave functions in the active shells. This is illustrated in Figure 4.2, where radial wave functions for the $4s$, $4p$, $4f$, and $6s$ shells of $_{58}$Ce, based on the Hartree–Fock–Slater method, are exhibited. One sees there that the 6s wave function is relatively exterior and it is this that accounts for the similar chemical properties through the rare-earth region from $_{57}$La through $_{58}$Ce, $_{59}$Pr, $\ldots$, $_{66}$Dy, $\ldots$, and on to $_{71}$Lu; these all share the features shown in Figure 4.2 and have their chemistry determined by outer electrons $(5d)^1(6s)^2$ while the successive addition of electrons is accommodated in the inner $4f$ shell. It is this that explains why the rare earths are rare or "dysprositous."

In general the application* of the Hartree–Fock method to atoms is quantitatively successful. As an example we show in Table 4.2 the ionization potentials for atoms through the $3p$ shell as derived on the basis of the Hartree–Fock approach and compared with observed values. The agree-ment is especially striking for the $3p^n$ configurations.

*An extensive discussion of methods and applications is given in C. Froese Fischer, *The Hartree–Fock Method for Atoms* (Wiley – Interscience, New York, 1977).

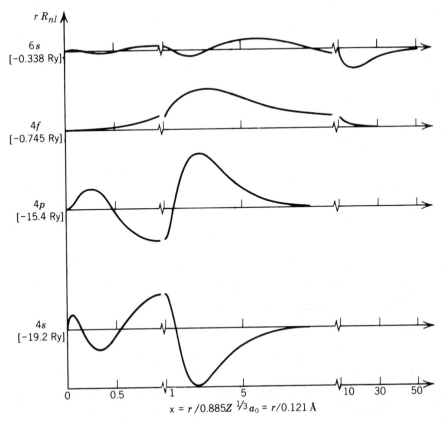

FIGURE 4.2 The radial wave functions for the 4s, 4p, 4f, and 6s single-electron states in $_{58}$Ce, normalized so that $\int_0^\infty R_{nl}^2 r^2\, dr = 1$. Binding energies in Rydberg units are given in brackets under the shell designation. [Plotted on the basis of results in F. Herman and S. Skillman, *Atomic Structure Calculations* (Prentice-Hall, Englewood Cliffs, N.J., 1963).]

4.5.2 Thomas–Fermi Methods and Atomic Systems

In Section 1.4 we have encountered the enormous simplification that results in the description of many-fermion systems if they can be treated as a Fermi gas. One could imagine that for a large atomic system the net effect of the self-consistent field is to provide a containment volume for a gas of electrons. Since the characteristics of the field vary over the atomic volume, one would have to suppose that there is a Fermi gas associated locally with the different regions of the atom. In other words, assuming that the system properties vary slowly over distance intervals comparable with an electron wavelength, at each point of the system we associate a Fermi gas with density equal to the density of the electrons at that point and then study the physics of the system in terms of the equivalent Fermi gas. This method is

TABLE 4.2   Comparison of Observed and Calculated
Ionization Potentials

Atom	Electron configuration	Observed ionization potential	Calculated ionization potential
Li	$2s$	0.198	0.196
Be	$2s^2$	1.012	0.962
B	$2p$	0.305	0.310
C	$2p^2$	1.310	1.357
N	$2p^3$	3.366	3.495
O	$2p^4$	6.657	6.855
F	$2p^5$	11.634	11.945
Ne	$2p^6$	18.669	19.113
Na	$3s$	0.189	0.182
Mg	$3s^2$	0.834	0.785
Al	$3p$	0.220	0.210
Si	$3p^2$	0.900	0.891
P	$3p^3$	2.234	2.229
S	$3p^4$	4.268	4.262
Cl	$3p^5$	7.277	7.269
Ar	$3p^6$	11.409	11.418

(From C. Froese Fischer, *The Hartree–Fock Method for Atoms*, Wiley–Interscience, New York, 1977). The ionization shown is for the removal of all the electrons in the configuration noted, and is given in units of me^4 ($\cong 27.21$ eV).

thus sometimes called the *local-density approximation*. We pursue this method here because it illustrates a technique that is often useful as a first approximation for complicated phenomena in many-fermion systems other than atoms. For the atomic case this approach has no hope, of course, of accounting for shell structure, as in Table 4.1 say, but should give the behavior of the atom after the shell features have been averaged out, and thus provides a kind of average of the Hartree or Hartree–Fock method.

Towards this end we consider the Fermi energy at each point to be

$$\epsilon_F = V(r) + \frac{k_F^2(r)}{2m} \, , \tag{56}$$

where $V(r)$ is the potential acting on an electron of mass m at radial distance r, and $k_F(r)$ is the local Fermi momentum derived from the electron density ρ at r according to Eq. 1.54, that is,

$$k_F(r) = [3\pi^2 \rho(r)]^{1/3} . \tag{57}$$

The Fermi energy ϵ_F is not a function of r, since if it were, electrons would migrate in the atom to the minimum point in energy ϵ_F, whereas we assume that the electrons are distributed with a given density $\rho(r)$ [that determines $k_F(r)$] that is consistent with the potential $V(r)$. It is then convenient to define the potential relative to the Fermi energy,

$$\mathcal{V}(r) \equiv V(r) - \epsilon_F ; \tag{58}$$

the consistency of this potential with electron density is contained in Poisson's equation relating the electrostatic potential to the charge density,

$$\nabla^2 \mathcal{V} = -4\pi e^2 \rho , \tag{59}$$

where e is the electron charge ($e^2 = \alpha$ in this section to conform with conventional atomic usage). Using Eqs. 56–59 as well as the Laplacian in spherical coordinates, we arrive at a differential equation for the potential,

$$\frac{1}{r^2} \frac{d}{dr} r^2 \frac{d}{dr} \mathcal{V} = \frac{1}{r} \frac{d^2}{dr^2} r\mathcal{V} = -\frac{4e^2}{3\pi} [-2m\mathcal{V}(r)]^{3/2} , \tag{60}$$

with boundary conditions

$$V \xrightarrow[r\to0]{} -\frac{Ze^2}{r} , \quad \text{or} \quad r\mathcal{V} \xrightarrow[r\to0]{} -Ze^2 , \tag{61}$$

and $\mathcal{V} \xrightarrow[r\to\infty]{} 0$ (see Eq. 56).

To eliminate the explicit Z-dependence we introduce variables

$$r = xb , \quad b \equiv \frac{(3\pi)^{2/3}}{2^{7/3}} \frac{1}{me^2} Z^{-1/3} = 0.885 a_0 Z^{-1/3} , \tag{62}$$

where $a_0 = (me^2)^{-1}$ is the hydrogen Bohr radius, and

$$r\mathcal{V}(r) = -Ze^2 \Phi(x) . \tag{63}$$

In terms of these universal quantities,

$$\frac{d^2\Phi}{dx^2} = \frac{1}{\sqrt{x}} \Phi^{3/2} , \quad \Phi(0) = 1 , \tag{64a, b}$$

and the length scaling is according to $Z^{-1/3}$ throughout the rest of the problem. Note that the cube root arises because of the basic Fermi-gas assumption, Eq. 57, and that the atoms become spatially smaller as Z gets larger, that is, the Coulomb attraction wins weakly over the exclusion principle as more electrons are added.

The solution to Eqs. 64 can be approximated near the origin by

$$\Phi(x) = 1 - ax + \tfrac{4}{3}x^{3/2} \, , \tag{65}$$

where $a > 1.589$ leads to a Φ that vanishes at some radius and corresponds to a free ion; $a = 1.589$ causes Φ to approach the axis asymptotically with no sharp boundary—a neutral atom but without a clear-cut radius; and $a < 1.589$ makes Φ diverge at large x, which is not of physical interest. An approximate solution to Eq. 64 for large x (but good to 10% even for $x < 10$) is

$$\Phi(x) = \left[1 + \left(\frac{x^3}{144} \right)^{\lambda/3} \right]^{-3/\lambda} \, , \qquad \lambda \equiv \tfrac{1}{2}(\sqrt{73} - 7) = 0.772 \, . \tag{66}$$

In Figure 4.3 we compare a numerical solution of Eqs. 64 with a Hartree

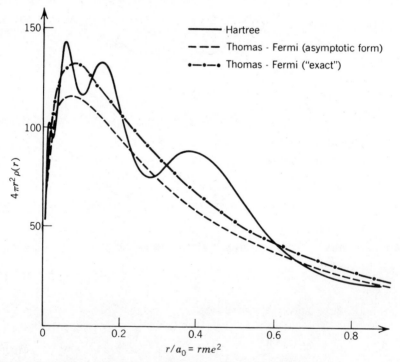

FIGURE 4.3 Spherical densities for Hg, normalized to $\int_0^\pi 4\pi r^2 \rho(r) \, dr = Z$, for a Hartree solution [plotted on the basis of D. R. Hartree and W. Hartree, Proc. Roy. Soc. A149, 210 (1935)], for the Thomas–Fermi solution [plotted on the basis of V. Bush and S. H. Caldwell, Phys Rev. 38, 1898 (1931)], and for the asymptotic expression of Eq. 66.

solution for the neutral Hg atom (and also with the form of Eq. 66); the comparison is made with a Hartree solution, and not Hartree–Fock, because the Thomas–Fermi method does not include exchange effects. As anticipated, the Thomas–Fermi solution averages over the shell structure of the Hartree result, but describes this average behavior very well.

To include exchange effects one may generalize the approach to what is called the Thomas–Fermi–Dirac method, which considers the exchange potential of Eq. 6d in the form

$$\mathcal{W}(\mathbf{r}, \mathbf{r}') = - \frac{e^2 \rho(\mathbf{r}, \mathbf{r}')}{|\mathbf{r} - \mathbf{r}'|} , \tag{67}$$

with the exchange density

$$\rho(\mathbf{r}, \mathbf{r}') = \sum_j \phi_j^*(\mathbf{r}') \phi_j(\mathbf{r}) \tag{68}$$

in terms of single-particle wave functions $\phi_j(\mathbf{r})$. These are here approximated by the Fermi gas form of Eq. 1.52, so that

$$\rho(\mathbf{r}, \mathbf{r}') = \frac{1}{\Omega} \sum_j e^{i\mathbf{k}_j \cdot (\mathbf{r} - \mathbf{r}')} = \int_{k \le k_F} e^{i\mathbf{k} \cdot (\mathbf{r} - \mathbf{r}')} \frac{d\mathbf{k}}{(2\pi)^3}$$

$$= \frac{1}{2\pi^2 |\mathbf{r} - \mathbf{r}'|^3} \left(\sin k_F |\mathbf{r} - \mathbf{r}'| - k_F |\mathbf{r} - \mathbf{r}'| \cos k_F |\mathbf{r} - \mathbf{r}'| \right) \tag{69}$$

for a system of volume Ω. Equation 69 is to be seen as valid for $\mathbf{r} \sim \mathbf{r}'$, whence the local Fermi momentum to be inserted is $k_F(r) \sim k_F(r')$. The effective exchange potential resulting from Eqs. 67–69 is defined so as to be of local form, again exploiting the Fermi-gas approximation:

$$V_{\text{eff}}(r) \equiv - \int \frac{e^2}{|\mathbf{r} - \mathbf{r}'|} \rho(\mathbf{r}, \mathbf{r}') \frac{\phi_j(\mathbf{r}')}{\phi_j(\mathbf{r})} d\mathbf{r}'$$

$$= - \int \frac{e^2}{|\mathbf{r} - \mathbf{r}'|} \rho(\mathbf{r}, \mathbf{r}') e^{i\mathbf{k}_j \cdot (\mathbf{r}' - \mathbf{r})} d\mathbf{r}'$$

$$= - \frac{2}{\pi} e^2 k_F F(\eta) , \tag{70a}$$

where

$$F(\eta) = \frac{1}{2} + \frac{1 - \eta^2}{4\eta} \log \frac{1 + \eta}{1 - \eta} , \qquad \eta \equiv \frac{k_j}{k_F} . \tag{70b}$$

Averaging over all electrons, this is, using Eqs. 69 and 70a,

$$\bar{V}_{\text{eff}} = \frac{\int_0^{k_F} V_{\text{eff}} \, d\mathbf{k}_j}{\int_0^{k_F} d\mathbf{k}_j} = -\frac{2e^2}{\rho} \int \frac{\rho^2(\mathbf{r}, \mathbf{r}')}{|\mathbf{r} - \mathbf{r}'|} \, d\mathbf{r}' = -\frac{3e^2}{2\pi} \, k_F \, . \tag{71}$$

When this term is included with the direct potential of Eq. 58, the Thomas–Fermi–Dirac equation results:

$$\frac{d^2 \Phi}{dx^2} = x\left(\sqrt{\frac{\Phi}{x}} + \beta \right)^3 , \qquad \beta = \frac{1}{\pi \sqrt{2}} \sqrt{\frac{b}{a_0 Z}} = 0.212 \, Z^{-2/3} , \tag{72}$$

as a replacement for Eq. 64a. In this case the atoms will have finite radii as well as the ions (see below Eq. 65). Having solved Eq. 64a or 72 for Φ and hence for an average approximation to the self-consistent potential, one can also solve the Schrödinger equation with this potential to generate approximate single-electron wave functions.

4.5.3 The Hartree–Fock Method for Nuclear Systems

In parallel to the extraction of atomic shell structure from the Hartree–Fock method for atoms, one can develop nuclear shell phenomena from a Hartree–Fock approach applied to the nucleon–nucleon force. As we have already noted, the atomic problem differs from the nuclear one in having an important central potential present even before the self-consistent field is developed. The Coulomb force is also more easily treated than nuclear forces, especially since the latter involve very strong repulsion at short distances, which makes the matrix elements of the interaction large and requires special techniques to carry out the Hartree–Fock program. Among other possible approaches, the Hartree–Fock method may be used as a point of departure to which further refinements are then attached. The Hartree–Fock theory gives the order of magnitude of level splittings and a first approximation for single-particle orbitals. Then, for instance, one may proceed with improvements patterned after the Brueckner treatment of infinite fermion systems discussed in Section 10.4.

The qualitative feature of the emergence of a shell structure is clear in the nuclear case just as it was for atoms. In the nuclear case the systematics of nuclear ground states and excitations shows that a strong spin–orbit force operates in the self-consistent field and acts to lower the energy of total-spin states with $j = l + \frac{1}{2}$ as compared with $j = l - \frac{1}{2}$ for the single-nucleon states. The population of these j states is then by $2j + 1$ protons, one for each of the substates $m_j = -j, \ldots, -\frac{1}{2}, \frac{1}{2}, \ldots, j$, and $2j + 1$ neutrons. Eventually the Coulomb repulsion between protons makes it advantageous for the system to add neutrons preferentially. The population scheme for light nuclei is shown in Table 4.3. The shells close at values of proton or neutron number that are called "magic numbers." Amongst the specialized ways in which

TABLE 4.3   Shell Structure of Nuclei for the Lighter Nuclei, where Proton and Neutron States Are Populated Approximately Equally

Shell number	Nucleon states	Number of states	Nuclei generated
1	$1s_{1/2}$	2	H ($= {}^{1,2,3}$H), He ($= {}^{3,4}$He)
2	$1p_{3/2}, 1p_{1/2}$	$4 + 2 = 6$	Li, Be, B, C; N, O
3	$1d_{5/2}, 2s_{1/2}, 1d_{3/2}$	$6 + 2 + 4 = 12$	F, Ne, Na, Mg, Al, Si; P, S; Cl, Ar, K, Ca
4	$1f_{7/2}, 2p_{3/2}, 2p_{1/2}, 1f_{5/2}$	$8 + 4 + 2 + 6 = 20$	Sc, Ti, V, . . .
5	$1g_{9/2}, 2d_{5/2}, 1g_{7/2}, 3s_{1/2}, 2d_{3/2}$	. . .	. . .

self-consistent-field methods may be applied to nuclei is an approach called the *restricted Hartree–Fock procedure*, in which one treats the last previously closed shell as inert and then proceeds to carry out the search for self-consistency with respect to the space of the shell currently being filled only. In this way one can study the competition between solutions possessing spherical symmetry and those preferring prolate or oblate spheroidal deformations or solutions with no axial symmetry at all. Deformed solutions are not of interest in the application of Hartree–Fock methods to atoms, because the central potential supplied by the Coulomb potential of the nucleus leads naturally to spherical shapes, but in the nuclear case the self-consistent potential is produced purely out of the average of the nucleon–nucleon interaction, and the resulting Hartree–Fock ground state is often deformed. These nonspherical solutions raise naturally the questions of symmetry breaking in the Hartree–Fock method that were touched upon in Section 4.4. A particularly dramatic case is that of ^{28}Si, done with the Hartree–Fock method restricted to the nuclear $2s$–$1d$ shell. One finds a strong competition between prolate, spherical, and oblate solutions for ^{28}Si (which in a spherical shell model would close the $1d_{5/2}$ subshell—the lowest of the three orbitals $1d_{5/2}$, $2s_{1/2}$, $1d_{3/2}$ in the shell when spin–orbit splitting acts). The three solutions yield total Hartree–Fock energies of -122.0, -114.2, and -123.0 MeV, respectively, relative to the ^{16}O core (the comparable experimental value being -134.4 MeV). Thus the lowest-energy solution is almost certainly deformed, but with which deformation shape is unclear. Equidensity surfaces for the two deformed cases are shown in Figure 4.4 and are obviously very different for the two situations.

4.5.4   The Hartree–Fock Method for Solids

The atomic Hartree–Fock solutions that we discussed in Subsection 4.5.1 pertain to a situation in which the atom is in isolation. In a metallic solid the atoms are so close together that the individual atomic Hartree–Fock states disappear (or inner, core electrons stay in somewhat modified states, while outer, valence electrons "merge"). They are then replaced by states that

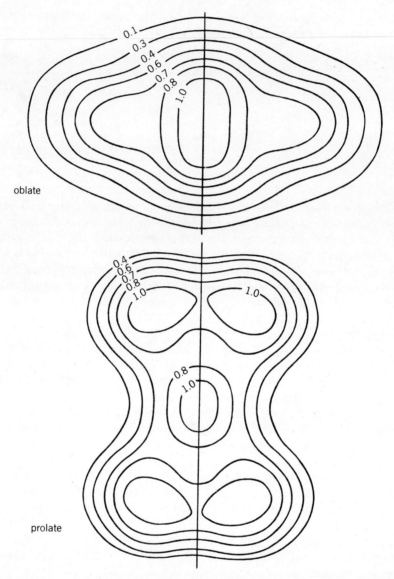

FIGURE 4.4 Equidensity surfaces of ^{28}Si for oblate and prolate Hartree–Fock solutions. The contour with highest density is normalized to unity, and the scale of length of the vertical axis is 12% smaller than for the horizontal axis. The plane of the cross section contains the symmetry axis (vertical line) of the solution. [From G. Ripka, in *Advances in Nuclear Physics, Vol. 1*, M. Baranger and E. Vogt, eds. (Plenum, New York, 1968), p. 183.]

extend through the entire solid, and one must attempt to address the self-consistency problem for this large system. (All of this has tacitly assumed that we take the background of positively charged nuclei as fixed during the discussion of the electron motion.)

The complexity of the problem involved in applying the Hartree–Fock method to electrons in a solid has required rather elaborate specialized approximation schemes,* often guided by empirical considerations, almost from the beginning of the procedure. It is common to begin such a program by noting that in a solid the positively charged ion cores may be viewed as merging into a uniform positive background charge. This is then neutralized by an approximately uniform distribution of electronic charge. (This system is the electron gas, which is discussed in more detail in Chapter 6.) That is to say, the valence electrons roughly form plane-wave states, and the Coulomb energy of their mutual repulsion cancels the energy of their attraction to the positive cores. The remaining energy term is then the exchange energy of the Hartree–Fock equations. One can show that these equations are consistent with plane-wave solutions for the electrons, as we have noted above. The resulting single-particle energy spectrum is

$$\epsilon_{\mathbf{k}} = \frac{k^2}{2m} - \frac{e^2 k_F}{2\pi} \left(2 + \frac{k_F^2 - k^2}{k k_F} \log \left| \frac{k_F + k}{k_F - k} \right| \right), \tag{73}$$

where m and e are the electron's mass and charge, $\mathbf{k}$ is its momentum in the plane-wave state, and k_F is the Fermi momentum or last occupied momentum for the system. The result for $\epsilon_{\mathbf{k}}$ has a singular slope at $k = k_F$, which leads for instance to an infinite electron velocity $\boldsymbol{v} = d\epsilon_{\mathbf{k}}/d\mathbf{k}$ at the Fermi surface. This unacceptable feature of the theory arises essentially from the infinite range of the Coulomb potential, and so has its cure in the screening of this potential. (Screening is sometimes introduced for this purpose in somewhat phenomenological ways; we shall not pursue this here, but do return to a more complete discussion of screening and electron–electron correlations in Sections 6.3, 10.5, and 10.6.)

Thus the Hartree–Fock theory is not a very useful point of departure for studying the behavior of electrons in solids. Since the offending term arises from the exchange energy, it has often been found convenient to use the Hartree method—with no exchange energy—instead. There one has neglected both the Pauli principle and the Coulomb correlations, and these two omissions tend to cancel, whereas in the Hartree–Fock case Coulomb correlations are ignored with no compensating, canceling error. The Hartree approximation of a free-electron gas gives helpful guidance in understanding

---

*The behavior of electrons in solids from the viewpoint of the Hartree–Fock method is discussed in many modern texts on solid-state physics, including N. W. Ashcroft and N. D. Mermin, *Solid State Physics* (McGraw-Hill, New York, 1970); J. D. Patterson, *Introduction to the Theory of Solid State Physics* (Addison-Wesley, Reading, Massachusetts, 1971).

experiments, but of course is inconsistent with the methods used for a free atom. To get around these difficulties of applying a successful self-consistent potential method to electrons in solids therefore requires refined, specialized, and often very complicated calculational techniques which take up a considerable part of texts on solid state physics (including those listed in the footnote near the beginning of this subsection), and we shall not pursue them further here.

4.6 THE RANDOM-PHASE APPROXIMATION

We consider an extension of the Hartree–Fock method that may improve the quality of the approximation for treating excitations in a many-body system. The new scheme also offers a formal solution—at this level of approximation—to some of the limitations of the Hartree–Fock approach with respect to the preservation of symmetry properties for excitations of the system and the fulfillment of energy-weighted sum rules. This extension is called the *random-phase approximation*. We treat it now through a fairly immediate approach involving linearization of the equations of motion. In Chapter 10 the random-phase approximation will be developed through diagrammatic techniques, and will prove useful in handling collective excitations of the electron gas.

The first basic notion that we require in order to develop the random-phase approximation—and that we shall need again in the next chapter in discussing pairing—is the blurring of the sharpness of the Fermi surface: Up to now it has been basic in the Hartree–Fock method, as expressed in Eq. 8 for our trial wave function, that within this approach we fully populate N levels for the N fermions and leave the remaining levels completely empty. We now envisage a more general situation in which the effects of the two-particle interaction may lead to a system wave function with partial occupation of the single-particle states, that is, we have a certain probability, not necessarily equal to zero or unity, to find a fermion in any given state. We shall attempt to make this generalization convincing by showing that it reduces to the Hartree–Fock result in the appropriate limit, so that we do not lose our mooring in the basic concepts of a self-consistent field.

For the random-phase approximation we consider a ground state $|\Psi\rangle$ which we expect to be different from the Hartree–Fock trial state of Eq. 8. It is to satisfy, for the Hamiltonian H of Eq. 7,

$$H|\Psi\rangle = E_0|\Psi\rangle , \qquad (74)$$

and excited states $|\Psi_n\rangle$ are to be generated by the action of the transition operator $Q_n^\dagger$ according to

$$|\Psi_n\rangle = Q_n^\dagger|\Psi\rangle . \qquad (75)$$

The introduction of $Q_n^\dagger$ at this point simply allows us to calculate directly with the ground state. The operator has the equation of motion

$$[H, Q_n^\dagger] = \Omega_n Q_n^\dagger , \tag{76}$$

whence the energy of the state $|\Psi_n\rangle$ is $E_n = E_0 + \Omega_n$ since

$$\begin{aligned}
H|\Psi_n\rangle &= HQ_n^\dagger|\Psi\rangle = [H, Q_n^\dagger]|\Psi\rangle + Q_n^\dagger H|\Psi\rangle \\
&= (E_0 + \Omega_n)Q_n^\dagger|\Psi\rangle = (E_0 + \Omega_n)|\Psi_n\rangle .
\end{aligned} \tag{77}$$

Now we make the simplifying assumption that the operator $Q_n^\dagger$ has the structure of a one-particle operator (higher-order forms could also be considered, leading to a more complex theory):

$$Q_n^\dagger = \sum_{\substack{\sigma > F \\ \lambda \leq F}} [x_{\sigma\lambda}^{(n)} a_\sigma^\dagger a_\lambda - y_{\sigma\lambda}^{(n)} a_\lambda^\dagger a_\sigma] , \tag{78}$$

where the first term is close in spirit to Hartree–Fock considerations for a particle-hole excitation, while the second introduces a new ingredient. The second term in Eq. 78 would vanish if applied to the Hartree–Fock trial state of Eq. 8, but escapes that fate here, since in generalizing to the state $|\Psi\rangle$ of Eq. 74 we have not ruled out the possibility of correlation in the ground state, that is, we do not demand perfect occupancy or vacancy for the single-particle states. Note, however, that in Eq. 78 and in the following, quantities enter that are defined with respect to a preexisting Hartree–Fock solution for the problem.

We have encountered commutators of the structure of Eq. 76 with operators like H and $Q_n^\dagger$ of Eq. 78 on several occasions (for example in Eq. 50), and we know that it will be impossible to make these forms compatible unless we use an approximation that eliminates higher orders in creation and annihilation operators. The linearizing approximation that generates the random-phase result is

$$\begin{aligned}
a_\alpha^\dagger a_\beta^\dagger a_\gamma a_\delta \rightarrow\ & a_\alpha^\dagger a_\delta \langle\Phi|a_\beta^\dagger a_\gamma|\Phi\rangle - a_\beta^\dagger a_\delta \langle\Phi|a_\alpha^\dagger a_\gamma|\Phi\rangle \\
& + a_\beta^\dagger a_\gamma \langle\Phi|a_\alpha^\dagger a_\delta|\Phi\rangle - a_\alpha^\dagger a_\gamma \langle\Phi|a_\beta^\dagger a_\delta|\Phi\rangle \\
= & \ a_\alpha^\dagger a_\delta \bar{\delta}_{\beta\gamma} - a_\beta^\dagger a_\delta \bar{\delta}_{\alpha\gamma} + a_\beta^\dagger a_\gamma \bar{\delta}_{\alpha\delta} - a_\alpha^\dagger a_\gamma \bar{\delta}_{\beta\delta} ,
\end{aligned} \tag{79}$$

in the notation of Eq. 12a. This form is linearized with respect to $a^\dagger a$, that is, terms of fourth order in the creation and annihilation operator are replaced with terms of the structure producing a single-particle excitation. We may expect this approximation to be valid if there is only a small admixture of particle–hole correlations in the ground state. The equation of motion 76 now closes algebraically and becomes

$$[H, Q_n^\dagger] = \sum_{\sigma\lambda} (\epsilon_\sigma - \epsilon_\lambda) a_\sigma^\dagger a_\lambda x_{\sigma\lambda}^{(n)} + \sum_{\alpha\gamma\sigma\lambda} \langle \alpha\lambda | v | \gamma\sigma \rangle_A a_\alpha^\dagger a_\gamma x_{\sigma\lambda}^{(n)}$$

$$+ \sum_{\sigma\lambda} (\epsilon_\sigma - \epsilon_\lambda) a_\lambda^\dagger a_\sigma y_{\sigma\lambda}^{(n)} - \sum_{\alpha\delta\sigma\lambda} \langle \alpha\sigma | v | \lambda\delta \rangle_A a_\alpha^\dagger a_\delta y_{\sigma\lambda}^{(n)}$$

$$= \Omega_n \sum_{\sigma\lambda} [a_\sigma^\dagger a_\lambda x_{\sigma\lambda}^{(n)} - a_\lambda^\dagger a_\sigma y_{\sigma\lambda}^{(n)}], \tag{80}$$

and identifying the corresponding coefficients of $a, a^\dagger$, we get

$$(\epsilon_\sigma - \epsilon_\lambda - \Omega_n) x_{\sigma\lambda}^{(n)}$$

$$+ \sum_{\substack{\tau > F \\ \mu \leq F}} [\langle \sigma\mu | v | \lambda\tau \rangle_A x_{\tau\mu}^{(n)} - \langle \sigma\tau | v | \mu\lambda \rangle_A y_{\tau\mu}^{(n)}] = 0 \tag{81a}$$

and

$$(\epsilon_\sigma - \epsilon_\lambda + \Omega_n) y_{\sigma\lambda}^{(n)}$$

$$- \sum_{\substack{\tau > F \\ \mu \leq F}} [\langle \lambda\tau | v | \mu\sigma \rangle_A y_{\tau\mu}^{(n)} - \langle \lambda\mu | v | \sigma\tau \rangle_A x_{\tau\mu}^{(n)}] = 0. \tag{81b}$$

If the correlation terms in $y_{\sigma\lambda}^{(n)}$ were dropped here, Eq. 81a would become the equation for diagonalizing the particle–hole secular matrix, so that this description of excitations would reduce to that of the Hartree–Fock.

Recalling the definitions in Eqs. 38 and the combination that appeared in Eq. 39,

$$M \equiv \begin{pmatrix} A & B \\ B* & A* \end{pmatrix}, \tag{82}$$

we see that the random-phase equations can be written in terms of column vectors

$$X_n \equiv \{x_{\sigma\lambda}^{(n)}\} \quad \text{and} \quad Y_n \equiv \{y_{\sigma\lambda}^{(n)}\} \tag{83a, b}$$

as

$$M\begin{pmatrix} X_n \\ Y_n \end{pmatrix} = \begin{pmatrix} A & B \\ B* & A* \end{pmatrix}\begin{pmatrix} X_n \\ Y_n \end{pmatrix} = \Omega_n \begin{pmatrix} X_n \\ -Y_n \end{pmatrix} \tag{84a}$$

or

$$\begin{pmatrix} A & B \\ -B* & -A* \end{pmatrix}\begin{pmatrix} X_n \\ Y_n \end{pmatrix} = \Omega_n \begin{pmatrix} X_n \\ Y_n \end{pmatrix}, \tag{84b}$$

so that the solution of these equations requires finding the eigenvalue for a non-Hermitian matrix. Note that if we take the complex conjugate of Eq. 84a we get, provided Ω_n is real,

$$\begin{pmatrix} A & B \\ B^* & A^* \end{pmatrix} \begin{pmatrix} Y_n^* \\ X_n^* \end{pmatrix} = -\Omega_n \begin{pmatrix} Y_n^* \\ -X_n^* \end{pmatrix}, \tag{84c}$$

whence we see that a negative eigenvalue will occur along with each positive Ω_n and will have the same magnitude as for the positive case. Since the matrices in Eqs. 84 are not Hermitian, we have no guarantee that the eigenvalues are real, but the discussion surrounding Eqs. 40 suggests that they will be if a stable Hartree–Fock basis has been used. Moreover, $X_n^\dagger X_n - Y_n^\dagger Y_n$ will then have the same sign as Ω_n, again on the grounds of requiring Hartree–Fock stability (Eqs. 39, 40).

To establish the orthogonality of the resulting vectors, we note from Eq. 84a that

$$(X_{n'}^\dagger, Y_{n'}^\dagger) M \begin{pmatrix} X_n \\ Y_n \end{pmatrix} = \Omega_n (X_{n'}^\dagger, Y_{n'}^\dagger) \begin{pmatrix} X_n \\ -Y_n \end{pmatrix}, \tag{85a}$$

while from the complex conjugate of that equation

$$(X_{n'}^\dagger, Y_{n'}^\dagger) \begin{pmatrix} A & B \\ B^* & A^* \end{pmatrix} = \Omega_{n'} (X_{n'}^\dagger, -Y_{n'}^\dagger), \tag{85b}$$

whence

$$(X_{n'}^\dagger, Y_{n'}^\dagger) M \begin{pmatrix} X_n \\ Y_n \end{pmatrix} = \Omega_{n'} (X_{n'}^\dagger, -Y_{n'}^\dagger) \begin{pmatrix} X_n \\ Y_n \end{pmatrix}, \tag{85c}$$

and subtracting 85c from 85a

$$(X_{n'}^\dagger, -Y_{n'}^\dagger) \begin{pmatrix} X_n \\ Y_n \end{pmatrix} = 0 \qquad \text{for} \quad \Omega_{n'} \neq \Omega_n. \tag{85d}$$

The normalization can be chosen with the same metric, namely such that

$$(X_{n'}^\dagger, -Y_{n'}^\dagger) \begin{pmatrix} X_n \\ Y_n \end{pmatrix} = \delta_{n'n} \, \text{sign} \, (\Omega_n). \tag{85e}$$

To complete the discussion of the orthonormality of the state vectors in the random-phase approximation we must generalize the trivial Hartree–Fock result

$$\langle \Phi | a_\lambda^\dagger a_\beta | \Phi \rangle = \delta_{\lambda\beta}, \tag{86a}$$

for $\lambda \leq F$, to a parallel random-phase combination

$$\langle \Psi | (\delta_{\sigma\alpha} a_\lambda^\dagger a_\beta - \delta_{\lambda\beta} a_\alpha^\dagger a_\sigma) | \Psi \rangle = \delta_{\sigma\alpha} \delta_{\lambda\beta}, \tag{86b}$$

for $\lambda \leq F$, $\sigma > F$. This represents a further approximation required to complete this method of linearizing the equations of motion, and can be seen to reduce to the usual Hartree–Fock situation—as before—when $|\Psi\rangle$ contains no particle–hole correlations, so that the action of second term in parentheses on $|\Psi\rangle$ yields zero.

This then leads to

$$\langle \Psi_n | \Psi_n \rangle = \langle \Psi | Q_n Q_n^\dagger | \Psi \rangle = \langle \Psi | [Q_n, Q_n^\dagger] | \Psi \rangle \,, \tag{87}$$

where we have used, in the transition to the commutator form,

$$Q_n | \Psi \rangle = 0 \,. \tag{88a}$$

This follows from the consideration of

$$HQ_n | \Psi \rangle = [H, Q_n] | \Psi \rangle + Q_n H | \Psi \rangle = (E_0 - \Omega_n) Q_n | \Psi \rangle \,, \tag{88b}$$

which violates that claim that $|\Psi\rangle$ is the ground state unless Eq. 88 is fulfilled. Then, calculating the commutator in Eq. 87, using Eqs. 86b and 85e, we get

$$\langle \Psi_n | \Psi_n \rangle = X_n^\dagger X_n - Y_n^\dagger Y_n = 1 \qquad \text{for} \quad \Omega_n > 0 \,. \tag{89}$$

Thus, while Eq. 85 gives the orthogonality of the RPA eigenvectors, Eq. 89 gives the normalization.

Transitions for one-particle operators as in Eq. 33 will be given, again using Eq. 88, by

$$\langle \Psi_n | F | \Psi \rangle = \langle \Psi | Q_n F | \Psi \rangle = \langle \Psi | [Q_n, F] | \Psi \rangle$$

$$= X_n^\dagger F_X + Y_n^\dagger F_Y \,, \tag{90}$$

where

$$\left(\frac{F_X}{F_Y} \right) = \begin{pmatrix} \langle \sigma | f | \lambda \rangle \\ \langle \rho | f | \lambda \rangle \\ \vdots \\ \text{---} : \text{---} \\ \vdots \\ \langle \lambda | f | \sigma \rangle \\ \langle \lambda | f | \rho \rangle \end{pmatrix} \,. \tag{91}$$

A central result for the random-phase approximation is Thouless's theorem, which states that

$$2 \sum_n |\langle \Psi_n | F | \Psi \rangle|^2 (E_n - E_0) = \langle \Phi | [F, [H, F]] | \Phi \rangle \,. \tag{92}$$

This has a structure very close to that of the sum rule in Eq. 55, except that the left-hand side refers to random-phase quantities and the right-hand side to the Hartree–Fock ground state. The proof of Eq. 92 proceeds by evaluating the double-commutator term directly and comparing it with the sum

$$2 \sum_{n=1}^{n_0} |\langle \Psi_n | F | \Psi \rangle|^2 |\Omega_n| = \sum_{n=1}^{n_0} |\langle \Psi_n | F | \Psi \rangle|^2 \Omega_n$$

$$- \sum_{n=n_0+1}^{2n_0} |\langle \Psi_n | F | \Psi \rangle|^2 \Omega_n, \qquad (93)$$

where we have explicitly introduced the positive- and negative-energy solutions of Eq. 84 into the summation. This is

$$\left(\sum_{n=1}^{n_0} - \sum_{n=n_0+1}^{2n_0} \right) \left| (X_n^\dagger, Y_n^\dagger) \binom{F_X}{F_Y} \right|^2 \Omega_n$$

$$= \left(\sum_{n=1}^{n_0} - \sum_{n=n_0+1}^{2n_0} \right) (F_X^\dagger, -F_Y^\dagger) \binom{X_n}{-Y_n} (X_n^\dagger, -Y_n^\dagger) \binom{F_X}{-F_Y} \Omega_n$$

$$= \left(\sum_{n=1}^{n_0} - \sum_{n=n_0+1}^{2n_0} \right) (F_X^\dagger, -F_Y^\dagger) M \binom{X_n}{Y_n} (X_n^\dagger, -Y_n^\dagger) \binom{F_X}{-F_Y}$$

$$= (F_X^\dagger, -F_Y^\dagger) M \binom{F_X}{-F_Y}, \qquad (94)$$

where the last step follows from the completeness of the random-phase solutions, and gives the right-hand side of Eq. 92 calculated along lines similar to those we used in Section 4.4.

Thouless's theorem, Eq. 92, in addition to providing a means for dealing with energy-weighted sum rules, also reduces the concern over symmetry breaking that arose in Section 4.4. This is because the very form of Eq. 92 implies that if the operator F is the generator of a symmetry of the Hamiltonian H (i.e., plays the role of G in Section 4.4 and has a vanishing commutator with H) then the states that it connects to the ground state must lie at zero energy. This comes about because the right-hand side of Eq. 92 then vanishes and the left-hand side involves a positive definite quantity if $\langle \Psi_n | F | \Psi \rangle \neq 0$ unless $E_n = E_0$. Thus these apparent excitations (sometimes called spurious excitations), which actually represent symmetry transformations of the ground state, can be picked out and discarded by exploiting the distinguishing characteristic that they lie at zero energy. The random-phase approximation thus provides a generalization of the Hartree–Fock method as applied to excitations that improves upon some of its formal properties and may or may not improve the accuracy of the approximate result as compared with the full solution of the many-particle problem, since the variational property of Hartree–Fock is lost. We shall also see in Chapter 10 that the random-phase approximation adds importantly to our understanding of collective excitations in the electron gas, of the electron–phonon interaction, and of correlation energy in metals.

In summary, we have seen how Hartree–Fock theory gives a first approximation to the ground state, and low excited states, of many-fermion systems. Often this is adequate for a semiquantitative understanding of the structure of atoms, of nuclei, and of crystalline solids. However, sometimes the Hartree–Fock description is simply inadequate, as we saw for the electron-gas model of metals, in Section 4.5.4 above. The random-phase approximation gives a method of improvement of Hartree–Fock theory for the particular circumstance that there are particle–hole excited states strongly coupled to the ground state of the system. With particularly strong coupling, the Hartree–Fock ground state may be unstable to a total change of character, described by a broken symmetry, as we saw. The Hartree–Fock approximation may break down for a number of other reasons, as we shall see in Chapters 5 and 6.

In the next chapter we turn to a generalization of the self-consistent-field approach for a problem in which part of the interaction must be dealt with nonperturbatively and so must be incorporated into the essential starting point of the method. There also, as for the random-phase approximation, an essential step in the generalization is the relaxation of the condition on the Hartree–Fock ground state that the lowest single-particle levels must be occupied with probability equal to unity and those above the Fermi surface must be empty.

EXERCISES

4.1 As a (highly unrealistic) model, consider a system of two equal-mass particles interacting through a potential, in which no statistics are obeyed. Take a trial wave function which is a simple product of single-particle wave functions.

  (a) Show in configuration space that the variational principle yields the Hartree equations.

  (b) Again in configuration space, but now for Fermi–Dirac statistics, show that the Hartree–Fock equations result.

  (c) Comment on the validity of these approximation for a two-particle system: how good (or bad) are the resulting ground-state energies and wave functions?

4.2 Consider a Fermi system with the following interaction:

$$H = \sum_i^N H_0(i) + \frac{1}{2} \sum_{i \neq j}^N V(i, j) \,,$$

with one-body part $H_0(i) = $ constant (all orbits degenerate), and two-body part $V(i, j) = \alpha X(i)X(j)$, where $X(i)$ is a (bounded) Hermitian operator on the ith particle, $i = 1, \ldots, N$.

(a) Write the Hartree–Fock equations explicitly for this case.

(b) Show that if one can find the one-body eigenstates of $X(i)$, one can solve the Hartree–Fock equations explicitly.

[This method is of particular interest if X has group properties.]

4.3 Consider the ground state $|\Phi\rangle$ of a bound, many-body system of fermions, and imagine the removal of a particle from the orbital λ, leaving the system in the state $|\Phi_n\rangle$. Define an energy for removing that particle,

$$\mathscr{E}_\lambda = \sum_n |\langle \Phi_n | a_\lambda | \Phi \rangle|^2 (E_0 - E_n) ,$$

where E_0 and E_n are the total energies for the respective systems and states.

(a) Show that

$$\mathscr{E}_\lambda = \langle \Phi | a_\lambda^\dagger [a_\lambda, H] | \Phi \rangle ,$$

where H is the system Hamiltonian.

(b) For H of Eq. 7 show that the initial ground-state energy may be written (exactly) as

$$E_0 = \langle \Phi | H | \Phi \rangle = \tfrac{1}{2} \langle \Phi | T | \Phi \rangle + \tfrac{1}{2} \sum_\lambda \mathscr{E}_\lambda .$$

(c) If $|\Phi\rangle$ is taken to be the Hartree–Fock result, find the relationship between $\mathscr{E}_\lambda$ and ϵ_λ of Eq. 15 for the occupied orbits $(\lambda \leq F)$. What happens for $\lambda > F$?

(d) What is the result equivalent to parts (a) and (b) for bosons? Explain.

4.4 Show Eq. 25 in detail.

4.5 (a) Use the Thomas–Fermi model to obtain an approximate expression for the energy of an atom in terms of the electronic density.

(b) Show that one obtains the Thomas–Fermi equation by minimizing this total energy with respect to variations in the density $\rho(r)$ while *the density is kept normalized*, $\int \rho(r)\, dr = $ constant.

(c) Using the result of (b) and the variational function

$$\rho_\lambda(r) = \lambda^3 \rho(\lambda r) ,$$

with variational parameter λ, prove the virial theory $2T = -V$ for the Thomas–Fermi model.

4.6 Consider the schematic model in which the particle–hole interaction matrix element of Eq. (25) is separable,

$$\langle \sigma v | v | \tau \mu \rangle_A = -v d_{(\sigma\mu)} d_{(\tau v)} .$$

Note that *each* particle–hole label $(\sigma\mu)$, (τv), ... functions here as a single array index.

(a) Find a simple expression from which the energy eigenvalues can be easily obtained. (Hint: Consider the sum $s = \sum_{\sigma\mu} d_{(\sigma\mu)} x_{(\sigma\mu)}$, which can be eliminated from the eigenvalue equations.)

(b) Find a simple approximate expression for the lowest (highest) eigenstate for v negative (positive) in the limit where the energy eigenvalue for this state is well removed from the other eigenvalues.

(c) Carry out these same studies for the random-phase approximation using this separable form in the secular equations 81–84, i.e. also assume $\langle \sigma\tau | v | \mu v \rangle_A = v d_{(\sigma\mu)} d_{(\tau v)}$.

4.7 Show Eq. 51 in detail.

4.8 **(a)** Consider the Hamiltonian for a perturbed harmonic oscillator: $H = A\alpha^\dagger \alpha + \frac{1}{2}(B\alpha^\dagger \alpha^\dagger + B^* \alpha\alpha)$ with $[\alpha, \alpha^\dagger] = 1$, $A^* = A$. The eigenvalue problem may be solved by a commutator method analogous to Eqs. 74–84, as follows. Let $Q^\dagger = x\alpha^\dagger - y\alpha$ such that $[H, Q^\dagger] = \Omega Q^\dagger$. Find a matrix equation (in A, B, B^*) which determines x, y, Ω, and compare to Eq. 84.

(b) Show that the Hamiltonian of a many-fermion system may be written in the Hartree–Fock basis in the form:

$$H = \sum_{a,b} A_{ab} \alpha_a^\dagger \alpha_b + \frac{1}{2}\sum_{a,b}(B_{ab}\alpha_a^\dagger \alpha_b^\dagger + B_{ab}^* \alpha_a \alpha_b) + \text{other terms} ,$$

where a, b label p-h states, e.g. $\alpha_a^\dagger = a_\sigma^\dagger a_\mu$, and A_{ab}, B_{ab} are given in Eq. 38.

(c) Show that the *quasiboson approximation*, for which we ignore the "other terms" in part (b), and assume that

$$[\alpha_a, \alpha_b^\dagger] = \delta_{a,b}$$

(with the method used for part (a)) leads to the random-phase approximation equations 84.

CHAPTER **5**

Pairing

In several systems of fermions two particles experience an attractive interaction when they occupy single-particle states near the Fermi surface. This attraction will have truly significant effects if it can act on many such pairs within the system. The system may achieve this by preferring to populate the single-particle orbitals in a pairwise fashion. Then each time two electrons interact, they go from one paired state to another. If the final states are occupied, the Pauli principle will of course prohibit this attractive scattering, but at least if the fermions are paired, then either *both* final single-particle states will be occupied or *neither* will be. This will minimize the obstructive effects of the exclusion principle in that otherwise the occupation of just *one* of the two final orbitals would keep the attraction from acting. Although other forces may act on or between the fermions, the pairing part of the interaction will be of special, favored importance and must receive special treatment, which is our purpose in this chapter.

A key example of this pairing phenomenon occurs in superconductivity in metals, where the pairing interaction is between an electron in a state with momentum and spin projection $\mathbf{k}$, μ and one in the time-reversed state $-\mathbf{k}$, $-\mu$. This interaction comes about here through interactions of the electrons with phonons (i.e., with vibrations in the lattice) that induce an attractive net interaction between the electrons. (Other mechanisms for producing such an interaction may exist in other systems.) Indeed as this book went to press superconducting materials were discovered with *much* higher transition temperatures than hitherto; the basic mechanisms involved are still unclear. The details of the mechanism producing the attraction are quite crucial in determining the characteristics of the superconductor and in particular the transition temperature below which superconducting properties exist—or even if there is to be a transition to a superconducting state. These considerations are rather more specialized than the general properties of a paired state, which are what will concern us here.

To illustrate the universality of the phenomena resulting from pairing, given that this interaction exists, we note that a behavior closely analogous to superconductivity is found in nuclei. There the attractive interaction arises from a short-range part of the nucleon–nucleon residual interaction. At the general level, the pairing phenomena that result can be treated just as for superconductors, and so the broad lines of the theory developed here will be applicable to nuclear systems as well. These common features which we shall explore include the appearance of an energy gap for the system, the generalization of the Hartree–Fock scheme to accommodate pairing, and the notion of a quasiparticle in this context. The methods we develop in this context allow for complete or nearly complete solutions of certain many-particle model systems and so give useful insight into the properties of such solutions. The most significant physical application of these results is to the phenomenon of superconductivity in metals, where one finds a vast number of related effects of both theoretical and practical importance.

5.1   ORIGINS OF AN ATTRACTIVE PAIRING INTERACTION

5.1.1   The Nuclear Case

Consider a model for the short-range force between two nucleons in which we approximate this as being of zero range compared to nuclear dimensions,

$$V = \tfrac{1}{2} V_0 \sum_{m_1 m_2} \sum_{m_1' m_2'} \sum_{IM, I'M'} \langle nl[jj]_{I'M'} | \delta(\mathbf{r} - \mathbf{r}') | nl[jj]_{IM} \rangle$$

$$\times (jjI'; m_1' m_2' M')(jjI; m_1 m_2 M)$$

$$\times a^\dagger_{jm_1'} a^\dagger_{jm_2'} a_{jm_2} a_{jm_1}. \tag{1}$$

We here assume that the nucleons participating in the pairing are the valence nucleons in a subshell with principal quantum number n, orbital angular momentum l, and total nucleon angular momentum j. We work in a coupled representation such that the two nucleons within this subshell and entering in the two-particle interaction matrix element are coupled to total angular momentum and projection I, M initially and I', M' finally. (These angular-momentum considerations are crucial to the pairing in nuclei.) The explicit Clebsch–Gordan coefficients appearing in Eq. 1 couple the creation or annihilation operators to states with good I, M or I', M'. The subsequent sum over those quantities together with the implicit Clebsch–Gordan coefficients that accomplish the coupling, denoted by $[jj]_{IM}$ or $[jj]_{I'M'}$, carries us back to the original representation with good projections of j which we have not made explicit in Eq. 1. The interaction strength V_0 is taken negative, consistent with the anticipation of a residual interaction that is on the average attractive.

Since the interaction appearing explicitly in Eq. 1 is a scalar under rotations, it will conserve total angular momentum and projection, $I' = I$ and $M' = M$, and its matrix elements are independent of M, so V may be rewritten as

$$V = \sum_{IM} C(I) A^\dagger_{IM} A_{IM} ,\qquad(2)$$

where

$$A_{IM} \equiv \sum_{m_1 m_2} (jjI; m_1 m_2 M) a_{jm_2} a_{jm_1}\qquad(3a)$$

and

$$A^\dagger_{IM} = \sum_{m_1 m_2} (jjI; m_1 m_2 M) a^\dagger_{jm_1} a^\dagger_{jm_2} .\qquad(3b)$$

The coefficients in Eq. 2 are

$$C(I) \equiv \tfrac{1}{2}V_0 \langle nl[jj]_{IM}|\delta(\mathbf{r} - \mathbf{r}')|nl[jj]_{IM}\rangle ,\qquad(4a)$$

where any value of M may be selected for the calculation of $C(I)$, which is independent of M. Some angular-momentum algebra, which is of no great interest for us here, allows a simplification of this expression to

$$C(I) = \frac{(2j+1)^2}{2I+1} (jjI; \tfrac{1}{2} - \tfrac{1}{2} 0)^2 F ,\qquad I \text{ even},\qquad(4b)$$

where F is the radial integral for the zero-range force,

$$F \equiv \frac{1}{4\pi} V_0 \int R^4_{nlj}(r) r^2\, dr < 0 ,\qquad(4c)$$

for the radial wave function $R_{nlj}(r)$.

The cases we consider involve large subshells, so that we may accommodate what is for nuclei a fairly large number of active fermions in the pairing process (say four or five). We therefore take, in a numerical example for Eq. 4b, the value $j = \tfrac{7}{2}$, and obtain for the relative contributions of the various total angular momenta I the results shown in Table 5.1. We see from this table that the $I = 0$ contribution dominates all others. This suggests that an adequate approximation results from restricting the interaction to that value alone,

**TABLE 5.1 Relative Contributions of Various
Total-Angular-Momentum Values for a Zero-Range
Nucleon–Nucleon Force in the $j = \frac{7}{2}$ Subshell**

I	$C(I)/F$
0	8.00
2	1.90
4	0.94
6	0.47

$$\mathscr{V} = -|G| \sum_{m,m'=1/2}^{j} a_{m'}^{\dagger} a_{-m'}^{\dagger} a_{-m} a_{m} , \qquad (5)$$

where a suitable phase choice has been taken for the angular-momentum states [such that $T|m\rangle = |-m\rangle$, where T is the time-reversal operator]. The positive coefficient $|G|$ contains the interaction strength, radial integrals, and necessary angular-momentum statistical factors. The interaction of Eq. 5 is structured so as to give an attractive contribution when a pair of time-reversed states m, $-m$ is occupied initially and the fermions in question scatter into a final time-reversed pair of states m', $-m'$. In simplest physical terms, the preferred population of the paired m, $-m$ states gives maximum overlap for the particle wave functions involved—which are degenerate for those orbitals and have identical densities— in order that the very short-range attractive force may operate maximally to lower the system energy. The fact that the force acts between fermions populating time-reversed doublets is the defining characteristic of a pairing force.

5.1.2 Superconductivity in Metals

The same general features of the pairing force may be found for the electrons in the conduction bands of metals. The electrons do not of course attract each other directly, since their Coulomb interaction is repulsive. But the positive lattice ions provide a background that on the average neutralizes the mutual repulsion. The attractive pairing force then results from lattice dynamics roughly as follows: An electron moving through the lattice will at a given moment pull the positive ions of the lattice slightly towards itself (see Figure 5.1), thus generating a momentary average excess of positive charge in its wake, towards which a second electron may be attracted. This interaction with lattice vibrations, or phonons, may be described by the boson-exchange mechanisms that we discussed in Section 2.3, depicted here in Figure 5.2. The matrix element in question is

$$\langle \mathbf{k} + \mathbf{K}, \mathbf{k}' - \mathbf{K}|v|\mathbf{k}, \mathbf{k}'\rangle = \frac{2\omega_q |M|^2}{[\epsilon(\mathbf{k}) - \epsilon(\mathbf{k} + \mathbf{K})]^2 - \omega_q^2} , \qquad (6)$$

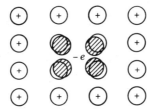

FIGURE 5.1   The passage of a negative electron distorts the ions of the lattice and creates a brief net positive charge in its wake, which attracts another electron.

where the initial and final electron momenta are $\mathbf{k}$, $\mathbf{k}'$ and $\mathbf{k} + \mathbf{K}$, $\mathbf{k}' - \mathbf{K}$ and the phonon momentum is $\mathbf{q}$. The corresponding electron and phonon energies are denoted by ϵ and ω, while M represents the electron–phonon coupling at each of the two vertices in the graph; this interaction is assumed to be of scalar form, and Eq. 6 results from simple considerations of second-order perturbation theory as noted in the context of Figure 2.3. In Eq. 6 the momentum of the phonon $\mathbf{q}$ is not necessarily equal to the momentum transfer between the electrons, $\mathbf{K}$. This is because the lattice may take up momentum here (by a unit $\mathbf{g}$, the reciprocal lattice vector), so that one does not always have $\mathbf{q} = \mathbf{K}$.

From Eq. 6 it is clear that the interaction matrix element is negative, or attractive, for a narrow range of energies* such that

$$|\epsilon(\mathbf{k}) - \epsilon(\mathbf{k} + \mathbf{K})| < \omega_q ,$$

where

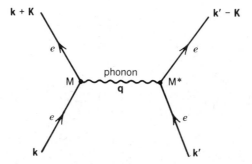

FIGURE 5.2   The electron–electron interaction generated by phonon exchange, i.e. by the lattice-vibration mechanism in the previous figure. For normal electron–phonon interactions $q = K$, but this need not be the case in umklapp processes, where conservation of momentum does not hold, since the electron can exchange momentum with the lattice (by a unit of the reciprocal lattice vector).

*The phonon frequency ω_q is limited by the Debye frequency of the lattice, which in turn induces a cutoff in momentum space for the effective pairing force.

$$v(\mathbf{k}, \mathbf{K}) \cong -2\frac{|M|^2}{\omega_q}.\tag{7}$$

This attraction can act most effectively for electrons in a shell near the surface of the Fermi sea, where the exclusion principle least inhibits the participation of many electrons in attractive interactions with minimal need for added energy to reach unoccupied levels. Furthermore, its effect for two electrons in the medium is maximal when the total electron momentum is zero—that is, when the electrons are initially in paired states $\mathbf{k}, -\mathbf{k}$ and finally in $\mathbf{k} + \mathbf{K}, -\mathbf{k} - \mathbf{K}$—as can be seen from a phase-space argument based on Figure 5.3: The initial momenta for the two electrons participating in the scattering must lie just below the Fermi surface k_F, so that the two fermions can be carried by the scattering from occupied to unoccupied levels. Furthermore the final momenta must add up to the same total momentum as the initial and must lie just outside k_F. Only very limited values of $\mathbf{K}$, leading to final momenta very close to the initial ones, fit both these criteria. The situation changes radically, however, if the total momentum is zero,

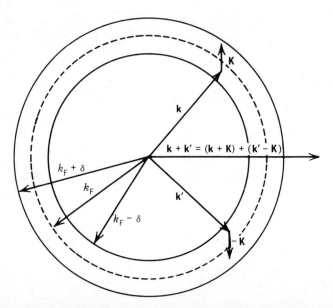

FIGURE 5.3 Momenta that contribute appreciably to the scattering of electron pairs, where by "scattering" we mean the appearance of the electrons in the matrix element $\langle \mathbf{k} + \mathbf{K}, \mathbf{k}' - \mathbf{K} | v | \mathbf{k}, \mathbf{k}' \rangle$. The initial momenta must lie just below the Fermi surface k_F and the final ones just above it to permit scattering to take place readily from filled to unfilled levels. Further, momentum must be conserved, i.e., the final momenta are constrained to add up to the horizontal vector in the figure. Within the shell restrictions this limits the available phase space severely, one legitimate choice of K being shown in the figure. If $\mathbf{k} = -\mathbf{k}'$, so that the total momentum vanishes, then the entire shell becomes available for scattering.

that is, $\mathbf{k}' = -\mathbf{k}$. Then the initial vectors lie along any diameter of an inner shell and the final ones are along any diameter of an outer shell: The entire vicinity of the Fermi surface becomes available for scattering. Thus the interaction of Eq. 7 is vastly more effective in the paired case $\mathbf{k}' = -\mathbf{k}$. This leads to a pairing interaction along the lines of Eq. 5, but now with electron states labeled by momentum and spin projection $\mathbf{k}$, μ (and the time-reversed states $-\mathbf{k}$, $-\mu$), which replace the index m there. (It is desirable to take pairs of opposite spin because exchange terms reduce the interaction for parallel spins; that is, since the attractive interaction has short range, it is expected to be most effective in antisymmetric spin states that permit symmetric—and hence s-wave—spatial states.)

To summarize these comments on the origins of pairing forces in fermion systems, we note that the detailed nature of the mechanism producing pairing is not crucial to the phenomena that result from it. It is only important that in some way there be produced a short-range, attractive force that acts near the Fermi surface. In the nuclear situation it is immediately clear that the pairing potential is of much shorter range than nuclear sizes; in the case of superconductivity in metals the electron–electron potential arising from phonon exchange is very nearly constant in momentum space and so essentially of zero range in configuration space, relative to the physical extent of the pair. We shall now assume that such an interaction exists as a rough model for the more realistic and more complex actual physical situations and explore the main phenomena that result. This may be viewed—much as for the Hartree or Hartree–Fock methods—as a point of departure on which one may base more refined and more specialized theories. The crucial consequence of the pairing situation will be encompassed within the approach, namely that in the presence of pairing normal assemblages of fermions are unstable against transitions to a superconducting phase.

5.2 THE SENIORITY MODEL

To gain some insight into effects produced by the pairing force in a case where exact solutions can be generated, in a lucid way, we consider the seniority model. This is based on a two-particle interaction given by the pairing form of Eq. 5,

$$\langle m_2 m_2' | v | m_1 m_1' \rangle = \begin{cases} -|G|, & m_1 = -m_1', \; m_2 = -m_2' \quad (m_1, m_2 > 0) \\ 0 & \text{otherwise}, \end{cases}$$

$$\text{(8a)}$$

where the notation refers to a conventional two-particle matrix element to be inserted into the form of Eq. 1.47a to produce the Fock-space interaction $\mathscr{V}$ of Eq. 5. This interaction is taken for particles all having the same single-particle energy ϵ. Thus Eq. 8a represents the full Hamiltonian matrix if we take the zero of energy at ϵ, that is, we choose $\epsilon = 0$ for convenience. (In our subsequent, more complete treatment of the pairing interaction we shall relax this special feature and admit differing single-particle unperturbed energies ϵ_k, as in superconductivity say, where one may have, approximately, $\epsilon_k = k^2/2m$.) Such a Hamiltonian is pertinent, for example, for a nucleus with several nucleons in a single subshell j with a constant pairing force acting between them. The explicit structure of the Hamiltonian is

$$
\mathcal{V} = -|G| \overbrace{\begin{pmatrix}
1 & 1 & 1 & \cdots & 1 & 0 & 0 & 0 & \cdots \\
1 & 1 & 1 & \cdots & 1 & 0 & 0 & 0 & \cdots \\
\vdots & \vdots & \vdots & & \vdots & \vdots & \vdots & \vdots \\
1 & 1 & 1 & \cdots & 1 & 0 & 0 & 0 & \cdots \\
0 & 0 & 0 & \cdots & 0 & 0 & 0 & 0 & \cdots \\
\vdots & \vdots & \vdots & & \vdots & \vdots & \vdots & \vdots
\end{pmatrix}}^{\Omega}, \tag{8b}
$$

where Ω is the size of the relevant space of pairs and we use a representation based on pair states $|m, -m\rangle$, or in Fock space $|1_m, 1_{-m}\rangle$. For two particles in this subshell the trivial diagonalization of Eq. 8b yields for the lowest eigenvector

$$
\psi_{N=2} = \frac{1}{\sqrt{\Omega}} \left.\begin{pmatrix} 1 \\ 1 \\ \vdots \\ 1 \\ 0 \\ \vdots \\ 0 \end{pmatrix}\right\} \Omega, \tag{9a}
$$

satisfying

$$
\mathcal{V}\psi_{N=2} = -\Omega|G|\psi_{N=2}. \tag{9b}
$$

The other eigenvectors lie in a space orthogonal to $\psi_{N=2}$ and all are degenerate with eigenvalue zero; the state $\psi_{N=2}$ exhausts the invariant trace of the matrix in Eq. 8b according to Eq. 9b. Since V is negative definite, once we have exhausted the trace we can be sure that all other eigenvalues lie at zero energy.

If there are N particles in the subshell, it is convenient to introduce, as in Eq. 3a with $I = 0$,

$$A^\dagger \equiv \sum_{m=1/2}^{j} a_m^\dagger a_{-m}^\dagger , \tag{10}$$

in terms of which

$$\mathcal{V} = -|G|A^\dagger A . \tag{11}$$

The particle number operator is

$$N \equiv \sum_{m=-j}^{j} a_m^\dagger a_m , \tag{12}$$

and the commutation relations that are satisfied by these operators are

$$[A^\dagger, N] = -2A^\dagger , \tag{13}$$

and for the interaction

$$[\mathcal{V}, A^\dagger] = -|G|(\Omega + 2 - N)A^\dagger \tag{14a}$$

and

$$[\mathcal{V}, (A^\dagger)^f] = [\mathcal{V}, (A^\dagger)^{f-1}]A^\dagger + (A^\dagger)^{f-1}[\mathcal{V}, A^\dagger]$$
$$= -|G|[f(\Omega + 2 - N) + f(f-1)](A^\dagger)^f , \tag{14b}$$

where the last result follows by induction. We shall see shortly that the operator $A^\dagger$ allows us to define a fully paired state $(A^\dagger)^{\frac{1}{2}N}|0\rangle$ whose energy is substantially lower, by $|G|\Omega$, than that of a state with even one broken pair.

As we noted in connection with Eqs. 8 and 9 for the $N=2$ case, it is convenient to know the trace of the many-particle Hamiltonian, which is invariant and thus provides guidance as to how to search for eigenstates. For N particles the trace is

$$\mathrm{Tr}_N \mathcal{V} = \sum_i \langle Ni|\mathcal{V}|Ni\rangle$$

$$= -|G| \sum_{m=1/2}^{j} \sum_i \langle Ni|a_m^\dagger a_{-m}^\dagger a_{-m} a_m|Ni\rangle , \tag{15}$$

where we have labeled states by the number of particles N and configuration i, summing over diagonal elements for the trace. In order to contribute to the sum in Eq. 15, a state $|Ni\rangle$ must be populated by pairs in the single-particle states, m, $-m$. For a given value of m this means that two particles occupy m and $-m$ and $N-2$ are left over to occupy the other $2\Omega-2$ states $m' \neq m$. Thus the number of such nonvanishing matrix elements is the combinatorial factor $\binom{2\Omega-2}{N-2}$, so that

$$\text{Tr}_N \mathcal{V} = -|G| \sum_{m=1/2}^{j} \binom{2\Omega-2}{N-2} = -|G|\Omega \frac{(2\Omega-2)!}{(2\Omega-N)!(N-2)!}. \qquad (16)$$

We now proceed progressively to build up solutions for the many-particle Hamiltonian. For $N=2$ these are already given in Eqs. 9. In the language of Eqs. 10–14 the eigenstate of lowest energy is $A^\dagger|0\rangle$, for which

$$\mathcal{V}A^\dagger|0\rangle = [\mathcal{V}, A^\dagger]|0\rangle = -|G|\Omega A^\dagger|0\rangle \qquad (17)$$

since (the trivial $N=0$ case)

$$\mathcal{V}|0\rangle = 0. \qquad (18)$$

Orthogonal to this we have states given by $B_i^{(2)}|0\rangle$ for which

$$\mathcal{V}B_i^{(2)}|0\rangle = 0, \qquad i=1,2,3,\ldots,t_2-1, \qquad (19)$$

where $t_2 = \binom{2\Omega}{2}$ is the number of states for $N=2$.

For $N=4$ the lowest eigenstate is $(A^\dagger)^2|0\rangle$, satisfying

$$\mathcal{V}(A^\dagger)^2|0\rangle = [\mathcal{V}, (A^\dagger)^2]|0\rangle = -2|G|(\Omega-1)(A^\dagger)^2|0\rangle, \qquad (20)$$

after which come the states $A^\dagger B_i^{(2)}$ fulfilling

$$\begin{aligned}
\mathcal{V}A^\dagger B_i^{(2)}|0\rangle &= [\mathcal{V}, A^\dagger]B_i^{(2)}|0\rangle \\
&= -|G|(\Omega-2)A^\dagger B_i^{(2)}|0\rangle, \qquad i=1,2,3,\ldots,t_2-1.
\end{aligned} \qquad (21)$$

These two sets of states exhaust the trace of $\mathcal{V}$, and the remaining states are constructed orthogonal to them and have eigenvalue zero:

$$B_i^{(4)}|0\rangle, \quad i=1,2,\ldots,t_4-t_2, \quad \text{with} \quad \mathcal{V}B_i^{(4)}|0\rangle = 0, \qquad (22)$$

where

$$t_4 = \binom{2\Omega}{4}$$

is the total number of states for $N=4$.

In order to classify general solutions it is useful to introduce the *seniority number S*, which is the number of *un*paired particles in the system. Then for N even the general structure of the eigenstates is

$$|NS\rangle = (A^\dagger)^{\frac{1}{2}(N-S)}B_i^{(S)}|0\rangle \,, \qquad N \le \Omega \,, \quad S = 0, 2, 4, \ldots, N \,, \quad (23)$$

where we have labeled the states by number of particles and seniority number S, and

$$B_i^{(0)} = 1 \,, \qquad \mathscr{V}B_i^{(S)}|0\rangle = 0 \,. \qquad\qquad\qquad (24a, b)$$

Then the eigenequation for these states is

$$\mathscr{V}|NS\rangle = E_S^{(N)}|NS\rangle \,, \qquad\qquad\qquad (25a)$$

with

$$E_S^{(N)} = -\tfrac{1}{4}|G|(N-S)(2\Omega + 2 - N - S) \,, \qquad\qquad (25b)$$

all of which may be compared with our previous specific results for $N = 2$ and 4.

A major conclusion from the seniority model—and one that provides guidance for more general considerations concerning pairing—is the existence of a gap between the ground state, $S = 0$, that results for the system of N particles with a pairing force, and the first excited state, $S = 2$, arising from the breaking of a pair. The breaking of a pair in the state with $S = 2$ is signaled by the appearance of one less operator $A^\dagger$ for $S = 2$ than for $S = 0$, where we have the ground state

$$|N0\rangle = (A^\dagger)^{\frac{1}{2}N}|0\rangle \qquad\qquad\qquad\qquad (26)$$

and all the fermions are paired. The energy difference between these states is

$$E_2^{(N)} - E_0^{(N)} = |G|\Omega \,; \qquad\qquad\qquad\qquad (27)$$

thus the energy required to break a pair is not measured merely by the magnitude of a single two-particle matrix element $\sim|G|$, but involves a large multiplicative factor Ω which is the number of single-particle states available for pairing. We have in fact already found this to be the case for only two particles $(N = 2)$, or a single pair, in Eq. 9b, where the energy gap for breaking the pair is of course also that given in Eq. 27.

5.3 THE BCS GROUND STATE

We now consider the general pairing Hamiltonian

$$H = \sum_k \epsilon_k^0 a_k^\dagger a_k - |G| \sum_{\substack{k>0 \\ k'>0}} a_k^\dagger a_{-k}^\dagger a_{-k'} a_{k'} \,, \qquad\qquad (28)$$

involving single-particle energies ϵ_k^0 for the fermion single-particle state k.*
The energies ϵ_k^0 are no longer necessarily all equal as in the seniority model.
This modification requires a quite different approach for the solution of the
problem. In Eq. 28 there appears the all-important selective operation of
the pairing force between the paired fermions in the substates $k', -k'$ that
are scattered into $k, -k$. The minus sign on the single-particle state label
refers to the time-reversed state. We assume that the single-particle energies
ϵ_k^0 are degenerate for k and $-k$, that is $\epsilon_k^0 = \epsilon_{-k}^0$, as will be the case for a
Hamiltonian possessing time-reversal invariance. This greatly simplifies the
solution of the pairing problem. If this degeneracy is sufficiently broken—as,
say, through the action of a magnetic field on the electrons in a metal—then
the pairing configuration may cease to be favored energetically and superconductivity is destroyed. A variational method for solving this Hamiltonian
was given by Bardeen, Cooper, and Schrieffer and is known as the *BCS
solution*. It assumes a trial wave function of the form

$$|\text{BCS}\rangle = \prod_{k>0}^{\infty} (u_k + v_k a_k^\dagger a_{-k}^\dagger)|0\rangle , \qquad (29a)$$

where u_k and v_k are the real parameters to be determined subject to the
normalization constraint

$$u_k^2 + v_k^2 = 1 , \qquad (29b)$$

which is motivated by an assumption that the various k states are independent.

Note that the state in Eqs. 29 does not have a fixed number of particles
(reminiscent of the situation for the coherent state of Section 3.3.5). This
sacrifice of the sharpness of the number of particles allows one to think of
the problem in terms of pairs that are essentially independent of each other
and thus greatly simplifies the structure of the solution. The relaxation of
the fixed number of particles is dealt with by a method of Lagrange
multipliers, introducing an extra term into the Hamiltonian,

$$H' = H - \lambda N , \qquad (30)$$

involving a multiplicative parameter λ and the number operator

$$N \equiv \sum_k a_k^\dagger a_k . \qquad (31)$$

This procedure is a common one when it is desirable to relax the condition

---

*Here k is a generic orbit label: for plane waves with spin $\frac{1}{2}$, $k = (\mathbf{k}, \mu)$, and $k > 0$ means all $\mathbf{k}$,
$\mu = \frac{1}{2}$, so that $-k$ means all $-\mathbf{k}$, $\mu = -\frac{1}{2}$. The restrictions $k, k' > 0$ are a convenience for
counting states; summation over $k, k' < 0$ would require the inclusion of a sign convention (see
Eq. 65) to avoid the vanishing of the interaction due to anticommutation.

that the number of particles in the system is fixed. It is well known, for example, in the transition from the canonical ensemble of statistical mechanics to the grand canonical ensemble, where just such a relaxation is introduced. The parameter λ plays the role of a chemical potential or removal energy for the system (see the discussion in Section 4.2). We fix λ by ultimately requiring that on the average the number of particles be appropriate,

$$\langle \text{BCS}|N|\text{BCS} \rangle = 2 \sum_{k>0} v_k^2 = N , \qquad (32)$$

where we intend by this last N the actual number of fermions in the system. Equation 32 suggests that v_k^2 represents the probability that a single-particle state k is occupied by a pair. Fluctuations in the number of particles are

$$(\Delta N)^2 = \langle \text{BCS}|N^2|\text{BCS} \rangle - \langle \text{BCS}|N|\text{BCS} \rangle^2$$

$$= 4 \sum_{k>0} u_k^2 v_k^2 , \qquad (33)$$

and are required to be small if the method of Lagrange multipliers is to be deemed adequate. (Indeed, for the degenerate case of the seniority model in the last section all the population probabilities v_k^2 are equal, as may be seen from Eq. 10, and have the value $\frac{1}{2}N/\Omega$, so that the relative fluctuations

$$\frac{\Delta N}{N} = \left(\frac{2 - N/\Omega}{N} \right)^{1/2}$$

are small.)

To solve the BCS variational problem we evaluate the expectation value of H' in Eq. 30 for the trial wave function of Eq. 29,

$$\langle \text{BCS}|H'|\text{BCS} \rangle = 2 \sum_{k>0} \epsilon_k v_k^2 - |G| \left[\sum_{k>0} u_k v_k \right]^2 , \qquad (34)$$

where we have measured single-particle energies with respect to the chemical potential,

$$\epsilon_k \equiv \epsilon_k^0 - \lambda . \qquad (35)$$

In Eq. 34 we have dropped a term

$$-|G| \sum_{k>0} v_k^4 , \qquad (36)$$

which is small if the occupation probabilities are not too large. (This can be checked easily in the seniority model, where the relative contribution of

Eq. 36 compared to the interaction term in Eq. 34 is $\sim \frac{1}{2} N / \Omega^2$ and is small for large Ω.) This term could be included as a correction of Hartree–Fock character in ϵ_k or could be treated in perturbation theory (see also the discussion in Section 5.4.2 and especially Eqs. 52 and 60 there).

The variational procedure for the quantity in Eq. 34 consists in changing v_k, recalling that this modifies u_k as well through Eq. 29b. Then we have

$$\left(\frac{\partial}{\partial v_k} + \frac{\partial u_k}{\partial v_k} \frac{\partial}{\partial u_k} \right) \langle \text{BCS} | H' | \text{BCS} \rangle$$

$$= \frac{2}{u_k} \left[2 \epsilon_k u_k v_k + |G| (v_k^2 - u_k^2) \sum_{k'} u_{k'} v_{k'} \right]$$

$$= 0 . \tag{37}$$

To see the consequences of requiring the quantity in brackets to vanish we introduce the *gap parameter*

$$\Delta \equiv |G| \sum_{k>0} u_k v_k , \tag{38}$$

whence Eq. 37 becomes what is called *the gap equation*,

$$2 \epsilon_k u_k v_k + (v_k^2 - u_k^2) \Delta = 0 . \tag{39}$$

This yields explicit solutions for the population terms,

$$u_k^2 = \frac{1}{2} \left[1 + \frac{\epsilon_k}{\sqrt{\epsilon_k^2 + \Delta^2}} \right] , \tag{40a}$$

$$v_k^2 = \frac{1}{2} \left[1 - \frac{\epsilon_k}{\sqrt{\epsilon_k^2 + \Delta^2}} \right] , \tag{40b}$$

where the choice of sign in solving the quadratic forms is made so that v_k^2 measures particle occupancy ($v_k^2 \to 1$ for $\epsilon_k \ll 0$) and u_k^2 reflects vacancy. These factors express a blurring of the sharp Fermi level found in Hartree–Fock solutions, as shown in Figure 5.4. The gap equation in terms of these explicit solutions is

$$\frac{1}{2} |G| \sum_{k>0} \frac{1}{\sqrt{\epsilon_k^2 + \Delta^2}} = 1 , \tag{41a}$$

while the number equation, Eq. 32, becomes

$$\sum_{k>0} \left[1 - \frac{\epsilon_k}{\sqrt{\epsilon_k^2 + \Delta^2}} \right] = N . \tag{41b}$$

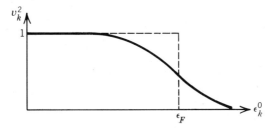

FIGURE 5.4 The population of single-particle levels in a Hartree–Fock solution (dashed line) and in the BCS pairing solution of Eqs. **40** (solid line). Note that in Eqs. **40** the zero of single-particle energies has been placed at λ (ϵ_F here), that is, $\epsilon_k = \epsilon_k^0 - \lambda$.

These must be solved together for Δ and λ, the gap and the chemical potential, given the interaction strength $|G|$, the single-particle spectrum $\{\epsilon_k = \epsilon_k^0 - \lambda\}$, and the number of particles, N.

5.4 QUASIPARTICLES AND THE SIGNIFICANCE OF THE GAP PARAMETER

5.4.1 The Bogolyubov Transformation

In order to explore the full significance of the population factors in Eqs. 40 and the gap parameter Δ in them, we must understand more fully the excitation spectrum of the paired system, which is largely the determining factor for the physics of pairing. We shall do this here through the use of the idea of a quasiparticle, which will arise in the present context from the use of a transformation due to Bogolyubov. This will at the same time provide a convenient framework in which to pull together the Hartree–Fock method and the treatment of pairing. In the course of doing this it is also easy to treat—at least formally—general interactions in which pairing is a dominant effect, but not the exclusive one as in Eq. 28.

We first recall the concept of a quasiparticle in the context of the Hartree–Fock solution introduced in Eqs. 4.17–4.20. In that case the notion of a quasiparticle was merely a convenient alternative notation for dealing with hole states in a manner that brought out their symmetry with particle states. For our present purposes it is convenient first to define along similar lines

$$\alpha_k \equiv \begin{cases} a_k , & k \text{ unoccupied}, \\ -a_{-k}^\dagger , & k \text{ occupied}, \end{cases} \tag{42a}$$

and

$$\alpha_k^\dagger \equiv \begin{cases} a_k^\dagger , & k \text{ unoccupied}, \\ -a_{-k} , & k \text{ occupied}, \end{cases} \tag{42b}$$

whereupon α_k annihilates the Hartree–Fock ground state $|\Phi\rangle$ of Eq. 4.8,

$$\alpha_k|\Phi\rangle = 0 , \tag{43}$$

which thus serves for it as a vacuum. The extension of this to the pairing situation involves a more complicated transformation (within each single-particle substate), known as the *Bogolyubov transformation*:

$$\alpha_k = u_k a_k - v_k a^\dagger_{\ k} \tag{44a}$$

and

$$\alpha^\dagger_k = u_k a^\dagger_k - v_k a_{-k} , \tag{44b}$$

with u_k and v_k real and satisfying, for unitarity or normalization,

$$u_k^2 + v_k^2 = 1 . \tag{44c}$$

Apart from this constraint u_k and v_k are arbitrary at this stage. Note that in Eqs. 44, using our earlier interpretation of v_k and u_k as occupation and vacancy probability amplitudes, we have simply weighted the operators in the same spirit as for the Hartree–Fock case in Eqs. 42. Indeed, in that limit we would have

$$u_k = 1 , \quad v_k = 0 , \quad k \text{ unoccupied} , \tag{45a}$$

and

$$u_k = 0 , \quad v_k = 1 , \quad k \text{ occupied} , \tag{45b}$$

thus regaining Eqs. 42. The relaxation of the sharpness of the Hartree–Fock occupancy factors is of course inevitable in a many-fermion ground-state wave function that proposes to go beyond the single Slater determinant of Eq. 4.8; thus we encountered a similar relaxation—though one that did not make explicit the form of the proposed ground state—in discussing the random-phase approximation in Section 4.6.

The new operators of Eqs. 44 fulfill anticommutation relations that, as a result of the structure of the transformation, are the same as for the old a's, namely

$$\{\alpha_k, \alpha_{k'}\} = 0 , \qquad \{\alpha^\dagger_k, \alpha^\dagger_{k'}\} = 0 , \qquad \{\alpha_k, \alpha^\dagger_{k'}\} = \delta_{kk'} ; \quad (46\text{a, b, c})$$

because the transformation of Eqs. 44 preserves the form of the anticommutation relations, it is referred to as canonical. In its definition we encounter, for the case of the paired, time-reversed companion to k, negative indices in the occupancy amplitudes, that is,

$$\alpha_{-k} = u_{-k}a_{-k} - v_{-k}a_k^\dagger . \qquad (47a)$$

In order to complete the anticommutation relations uniformly for this case as well, we identify these by setting

$$u_{-k} = u_k \quad \text{and} \quad v_{-k} = -v_k , \qquad (47b)$$

whence

$$\alpha_{-k} = u_k a_{-k} + v_k a_k^\dagger , \qquad \alpha_{-k}^\dagger = u_k a_{-k}^\dagger + v_k a_k , \qquad (47c, d)$$

and the resulting anticommutators are, as expected,

$$\{\alpha_k, \alpha_{-k'}\} = 0 , \qquad \{\alpha_k^\dagger, \alpha_{-k'}^\dagger\} = 0 , \qquad \{\alpha_k, \alpha_{-k'}^\dagger\} = \delta_{k, -k'} . \qquad (48a, b, c)$$

Since the single-particle energies ϵ_k are degenerate, $\epsilon_k = \epsilon_{-k}$, we have in Eq. 47b built in our expectation that the transformations have the same structure for $-k$ as for k up to a phase. (See footnote after Eq. 28, Section 5.3.)

The central point is that the relationship of the *quasiparticle* creation and annihilation operators $\alpha_k^\dagger$ and α_k to the BCS ground state is parallel to that of the original operators $a_k^\dagger$ and a_k to the Hartree–Fock ground state. Thus by construction α_k annihilates a state with the structure of the BCS ground state of Eqs. 29 in analogy to Eq. 43, namely, for any u_k, v_k ,

$$\alpha_k|\text{BCS}\rangle = (u_k a_k - v_k a_{-k}^\dagger) \prod_{k'>0} (u_{k'} + v_{k'}a_{k'}^\dagger a_{-k'}^\dagger)|0\rangle$$

$$= (u_k v_k - v_k u_k)a_{-k}^\dagger \sum_{\substack{k'\neq k \\ k'>0}} (u_{k'} + v_{k'}a_{k'}^\dagger a_{-k'}^\dagger)|0\rangle$$

$$= 0 . \qquad (49)$$

The general notion behind the Bogolyubov transformation is that in the new basis the quasiparticle behaves similarly to the original particle but has incorporated some part of the interaction into its fundamental definition—as it were, it drags along its own dressing. [Note that the quasiparticle $\alpha_k^\dagger$ also possesses the same quantum numbers k as the particle $a_k^\dagger$ (and hole a_{-k}) from which it was made.] There will still remain (presumably small) residual interactions between the quasiparticles, but the bulk of the interaction will have been dealt with in the new representation. The quasiparticles involved in pairing require the generalization noted in going from Eqs. 42 to Eqs. 44 in order to deal with the paired state $|\text{BCS}\rangle$.

5.4.2 The Bogolyubov Transformation for the Hamiltonian

If we take the Hamiltonian of one- and two-particle character in Eq. 4.7, using our current notation,

$$H = \sum_{kk'} \langle k|t|k' \rangle a_k^{\dagger} a_{k'} + \frac{1}{2} \sum_{\substack{k_1 k_2 \\ k_1' k_2'}} \langle k_1 k_2 | v | k_1' k_2' \rangle a_{k_1}^{\dagger} a_{k_2}^{\dagger} a_{k_2'} a_{k_1'} , \qquad (50)$$

and carry out the canonical transformation in Eqs. 44 and 47 for arbitrary u_k and v_k consistent with 44c, we obtain

$$\tilde{H} = U + H_{11} + H_{20} + H_{\text{int}} , \qquad (51)$$

where the numerical subscripts indicate the number of quasiparticle creation and annihilation operators in each piece of the Hamiltonian. We shall also include in $\tilde{H}$ the transformed side condition for particle-number conservation of Eq. 30 (which we shall see in a moment is necessary), and require that Eq. 51 be in "normal order", that is, annihilation operators α are to stand to the right of creation operators $\alpha^{\dagger}$. The terms in $\tilde{H}$ first involve a part with no such operators, the constant

$$U = \sum_{k} \left\{ [\langle k|t|k \rangle - \lambda] + \frac{1}{2} \sum_{k'} \langle kk'|v|kk' \rangle_A v_{k'}^2 \right\} v_k^2$$

$$+ \frac{1}{2} \sum_{k,k'} \langle k, -k|v|k', -k' \rangle u_k v_k u_{k'} v_{k'} , \qquad (52)$$

and then a piece that annihilates and creates one quasiparticle,

$$H_{11} = \sum_{k,k'} \left\{ \left[\langle k|t|k' \rangle - \lambda \delta_{kk'} + \sum_{k''} \langle kk''|v|k'k'' \rangle_A v_{k''}^2 \right] \right.$$

$$\times (u_k u_{k'} \alpha_k^{\dagger} \alpha_{k'} - v_k v_{k'} \alpha_{-k'}^{\dagger} \alpha_{-k})$$

$$- \sum_{k''} \langle k, -k'|v|k'', -k'' \rangle$$

$$\left. \times u_{k''} v_{k''} u_k v_{k'} (\alpha_k^{\dagger} \alpha_{k'} + \alpha_{-k'}^{\dagger} \alpha_{-k}) \right\} . \qquad (53)$$

There follows a term that changes the number of quasiparticles by two,

$$H_{20} = \sum_{k,k'} \left\{ \left[\langle k|t|k' \rangle - \lambda \delta_{kk'} + \sum_{k''} \langle kk''|v|k'k'' \rangle_A v_{k''}^2 \right] \right.$$

$$\times (u_k v_{k'} \alpha_k^{\dagger} \alpha_{-k'}^{\dagger} + u_k v_k \alpha_{-k} \alpha_{k'})$$

$$+ \frac{1}{2} \sum_{k''} \langle k, -k'|v|k'', -k'' \rangle u_{k''} v_{k''}$$

$$\left. \times (u_k u_{k'} - v_k v_{k'})(\alpha_k^{\dagger} \alpha_{-k'}^{\dagger} + \alpha_{-k} \alpha_{k'}) \right\} ; \qquad (54)$$

this is clearly a complicating feature for any simple description in terms of the new constructs. Still worse is the last term in Eq. 51, which has the structure

$$H_{\text{int}} = H_{40} + H_{31} + H_{22} \tag{55}$$

and involves in the three respective pieces $\alpha^\dagger\alpha^\dagger\alpha^\dagger\alpha^\dagger$, $\alpha\alpha\alpha\alpha$ or $\alpha^\dagger\alpha^\dagger\alpha^\dagger\alpha$, $\alpha^\dagger\alpha\alpha\alpha$ or $\alpha^\dagger\alpha^\dagger\alpha\alpha$. We ignore this term at this stage, leaving it for subsequent perturbative treatment if necessary. As will be more apparent shortly, it represents an interaction between quasiparticles after the main part of the pairing features in the Hamiltonian have been treated in the transformation to quasiparticles. It is as a result of dropping H_{int} that the Hamiltonian ceases to conserve particle number—for a different technical reason than in the BCS approach—thus necessitating the use of the Lagrange multiplier λ that appears in U, H_{11}, and H_{20}. Naturally this entire classification and treatment assumes that pairing dominates the physical system in question, in which case the resulting phenomena will have been tended to in the main and the terms neglected so far will be minor corrections to be attended to subsequently; if pairing is not dominant, then the present approach is not a good point of departure.

The results of Eqs. 51–55 are obtained in a completely straightforward—but rather tedious—fashion by using the anticommutation relations Eqs. 46 and 48 of the operators $\alpha^\dagger$ and α. They also exploit the symmetry based on the definition of the two-particle matrix element,

$$\langle k_2 k_1 | v | k_4 k_3 \rangle = \langle k_1 k_2 | v | k_3 k_4 \rangle , \tag{56a}$$

and a further relationship that we assume holds for the systems of interest to us on grounds of time-reversal invariance,

$$\langle -k_1, -k_2 | v | -k_3, -k_4 \rangle = \langle k_1 k_2 | v | k_3 k_4 \rangle , \tag{56b}$$

as well as the property of Eq. 47b.

Once we are working with quasiparticles as the basic theoretical entity, we wish also to consider the number operator for quasiparticles,

$$\mathcal{N} = \sum_k \alpha_k^\dagger \alpha_k . \tag{57}$$

To complete the definition of the new representation, we introduce the natural assumption for simplicity that the Hamiltonian $\tilde{H}$ in terms of quasiparticles is diagonal with respect to quasiparticle number (this, of course, after having dropped H_{int} of Eq. 55). Thus the same number of quasiparticles is always to be present in the system, and so we must require

$$H_{20} = 0 , \tag{58}$$

having noted that $U + H_{11}$ is already diagonal in quasiparticle number,

$$[U + H_{11}, \mathcal{N}] = 0. \tag{59}$$

In examining the consequences of the condition of Eq. 58 we first assume that we are working in a basis of single-particle states that diagonalizes the expression in braces in Eq. 54, so that

$$\langle k|t|k'\rangle - \lambda\delta_{kk'} + \sum_{k''} \langle kk''|v|k'k''\rangle_A v_{k''}^2 = (\epsilon_k^0 - \lambda)\delta_{kk'} = \epsilon_k \delta_{kk'}, \tag{60}$$

which is an obvious generalization of the Hartree–Fock condition of Eqs. 4.13–4.15 to the case where the single-particle occupancy is not 0 or 1 but rather is measured by v_k^2. Then Eq. 54 becomes

$$H_{20} = \sum_{k,k'} \left\{ \epsilon_k u_k v_k \delta_{kk'} \right.$$

$$+ \frac{1}{2} \sum_{k''} \langle k, -k'|v|k'', -k''\rangle u_{k''} v_{k''} (u_k u_{k'} - v_k v_{k'}) \Big\}$$

$$\times (\alpha_k^\dagger \alpha_{-k'}^\dagger + \alpha_{-k} \alpha_{k'}), \tag{61}$$

and the requirement that this vanish is the *Hartree–Fock–Bogolyubov equation*

$$2\epsilon_k u_k v_k \delta_{kk'} + \sum_{k''} \langle k, -k'|v|k'', -k''\rangle u_{k''} v_{k''} (u_k u_{k'} - v_k v_{k'}) = 0. \tag{62}$$

For a sharp Fermi surface,

$$u_k = 0, \qquad v_k = 1, \qquad k \leq k_F \tag{63a}$$

and

$$u_k = 1, \qquad v_k = 0, \qquad k > k_F, \tag{63b}$$

this equation becomes trivial. The restriction of Eq. 62 is to be viewed as a condition on u_k or v_k (which are related by Eq. 44c, of course), while the prediagonalization of Eq. 60 restricts the single-particle basis. This represents an exceedingly complicated problem in self-consistency if it is to be carried out in full generality.

Our formalism up to this point is for a general interaction V in Eq. 50, with the background supposition that a large part of this interaction has the character of a pairing force and will thus be dealt with by the Bogolyubov transformation of Eqs. 44. To see how this comes about, we specialize now to the pairing force of Eq. 28. Note the restriction to positive values of the quantum labels in Eq. 28 (k and $k' > 0$), which prevents the interaction from vanishing by reason of the anticommutation relations if the terms are "double-counted"; the restriction $k' > 0$, for example, prohibits the appearance of $a_{-k'}a_{k'} + a_{k'}a_{-k'} = 0$. (This is incorporated into Eq. 65 below through the device of a sign reversal for the redundant terms.) In our present language this implies a force such that

$$\langle k, -k' | v | k'', -k''' \rangle = \delta_{kk'} \delta_{k''k'''} \langle k, -k | v | k'', -k'' \rangle \,. \tag{64}$$

For the especially simple case of the constant pairing force in Eq. 28, and to be consistent with the symmetry conditions (Eqs. 56) that we have used, we have

$$\langle k, -k | v | k'', -k'' \rangle = -\tfrac{1}{2} \, \text{sign}(k) \, \text{sign}(k'') \, |G| \,. \tag{65}$$

The form of the summations in Eqs. 52–54 or 61 and 62 motivates the definition (compare Eq. 38)

$$\Delta_k \equiv -\sum_{k''} \langle k, -k | v | k'', -k'' \rangle u_{k''} v_{k''} \,, \tag{66}$$

where we here allow for dependence of the gap parameter on the pertinent single-particle state. If the pairing force of Eq. 64 is generally attractive, then $\Delta_k > 0$. In terms of the generalized gap parameter of Eq. 66, the Hartree–Fock–Bogolyubov equation 62 is

$$2\epsilon_k u_k v_k - \Delta_k (u_k^2 - v_k^2) = 0 \,, \tag{67}$$

as in Eq. 39 (where the gap parameter is independent of k however). Together with Eq. 44c, this yields solutions parallel to those of Eqs. 40 (again choosing signs such that $v_k^2 \to 1$ and $u_k^2 \to 0$ when $\epsilon_k \ll 0$),

$$u_k^2 = \frac{1}{2} \left[1 + \frac{\epsilon_k}{\sqrt{\epsilon_k^2 + \Delta_k^2}} \right], \qquad v_k^2 = \frac{1}{2} \left[1 - \frac{\epsilon_k}{\sqrt{\epsilon_k^2 + \Delta_k^2}} \right]. \tag{68a, b}$$

Inserting these into the definition of Eq. 66, the gap equation is

$$\Delta_k = -\sum_{k''>0} \langle k, -k | v | k'', -k'' \rangle \frac{\Delta_{k''}}{[\epsilon_{k''}^2 + \Delta_{k''}^2]^{1/2}} \,, \qquad k > 0 \,, \tag{69}$$

which again reduces to Eq. 41a for a simple constant pairing force, Eq. 65. (Note that the summations in Eq. 69 and above are restricted to positive values if and only if this is explicitly indicated. See footnote after Eq. 28, Section 5.3.)

For the pairing-force case of Eqs. 64 and 66–69, the quasiparticle Hamiltonian becomes, now that we have dropped H_{int} and required $H_{20} = 0$,

$$\tilde{H} \Rightarrow H_0 = U + H_{11}$$

$$= U + \sum_{k>0} [\epsilon_k(u_k^2 - v_k^2) + 2\Delta_k u_k v_k](\alpha_k^\dagger \alpha_k + \alpha_{-k}^\dagger \alpha_{-k})$$

$$= U + \sum_{k>0} [\epsilon_k^2 + \Delta_k^2]^{1/2}(\alpha_k^\dagger \alpha_k + \alpha_{-k}^\dagger \alpha_{-k}) . \tag{70}$$

From this we can calculate the spectrum for quasiparticle excitation: For the BCS ground state of Eqs. 29 the calculation is especially simple, since we know from Eq. 49 that it serves as the vacuum for the quasiparticles and since the rightmost operator in the nonconstant part of the Hamiltonian H_0 of Eq. 70 is always an annihilation operator α. (This was the purpose of the "normal ordering".) Thus, ignoring the constant energy U,

$$H_0|\text{BCS}\rangle = 0 . \tag{71}$$

For one quasiparticle present the energy is

$$H_0 \alpha_k^\dagger|\text{BCS}\rangle = E_k \alpha_k^\dagger|\text{BCS}\rangle , \tag{72}$$

where

$$E_k \equiv [\epsilon_k^2 + \Delta_k^2]^{1/2} \geq \Delta_k \tag{73}$$

is the quasiparticle energy. In the state of Eq. 72 the odd number of particles present are all paired except for the last one labeled by k. The spectrum of excitations is continuous, as is seen from Eq. 73. For two quasiparticles present—again all other fermions are paired—we have again a simple counting procedure for the energy, and

$$H_0 \alpha_{k_1}^\dagger \alpha_{k_2}^\dagger|\text{BCS}\rangle = (E_{k_1} + E_{k_2})\alpha_{k_1}^\dagger \alpha_{k_2}^\dagger|\text{BCS}\rangle . \tag{74}$$

The energy eigenvalue here is always greater than or equal to 2Δ (for a constant pairing force as in Eq. 65 so that $\Delta_k \equiv \Delta$). For this even number of particles the ground state is $|\text{BCS}\rangle$, in which all particles are paired and the lowest excitation involves breaking a pair as in Eq. 74; this costs a minimum energy of 2Δ. Thus 2Δ is the energy gap imposed upon the fully paired system by the pairing force.

5.4.3 Corrections to Independent Quasiparticles

In exploring the consequences of the Bogolyubov transformation we dropped the term H_{int} in Eq. 55. It is now clear that within this transformation we arrive at a new Hamiltonian in which we have a constant term U and a one-(quasi)particle operator H_{11}. The term H_{int} then represents a kind of residual interaction between the quasiparticles. By dropping it we brought about a situation in which the conservation of fermion number was violated, and so we introduced the chemical potential λ as a Lagrange multiplier to deal with it. At the stage in which we ignored H_{int} we consoled ourselves with the thought that we could later incorporate its effects approximately by using it to improve our lowest-order results perturbatively. This is rather easily done for the ground state because H_{31} and H_{22} in H_{int} both contain an annihilation operator α as their rightmost operator by definition, and thus both annihilate the state of full pairing, $|BCS\rangle$. Only H_{40} connects to this state, and it is easily evaluated from Eq. 50. For a pure, constant pairing force we have

$$H_{40} = |G| \sum_{\substack{k>0 \\ k'>0}} u_k^2 v_{k'}^2 (\alpha_k^\dagger \alpha_{-k}^\dagger \alpha_{-k'}^\dagger \alpha_{k'}^\dagger + \alpha_{k'} \alpha_{-k'} \alpha_{-k} \alpha_k) . \tag{75}$$

This leads to a corrected ground state in perturbation theory that differs from the fully paired state by the breaking of two pairs (i.e. the creation of four quasiparticles),

$$|BCS\rangle' = \left\{ 1 - |G| \sum_{\substack{k>0 \\ k'>0}} \frac{1}{2(E_k + E_{k'})} u_k^2 v_{k'}^2 \alpha_k^\dagger \alpha_{-k}^\dagger \alpha_{-k'}^\dagger \alpha_{k'}^\dagger \right\} |BCS\rangle .$$

$$\tag{76}$$

The correction admixed is thus $\leqslant |G|/4\Delta$, which is expected to be small, as can be seen from Eq. 84 for example. Thus we expect that the effects of quasiparticle interaction do not destroy the gap properties of the pairing solution that emerges from the quasiparticle transformation in the presence of a pairing force.

5.5 HEURISTIC SOLUTIONS OF THE GAP EQUATION

To gain further insight into the properties of pairing we consider a heuristic version of the problem for constant pairing force. We assume that the gap Δ is sufficiently large to contain a sizable range of single-particle unperturbed energies. This allows the sums in the gap and number equations 41 to be replaced by integrals,

$$\tfrac{1}{2}|G| \int_a^b \frac{\rho(\epsilon)\, d\epsilon}{(\epsilon^2 + \Delta^2)^{1/2}} = 1 \tag{77a}$$

and

$$\int_a^b \left[1 - \frac{1}{(\epsilon^2 + \Delta^2)^{1/2}} \right] \rho(\epsilon)\, d\epsilon = N , \tag{77b}$$

where $\rho(\epsilon)$ is the density of the original single-particle states and $[a, b]$ is the range of energies covered by the pairing. Since we are measuring energies relative to the chemical potential λ,

$$\epsilon = \epsilon^0 - \lambda, \qquad a = \epsilon' - \lambda, \qquad b = \epsilon'' - \lambda, \tag{78}$$

the energy domain of relevance in the integrals of Eqs. 77, which involves $|\epsilon| \lesssim \Delta$, is for energies ϵ^0 near λ. We further assume that the density of states is constant over the relevant domain,

$$\rho(\epsilon) = \bar\rho \qquad \text{for} \quad a \le \epsilon \le b. \tag{79}$$

Then Eq. 77a gives

$$\operatorname{arcsinh} \frac{b}{\Delta} - \operatorname{arcsinh} \frac{a}{\Delta} = 2\eta, \qquad \eta \equiv (\bar\rho|G|)^{-1}, \tag{80}$$

or

$$\Delta = \frac{(b^2 + a^2 - 2ab \cosh 2\eta)^{1/2}}{\sinh 2\eta}. \tag{81}$$

In terms of the total number of pairing states

$$\Omega \equiv \bar\rho(\epsilon'' - \epsilon') = \bar\rho(b - a) \tag{82a}$$

we define an occupation factor

$$\chi_N \equiv 1 - \frac{N}{\Omega}, \tag{82b}$$

and the number equation 77b then yields for the chemical potential

$$\lambda = \tfrac{1}{2}(\epsilon'' + \epsilon') - \tfrac{1}{2}(\epsilon'' - \epsilon')\chi_N \coth \eta, \tag{83a}$$

while the gap parameter in these terms is

$$\Delta = \frac{1}{2} \frac{(\epsilon'' - \epsilon')(1 - \chi_N^2)^{1/2}}{\sinh \eta}. \tag{83b}$$

If the interaction is weak ($\bar\rho|G| \ll 1$), the gap parameter is approximately

$$\Delta \cong (\epsilon'' - \epsilon')e^{-1/\bar\rho|G|}. \tag{84}$$

From this expression it is clear that no satisfactory perturbation solution for the pairing problem is possible, since this would imply expansion in the small parameter $\bar\rho|G|$, but that is an expansion about an essential singularity for the central quantity of the theory—the gap parameter.

Thus we can now see that to include pairing properly we required a nonperturbative technique such as the BCS variational method or the Bogolyubov transformation. These techniques generalize the characteristic feature of the Hartree–Fock method: That method is based on a concept of independent-particle motion (though in accord with the Pauli principle and with self-consistent orbitals), while the approaches to paired systems assume the independence of these pairs. The pairing theories lead to the notion of quasiparticles in which the pairing part of the interaction is treated nonperturbatively—on the assumption that it dominates—leaving the rest to be included subsequently as corrections. The quasiparticle spectrum, and hence the spectrum of the whole paired system, is characterized by an energy gap. This spectrum and this gap are the central quantities in determining the physical properties of the paired system, as we shall now illustrate in the case of superconducting electrons in a metal.

5.6 APPLICATIONS: TUNNELING AND THE JOSEPHSON EFFECT

5.6.1 Tunneling between a Superconductor and a Normal Metal

Consider a physical situation, depicted roughly in Figure 5.5, in which a superconductor is separated from a normal metal by a very thin nonconducting (oxide) layer. We wish to measure the value of the gap parameter by applying a voltage V, which we increase until electrons can overcome the gap difference and begin to flow so that they are detected as a current. The details of this effect depend primarily on the densities of states ρ_N and ρ_S for the normal and the superconducting systems; these quantities are also sketched in the figure. For the normal system we approximate the electrons by a Fermi gas. Though we could treat temperature effects by using a Fermi distribution for this gas, these effects are relatively unimportant for the understanding of the superconducting aspects of the problem. We thus take $T = 0$ K, and $\rho_N(E)$ is the usual density of states for a Fermi gas, while the occupancy probability for the gas is the corresponding flat distribution

$$f(E) = \begin{Bmatrix} 1, & E \leq 0 \\ 0, & E > 0 \end{Bmatrix} = \theta(-E) . \tag{85}$$

The density of states for the superconductor is

$$\rho_S(E_k) \, dE_k = \frac{2 \times 4\pi k^2}{(2\pi)^3} \frac{dE_k}{dE_k/dk} , \tag{86}$$

where the factor of 2 takes account of the spin $\pm \frac{1}{2}$ degeneracy, and 4π represents the integration over solid angle. The energy is that of the quasiparticles

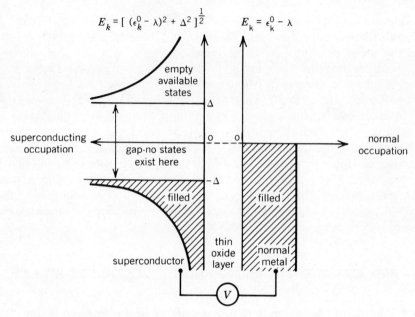

FIGURE 5.5 Representation of a tunneling situation for a thin nonconducting layer between a superconductor and a normal metal. The Fermi levels λ in the two systems are depicted as being at the same energy in the sense that far beyond the gap $E_k \equiv [(\epsilon_k^0 - \lambda)^2 + \Delta^2]^{1/2} \rightarrow \epsilon_k^0 - \lambda$, which is the same energy as in the normal metal. This will be the case for zero applied voltage $V = 0$. As V changes, the entire scale of energy of one system shifts correspondingly relative to the other.

$$E_k = [(\epsilon_k^0 - \lambda)^2 + \Delta^2]^{1/2} , \qquad (87)$$

where we measure relative to the chemical potential or Fermi level λ, as shown explicitly. We further assume that the original (i.e. normal) particle spectrum of the superconductor is well represented by the Fermi gas, $\epsilon_k^0 \cong k^2/2m$. Then

$$\rho_S(E_k)dE_k = \frac{mk_F}{\pi^2} \frac{E_k}{|\epsilon_k^0 - \lambda|} dE_k$$

$$= \frac{mk_F}{\pi^2} \frac{E_k}{(E_k^2 - \Delta^2)^{1/2}} dE_k , \qquad (88)$$

and the superconducting density of states exhibits a square-root divergence at $|E_k| = \Delta$.

In order to calculate the current flowing in the system we take the transition rate at which electrons tunnel through the barrier (electrons per unit time) and multiply it by the charge $-e$ on each electron. The transition rate is given by Fermi's "golden rule", so that the current is

$$I = 2\pi(-e)|T|^2 \int dE_k \; \rho_S(E_k)\rho_N(E_k + eV)$$

$$\times f(E_k)[1 - f(E_k + eV)] \,. \tag{89}$$

In writing this we have displaced the relative energies of the electrons in the normal metal by the energy value eV supplied by the applied voltage V. An important and complicated quantity in the problem is the tunneling probability amplitude or matrix element, which we denote by T. This quantity is crucial if we require the actual value of the current, but we do not need to know it (beyond an assumption here that it is roughly constant near the Fermi surface) to understand considerations of energetics, that is, the dependence on voltage. This is just as well, since T will depend on detailed information of the features of the oxide layer, such as its geometry and surface properties, and these are not likely to be well controlled. The remainder of the expression in Eq. 89 represents the fact that the transition probability is proportional to the density of available states in the two systems and to the probabilities that these states are occupied for the superconductor $[f(E_k)]$ and unoccupied for the normal metal $[1 - f(E_k + eV)]$, so that electrons can flow from the former to the latter.

We now consider the change in the current when the voltage is varied, dI/dV, and note from Eq. 85.

$$\frac{df(E_k + eV)}{dV} = -e\delta(E_k + eV) \,,$$

while the density of states varies relatively weakly with energy. Then

$$\left|\frac{dI}{dV}\right| = 2\pi e^2 |T|^2 \rho_N(\epsilon_k^0 = \lambda)\rho_S(E_k = -eV) \,, \tag{90}$$

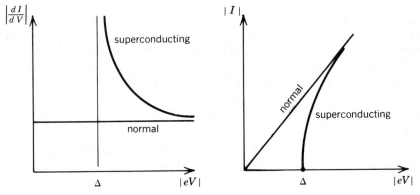

FIGURE 5.6 Current flow in tunneling between a normal metal and a superconductor begins when the applied voltage overcomes the superconducting gap.

and we can map the density of quasiparticle states in the superconductor explicitly by varying V. As we noted in Eq. 88, this diverges at $|E_k| = |eV| = \Delta$, so that the qualitative behavior of these quantities is that shown in Figure 5.6. Current flows only after the applied voltage can offset the gap of the superconductor. As higher voltage is imposed, the current eventually joins the conventional behavior for normal metals.

5.6.2 The Josephson Effect

We again consider two metals separated by a thin insulating layer, but this time both metals are superconducting, so that the anticipated tunneling is such that pairs on one side of the barrier are again found after the tunneling process as pairs on the other side, although the actual transfer takes place electron by electron. Our treatment of this system is somewhat intuitive [and is based on an interpretation originally due to R. A. Ferrel and R. E. Prange, Phys. Rev. Lett. **10**, 479 (1963) and further evolved in P. G. de Gennes, *Superconductivity of Metals and Alloys* (Benjamin, New York, 1966)]. We assume a Hamiltonian for the system of the form

$$H = H_S + H_{S'} + H_T , \qquad (91)$$

where H_S and $H_{S'}$ refer to the two superconductors and describe the pairing situation that has been the subject of this chapter. For the moment we only use the fact that the two systems are fully paired, and we characterize the situation by the number of pairs ν in system S say, so

$$(H_S + H_{S'})\psi_\nu = E_\nu \psi_\nu , \qquad (92)$$

where

$$\psi_\nu = \phi'_{2N-2\nu}\phi_{2\nu} \qquad (93)$$

for a constant total of $2N$ electrons, and where we assume $N \gg \nu$. The function $\phi_{2\nu}$ describes the system S, and $\phi'_{2N-2\nu}$ the system S'. The tunneling part of the Hamiltonian is described by a one-electron operator (as for normal metals) with tunneling coefficients T_{kl} and T^*_{kl} whose full properties are complicated but are not required for our purposes. Thus

$$H_T = \sum_{kl} (T_{kl}a^\dagger_k a'_l + T^*_{kl}a'^\dagger_l a_k) , \qquad (94)$$

where $a^\dagger_k, a_k$ refer to system S and $a'^\dagger_l, a'_l$ to system S', the electrons tunneling—one by one we presume at this stage—from one system to the other.

If the pair count ν is increased by one unit to $\nu + 1$, this means that two electrons were removed from S' and added to S, at an energy cost

$$E_{\nu+1} - E_\nu = 2(E_F - E_F') \,. \tag{95}$$

Here E_F and E_F' are the Fermi energies for S and S' and are equal if there is no applied voltage between the superconductors. The unperturbed solutions ψ_ν are then degenerate in ν, a degeneracy that is removed by the tunneling interaction H_T. Because of this interaction the second-order matrix element between ψ_ν and $\psi_{\nu+1}$ is

$$J_0 \equiv \sum_i \sum_{kl\bar{k}\bar{l}} \langle \nu + 1 | T_{kl} a_k^\dagger a_l' | i \rangle \frac{1}{E - E_i} \langle i | T_{\bar{k}\bar{l}} a_{\bar{k}}^\dagger a_{\bar{l}}' | \nu \rangle \,. \tag{96}$$

The intermediate state $|i\rangle$ has $2\nu + 1$ electrons in S and $2N - 2\nu - 1$ in S', that is, there is one unpaired electron in each. Since the systems are fully paired initially and finally, we must have $k = -\bar{k}$ and $l = -\bar{l}$; we also assume the tunneling interaction to be time-reversal invariant, so that $T_{-k,-l} = T_{kl}^*$. Then

$$J_0 = -4 \sum_{kl} |T_{kl}|^2 \frac{u_k v_k u_l v_l}{E_k + E_l} \,, \tag{97}$$

where we have used the properties of the superconducting ground state $|\text{BCS}\rangle$ of Eqs. 29 in calculating the numerator. The denominator in Eq. 97 represents the energy of two quasiparticles, and the factor of 4 includes the various matching combinations of $k, \bar{k}, l, \bar{l}$. Note that for normal states (i.e. no superconductivity) one has $u_i v_i = 0$ and thus $J_0 = 0$.

Now J_0 incorporates the superconductive feature which we require in our description, and so we may proceed to a solution for the full Hamiltonian, for which we have

$$H\psi_\nu = E\psi_\nu + J_0(\psi_{\nu+1} + \psi_{\nu-1}) \,, \tag{98}$$

where $E = E_\nu + J_0$ includes the diagonal part arising from J_0. Note that J_0 is independent of ν. The correct wave function is a linear superposition of ψ_ν, and, using the "translational invariance" of Eq. 98 in ν, in analogy with the discrete displacement features of a crystal lattice, we introduce

$$\psi(k) \equiv \sum_\nu e^{-ik\nu} \psi_\nu \,, \tag{99}$$

where we shall not restrict the summation, since $N \gg \nu$ (see Eq. 93). This is to be an eigenstate of H, and so we have

$$H\psi(k) = E(k)\psi(k) \,, \tag{100a}$$

or

$$\sum_{\nu} e^{-ik\nu}[E\psi_\nu + J_0(\psi_{\nu+1} + \psi_{\nu-1})] = E(k)\sum_{\nu} e^{-ik\nu}\psi_\nu , \qquad (100b)$$

whence

$$E(k) = E + 2J_0 \cos k . \qquad (101)$$

(In practice, since the experimental circuit containing the Josephson junction has resistance, a generator must be included in the circuit to balance the *IR* drop and produce zero voltage across the barrier layer. In the steady state this will fix the phase parameter k, which reflects details of the experimental setup.)

In the superconducting systems the number of pairs is very large, $\nu \sim 10^{20}$, and the fluctuations in this number are relatively small, $\Delta\nu \sim \sqrt{\nu} \sim 10^{10}$, whence $\Delta\nu/\nu \sim 10^{-10}$. The uncertainty in the corresponding variable is thus also small, $\Delta k \sim 1/\Delta\nu \sim 10^{-10}$. As a consequence we may specify ν and k simultaneously quite sharply. In other words the characterization of the pair number for each subsystem is a classical problem. If we invert the Fourier series in Eq. 99,

$$\psi_\nu = \int e^{ik\nu}\psi(k)\frac{dk}{2\pi} , \qquad (102)$$

and define a wave packet for the electron number,

$$\langle \nu \rangle \equiv \sum_{\nu} \psi_\nu^* \nu \psi_\nu$$

$$= \sum_{\nu} \int e^{-ik'\nu}\psi^*(k')\nu e^{ik\nu}\psi(k)\frac{dk\, dk'}{(2\pi)^2} , \qquad (103)$$

then the usual time dependence $\psi(k) \Rightarrow e^{-iE(k)t}\phi(k)$ implies an equation of motion for a wave packet

$$\frac{d\langle \nu \rangle}{dt} = \left\langle \frac{\partial E(k)}{\partial k} \right\rangle = -2J_0 \sin k , \qquad (104)$$

where the spread in k is small, as we have already noted. (This is all in precise analogy to the considerations of group velocity in conventional optics, for instance.) The current carried by this change in ν is

$$I = 2e\frac{d\langle \nu \rangle}{dt} = -4eJ_0 \sin k , \qquad (105)$$

and this current flows even in the complete *absence* of a voltage. For a nonsuperconducting system, of course, $J_0 = 0$ and this current vanishes.

If a voltage is applied across the systems S and S', then from Eq. 95 the energy cost for a pair to tunnel from one system to the other is

$$E_{\nu+1} - E_\nu = -2eV \tag{106}$$

and the wave packet obeys the force equation

$$\frac{d\langle k \rangle}{dt} = \left\langle -\frac{\Delta E}{\Delta \nu} \right\rangle = \langle E_\nu - E_{\nu+1} \rangle = 2eV \ . \tag{107}$$

If V is constant, then $k \propto t$ and I is an alternating current of frequency $2eV$,

$$I = -4eJ_0 \sin (2eVt + k_0) \ , \tag{108}$$

where k_0 is an arbitrary constant. This averages to zero if $V \neq 0$ because of its oscillatory behavior, and is paradoxically different from zero only when $V = 0$, as we saw for Eq. 105. This behavior is known as the *DC Josephson effect*, since it relates to a voltage that is constant in time.

The applied voltage may itself have an oscillatory component of frequency ω and amplitude V_1,

$$V = V_0 + V_1 \cos \omega t \ , \tag{109}$$

leading to the *AC Josephson effect*. Then the solution of Eq. 107 is

$$k = k_0 + 2eV_0 t + 2e \frac{V_1}{\omega} \sin \omega t \ , \tag{110}$$

and the resulting current is

$$I = -4eJ_0 \sin \left[k_0 + 2eV_0 t + 2e \frac{V_1}{\omega} \sin \omega t \right] . \tag{111}$$

For small V_1 this may be expanded as

$$I \cong -4eJ_0 \sin (2eV_0 t + k_0)$$
$$-8e^2 J_0 \frac{V_1}{\omega} \sin \omega t \cos (2eV_0 t + k_0) \ , \tag{112}$$

of which the first term is the DC Josephson effect and averages to zero for $V_0 \neq 0$, and the second term is the AC effect, which survives if we tune $\omega = 2eV_0$. The sensitivity to the electron charge in this tuning allows the exploitation of the AC Josephson effect for extremely precise measurement of the fine-structure constant. This is an unexpected dividend that we derive from the understanding of pairing phenomena.

EXERCISES

5.1 **(a)** Show that the expression in Eq. 9a satisfies Eq. 9b.
   (b) Show Eq. 14b in detail.
   (c) Calculate the energy of the first excited state in the seniority model for N even and $\leq \Omega$ (the energy gap).
   (d) Show that the degree of degeneracy of the levels with seniority S is independent of particle number N.

5.2 Define the three quasispin operators for a system of a finite number of fermions,

$$A_- = \sum_{m>0} a_m a_{-m}, \qquad A_+ = \sum_{m>0} a^\dagger_{-m} a^\dagger_m,$$

$$A_0 = \frac{1}{2} \left[\sum_{m>0} (a^\dagger_m a_m + a^\dagger_{-m} a_{-m}) - \frac{1}{2} S_0 \right],$$

where S_0 is the number of single-particle orbits. These can be used to solve the Hamiltonian of Eqs. 10 and 11.

   (a) Calculate the commutators of A_+, A_-, A_0, and show they are the same as for a spin operator $\mathbf{S}$. Show also that

$$\mathbf{A}^2 = \tfrac{1}{2}(A_+ A_- + A_- A_+) + A_0^2$$

   commutes with $A_\pm$, A_0.

   (b) Show that the fully paired state of $2m$ fermions,

$$(A_+)^m |0\rangle ,$$

   can be classified by eigenvalues of $\mathbf{A}^2$ and A_0 simultaneously. Give the eigenvalues for the vacuum, and for the state with all orbits filled.

   (c) Show that all other states of the system, with $0 \leq n \leq N_{max}$ particles, may also be classified by eigenvalues of $\mathbf{A}^2$ and A_0. What does A_0 denote? How can you interpret states orthogonal to those in part (b)?

   (d) Give formulas for the energies of the ground state and first excited state (for an even number of particles) in terms of $|G|$, S_0, and N.

5.3 Using the BCS trial function of Eq. 29 and the standard form for the number operator in Eq. 31, verify the expression for the fluctuation in the value of the number of particles in Eq. 33.

5.4 Use the BCS trial function of Eq. 29 to obtain the expression of Eq. 34 for the BCS system energy on the basis of the Hamiltonian proposed in Eqs. 28 and 30.

5.5 Using the inverse of the transformations defined in Eqs. 44 and the general Hamiltonian for a one- and two-particle interaction of Eq. 50, carry out the (rather lengthy) procedure of normal ordering in order to derive the expressions of Eqs. 51–55 for the Bogolyubov Hamiltonian in terms of quasiparticles.

5.6 From standard perturbation theory applied to the four-quasiparticle interaction of Eq. 75, derive the correction shown in Eq. 76 for the BCS trial function for the case where this part of H_{int} is not neglected.

5.7 For the heuristic model of section 5.5, calculate the full system ground-state energy and the number fluctuations, $\langle BCS|H|BCS\rangle$ and $\langle BCS|N^2|BCS\rangle$, of Eqs. 28, 33, and 34.

5.8 With the help of the usual techniques for treating group velocities in optics, derive Eq. 104 for the change in time of the pair partition between the two superconductors in the Josephson effect.

5.9 **(a)** Solve the BCS problem for a set of degenerate single-particle orbits: i.e. in Eq. 28, $\epsilon_k^0 = \epsilon$, with $\Sigma_{k>0} = \Omega$, for an average number $N < \Omega$. Find the ground-state energy and v_k^2.

   (b) Compare the BCS ground state wave function and energy to those for the seniority ground state solution (exact) for the same interaction: Eqs. 23 and 25b, with $S = 0$. Interpret v_k^2 (BCS) for the $S = 0$ exact solution.

   (c) Calculate the number fluctuation, Eq. 33; under what conditions is it small?

   (d) Explain the differences between the BCS solution and the exact solution.

5.10 Consider a Hamiltonian

$$H = \frac{1}{2} \sum_{k>0} [(\epsilon_k^0 + |G|)(a_k^\dagger a_k + a_{-k}^\dagger a_{-k}) + |G|(a_k^\dagger a_{-k}^\dagger + a_k a_{-k})] ,$$

where $a_k^\dagger$ and a_k create and annihilate *bosons* in a single-particle state k. Define the Bogolyubov transformation to quasiparticles according to

$$\alpha_k = u_k a_k - v_k a_{-k}^\dagger \quad \text{and} \quad \alpha_k^\dagger = u_k a_k^\dagger - v_k a_{-k},$$

where $u_{-k} = u_k$ and $v_{-k} = v_k$.

   (a) Show that canonical commutation relations require that

$$u_k^2 - v_k^2 = 1 ,$$

where by "canonical" we here mean that the α's and $\alpha^\dagger$'s are to obey the same commutation relations as the original boson operators a and $a^\dagger$.

(b) Transform the Hamiltonian H to quasiparticle variables and bring it to normal-ordered form, i.e., write it so that all quasiparticle creation operators $\alpha^\dagger$ stand to the left of all quasiparticle annihilation operators α.

(c) Require that the transformed Hamiltonian conserve the number of quasiparticles, so that no terms in which the number of quasiparticles is changed may be present. What condition does this give for the u_k and v_k?

(d) Write the Hamiltonian in terms of these u_ks and v_ks. What is the energy spectrum of the quasiparticles? (See Eq. 10.29.)

CHAPTER 6

The Interacting Fermi Gas: Hard Spheres and Electrons

In the previous two chapters we have considered interacting Fermi systems in two interesting and simple limits: First, when independent particle motion in an average potential is an adequate description, we can apply the Hartree–Fock methods of Chapter 4. Second, when a weakly interacting Fermi gas experiences an attractive interaction between particles in orbits near the Fermi surface, the phenomena associated with pairing occur, and are treated by the methods developed in Chapter 5. In the present chapter we consider two examples of interacting Fermi systems for which we shall need entirely new techniques to describe their dynamics and to obtain properties such as their ground-state energies. The examples are the hard-sphere gas and the electron gas, which we shall define shortly. The theoretical tools to treat these systems will be those of diagrammatic perturbation theory for many-body systems, which is the subject of Chapters 8 and 9.

The hard-sphere gas and the electron gas have in common the property of being *normal* Fermi systems in the ground state, *normal* meaning that it is possible to define a convergent perturbation expansion for the ground state in terms of the interaction among particles. Such expansions, as we shall see, are not necessarily simply in powers of the strength of the interaction, as given by Rayleigh–Schrödinger perturbation theory. But the expansions do imply that the ground-state properties are controlled by some physical parameter (or parameters) which may be taken to be zero for the limiting case of the noninteracting Fermi gas, and may be developed as power series for at least some domain of the parameter. This is not so for the case of pairing (for an infinite system of particles) as we noted in the discussion following Eq. 5.84 in the previous chapter: for an attractive pairing interaction, the ground state is not analytically expressible in terms of the interaction strength, no matter how weak the interaction. (The distinction of *normal* ground states from other possibilities will be discussed further in Section 7.4 in connection with the concept of *adiabatic switching* of an interaction in time.)

On the other hand, both examples we have chosen to discuss exhibit difficulties which make a straightforward application of ordinary perturbation expansions (e.g. Rayleigh–Schrödinger) impossible, as we shall see in this chapter. First, we shall find out what these difficulties are; then in the following chapters we shall develop a general method of treating such problems, and return in Chapter 10 to complete our discussion of the interacting Fermi gas.

6.1 HARD-SPHERE GAS AND ELECTRON GAS

6.1.1 Hard-Sphere Gas

Imagine a physical system consisting of a large number of identical particles moving freely in a large container, interacting with each other as hard or impenetrable spheres of diameter d. The interaction of hard sphere particles is equivalent to a constraint that no two particles can be separated by a distance less than d between their centers. Described as a potential, the two-particle interaction is infinitely repulsive for relative distances $r < d$, and zero otherwise:

$$V(r) = \begin{cases} +\infty & r < d \\ 0, & r > d \end{cases}, \tag{1}$$

as illustrated in Figure 6.1. This potential is an idealization of the repulsive interaction between atoms at $r < d \sim$ (atomic diameter), or that between nucleons for $r < d \sim 0.5$ fm. In both these physical cases, there is an attractive interaction of finite range at distances outside the repulsive "core" region, as shown in Figure 6.2. In atoms the attraction comes from electron exchange or mutual polarization (Van der Waals) and gives the binding potential for molecules. For nucleons the attraction is understood to be from meson exchange and/or mutual polarization of the quark structure, and leads to the binding potential for nuclei. Without the attraction, the atomic or nuclear systems would remain unbound, even at zero temperature, behaving as a gas. This is a somewhat simpler problem to treat than that of the full molecular or nuclear interaction. We shall study the purely repulsive interaction first, returning to some aspects of the more general problem in Section 10.4.

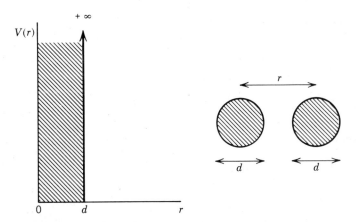

FIGURE 6.1 Potential describing hard-sphere repulsive interaction between two impenetrable spheres of diameter d.

The hard-sphere gas is a nontrivial system, since the interaction is very strong within the interaction range d. As we shall see, perturbation theory in any finite order is useless. However, it seems intuitively clear that for low particle density the effect of the repulsion on bulk properties of the gas, like the energy, should be small in some sense. By low particle density we mean that the mean separation r_{mean} between particles is large compared with d. We should expect that the added energy due to repulsion will vanish with the ratio d/r_{mean}, possibly as some power. This has, in fact, been shown for the ground-state energy of the hard-sphere gas, for both Fermi and Bose statistics. The study of the expansion in d has provided a classic problem for the development of a variety of techniques in many-body theory over many decades, and a number of exact results have been found.

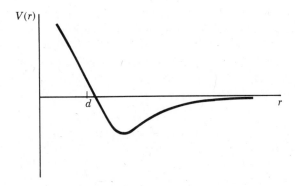

FIGURE 6.2 Schematic potential with finite repulsion and finite-range attraction, typical of atom–atom interaction ($d \sim 1$ Å) or nucleon–nucleon interaction (even l; $d \sim 0.5$ fm).

6.1.2 Electron Gas

The *electron gas* is defined as a system of electrons moving in the presence of a fixed uniform positive charge density of infinite extent. The electrons interact with each other through their mutually repulsive Coulomb interaction, as well as with fixed positive charge density through an attractive electrostatic interaction. The system so defined may serve as a model of the conduction electrons in a metal, where one ignores both the nonuniformity of the positive charge density of the metal ions, which is actually concentrated around sites in a crystal lattice, and the degrees of freedom of the lattice of ions, which can vibrate (phonon excitation). The neglect of these physical effects allows one to single out some specific effects of the electrons alone, which we consider in this chapter. Similarly, our idealized system can also represent some aspects of a gaseous plasma in which the degrees of freedom of the (heavier) positive ions are neglected, compared to those of the (lighter) electrons.

The features of special interest for the electron gas are related to the long range of the Coulomb interaction, in contrast to the short-range aspects of the hard-sphere gas, which we have just introduced.

6.2 PERTURBATION THEORY FOR A SHORT-RANGE REPULSIVE INTERACTION

We start with a uniform Fermi gas without interaction, at a fixed particle density ρ_0. The ground state is given by the antisymmetrized product wave function of Eq. 1.53:

$$|\Phi_0\rangle = \prod_{\substack{k \le k_F \\ \mu}} a_\mu^\dagger(\mathbf{k})|0\rangle , \qquad (2)$$

where the orbits are plane waves normalized in a box of volume Ω (which will later be taken in the limit $\Omega \to \infty$), obeying periodic boundary conditions on the surface, as in Eq. 1.52. Here we shall allow for internal degrees of freedom, as well as spin, in the index μ (which replaces s in Eq. 1.53), so that the degeneracy factor per particle is given by the even integer $g = \Sigma_\mu 1$. The density was given in Eq. 1.55 in terms of the Fermi momentum k_F:

$$\rho_0 = g k_F^3 / 6\pi^2 . \qquad (3)$$

The energy density of the noninteracting system is given by Eq. 1.57:

$$\frac{E_0}{\Omega} = \tfrac{3}{5}\epsilon_F \rho_0 , \qquad \epsilon_F = \frac{k_F^2}{2m} , \qquad (4)$$

and will be the unperturbed energy density of the interacting system.

Let us represent the interaction between particles by the repulsive square potential

$$v(r) = \begin{cases} V_0, & r < d, \\ 0, & r > d, \end{cases} \tag{5}$$

with strength $V_0 > 0$ and range d, illustrated in Figure 6.3. This becomes the hard-sphere potential in the limit $V_0 \to +\infty$. The first-order contribution to the energy density is given by the integral

$$\frac{\Delta E^{(1)}}{\Omega} = \frac{1}{2(2\pi)^6} \sum_{\mu_1,\mu_2} \int d\mathbf{k}_1 \, d\mathbf{k}_2 \; \theta(k_F - k_1)\theta(k_F - k_2)$$

$$\times \int d\mathbf{r} \, v(r)\{1 - \delta_{\mu_1,\mu_2} e^{-i(\mathbf{k}_1 - \mathbf{k}_2)\cdot\mathbf{r}}\}, \tag{6}$$

following the calculation of Eqs. 1.58–1.60, where we have replaced sums on $\mathbf{k}_1$, $\mathbf{k}_2$ by integrals, taking the limit of large Ω.

Now we shall explicitly assume that the interaction range d is small compared with the mean spacing between particles, at the given density ρ_0, by requiring that $k_F d \ll 1$. (Note that short range at fixed density is equivalent to low density for a given range.) In this limit, the exponential under the integral, which represents the exchange contribution to Eq. 6, may be approximated: $e^{i(\mathbf{k}_1 - \mathbf{k}_2)\cdot\mathbf{r}} \cong 1$. The spatial integral is then independent of momentum, given the volume integral of the potential, which for $v(r)$ of Eq. 5 becomes

$$\int d\mathbf{r} \, v(r) = \frac{4\pi}{3} V_0 d^3. \tag{7}$$

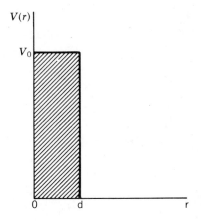

FIGURE 6.3 Square repulsive potential: strength V_0, range d.

For any finite-range potential, the volume integral is proportional to the scattering length in the Born approximation:

$$a_B = \frac{m}{4\pi} \int d\mathbf{r}\, v(r) , \qquad (8)$$

where $m/2$ is the reduced mass for scattering of two particles. In general, the scattering length is defined by the zero-energy limit of the scattering amplitude,

$$a = - \lim_{E\to 0} f(E, \theta) \qquad (9)$$

(which is independent of θ). For the hard-sphere potential, the Born approximation (Eq. 8) diverges, but the *exact* scattering length is given by

$$a = d \qquad \text{(hard sphere)} . \qquad (10)$$

(The radial wave function must vanish inside the infinite repulsive potential. The resulting s-wave phase shift is then easily shown to be $\delta_0 = -kd$, and $a = -\lim_{k\to 0} (\sin \delta_0)/k = d$.)

The momentum integrals in (6) may be done separately in the short-range limit. The sum on μ_1, μ_2 gives

$$\sum_{\mu_1,\mu_2} (1 - \delta_{\mu_1,\mu_2}) = g(g-1) \qquad (11)$$

and the first-order energy density may be written

$$\frac{\Delta E^{(1)}}{\Omega} \cong \frac{g(g-1)}{2} \left(\frac{k_F^3}{6\pi^2} \right)^2 \frac{4\pi a_B}{m} = \frac{2(g-1)}{3\pi} \epsilon_F \rho_0 (k_F a_B) . \qquad (12)$$

Comparing Eq. 12 with Eq. 4, we see that the first-order correction is small compared to the unperturbed energy density for

$$k_F |a_B| \ll 1 , \qquad (13a)$$

or, using Eqs. 7, 8,

$$\frac{m}{3} |V_0| k_F d^3 \ll 1 . \qquad (13b)$$

Since for low density we have $k_F d \ll 1$, the inequalities will hold for a sufficiently weak potential, i.e.

$$\frac{m}{3} |V_0| d^2 \lesssim 1 . \qquad (13c)$$

The energy shift is upward, as would be expected for a repulsive interaction. The perturbation formula Eq. 12 also obtains for a purely attractive potential $(a_B < 0)$, but perturbation theory is inapplicable even for a *weak* attractive potential, since the ground state in this case would be of the superconducting form (see Chapter 5).

For a weak repulsive potential, we might expect Eq. 12 to give a good estimate of the energy, but for a strong repulsion this need not be true. In the hard-sphere limit, $a_B \to +\infty$, and the perturbation expansion does not seem appropriate. It is interesting, however, that the exact scattering length is well defined in this limit (see Eq. 10) and "small", i.e. $k_F a \ll 1$, even though a_B diverges. This suggests the possibility of an expansion in a well-defined parameter like $k_F a$, rather than in a potential strength V_0.

To see how this might come about, let us first see what happens in second order (and higher order) in perturbation theory. The second-order energy density can be calculated from the Rayleigh–Schrödinger expression

$$\Delta E^{(2)} = \sum_{m \neq 0} \frac{|\langle \Phi_0 | V | \Phi_m \rangle|^2}{E_0 - E_m} , \tag{14}$$

where Φ_m are all unperturbed excited states of the Fermi gas, with kinetic energy E_m, and Φ_0, E_0 are given in Eqs. 2 and 4. Since V is a two-body interaction, only Φ_m with one or two particles excited out of the Fermi sea can contribute to the sum in Eq. 14. Further, since the system is uniform, and the interaction (Eq. 5) is translationally invariant, conserving two-body momentum, it is not possible to excite one particle alone out of the Fermi sea. Therefore the intermediate states Φ_m will have two particles above the Fermi level (labeled by momenta $\mathbf{p}_1$, $\mathbf{p}_2$) and two holes below (labeled by $\mathbf{k}_1$, $\mathbf{k}_2$); the energy denominator for Φ_m is given by

$$E_0 - E_m = \epsilon(k_1) + \epsilon(k_2) - \epsilon(p_1) - \epsilon(p_2) . \tag{15}$$

The second-order energy density is then easily obtained:

$$\frac{\Delta E^{(2)}}{\Omega} = \frac{1}{4\Omega} \sum_{\substack{k_1, k_2 \leq k_F \\ p_1, p_2 > k_F}} \frac{|\langle \mathbf{k}_1, \mathbf{k}_2 | v | \mathbf{p}_1, \mathbf{p}_2 \rangle_A|^2}{\epsilon(k_1) + \epsilon(k_2) - \epsilon(p_1) - \epsilon(p_2)} , \tag{16}$$

in terms of the antisymmetrized two-body matrix element (see Eq. 1.50) of the interaction of Eq. 5 (Exercise 6.3). We have suppressed the sums on the spin labels μ, which will give an integral factor, of order g^2. Let us make a *closure* estimate of the sum in (16), replacing the denominator by an average value D, as follows. The potential should have its major contribution for momentum transfers varying inversely with the range $q \sim d^{-1}$. Since by assumption $k_F d \ll 1$, the excited momenta reached from $k_1, k_2 < k_F$ should be $p_1, p_2 \sim d^{-1} \gg k_F$. The average energy denominator for this excitation is approximately

$$D \sim -2 \times \frac{q^2}{2m} \sim -(md^2)^{-1} . \tag{17}$$

(This may overestimate D, giving an underestimate of $|\Delta E^{(2)}|/\Omega$.) The numerator of Eq. 16 may be summed over $\mathbf{p}_1$, $\mathbf{p}_2$ (ignoring the k_F restriction) by closure:

$$\sum_{\mathbf{p}_1,\mathbf{p}_2} |\langle \mathbf{k}_1, \mathbf{k}_2|v|\mathbf{p}_1, \mathbf{p}_2\rangle_A|^2 = 2 \sum_{\mathbf{p}_1,\mathbf{p}_2} \{\langle \mathbf{k}_1, \mathbf{k}_2|v|\mathbf{p}_1, \mathbf{p}_2\rangle\langle \mathbf{p}_1, \mathbf{p}_2|v|\mathbf{k}_1, \mathbf{k}_2\rangle$$

$$- \langle \mathbf{k}_1, \mathbf{k}_2|v|\mathbf{p}_1, \mathbf{p}_2\rangle\langle \mathbf{p}_1, \mathbf{p}_2|v|\mathbf{k}_2, \mathbf{k}_1\rangle\}$$

$$= 2\{\langle \mathbf{k}_1, \mathbf{k}_2|v^2|\mathbf{k}_1, \mathbf{k}_2\rangle - \langle \mathbf{k}_1, \mathbf{k}_2|v^2|\mathbf{k}_2, \mathbf{k}_1\rangle\}$$

$$= 2\langle \mathbf{k}_1, \mathbf{k}_2|\epsilon^2|\mathbf{k}_1, \mathbf{k}_2\rangle_A , \tag{18}$$

which is the antisymmetrized matrix element of the "potential" $2v^2(r)$. (We have combined terms with $\mathbf{p}_1 \leftrightarrow \mathbf{p}_2$ to obtain the factor of 2 in Eq. 18.) The closure estimate may be written in the form

$$\frac{\Delta E^{(2)}}{\Omega} \cong \frac{1}{2\Omega D} \sum_{k_1,k_2 \le k_F} \langle \mathbf{k}_1, \mathbf{k}_2|v^2|\mathbf{k}_1, \mathbf{k}_2\rangle_A , \tag{19}$$

where again μ-sums are suppressed. The evaluation of Eq. 19 is then similar to that of Eq. 6 in the short-range (or low-density) limit, yielding the estimate

$$\frac{\Delta E^{(2)}}{\Omega} \cong - \frac{g(g-1)md^2}{2} \left(\frac{k_F^3}{6\pi^2}\right)^2 \int d\mathbf{r}\, v^2(r)$$

$$\cong - \frac{2(g-1)}{3\pi} \epsilon_F \rho_0(k_F a_B)md^2 V_0 . \tag{20}$$

Comparing the last form with Eq. 12 gives us an estimate (possibly an underestimate) of the ratio

$$\frac{\Delta E^{(2)}}{\Delta E^{(1)}} \cong -md^2 V_0 . \tag{21}$$

This means one can expect Eq. 12 to give a good estimate only for a potential considerably weaker than the limits given in Eq. 13. Clearly, for a strongly repulsive interaction, i.e., $m|V_0|d^2 \gg 1$, the fractional error (21) grows very large.

The same method can be used to make a crude estimate of a selected set of higher-order perturbation terms which involve a sequence of intermediate states with two particles and two holes, as in second order. These terms will be defined in Chapter 10.1, after we have developed the Goldstone perturbation expansion, and will be denoted as those corresponding to *ladder diagrams*. Following a procedure similar to that of Eqs. 16–21, it is possible to find an order-of-magnitude estimate of the *n*th-order ladder diagrams in the form

$$\frac{\Delta E^{(n)}_{\text{ladder}}}{\Delta E^{(1)}} \cong (-md^2V_0)^{n-1} . \tag{22}$$

The behavior for strong repulsion, with $m|V_0|d^2 \gg 1$, grows worse with increasing order n, and clearly the perturbation series diverges although the signs alternate. (Of course, for the hard-sphere limit, $V_0 \to \infty$, and *each* term diverges.) Clearly, the conventional expansion in powers of the interaction strength V_0 is not useful. Yet, as we noted earlier, we do expect that for a low-density gas the effect of the repulsive potential on the ground state energy will be small, based on volume considerations. The relevant dimensionless measure of density is $k_F d$, and we might expect that the ground-state energy can be expressed as an expansion in this parameter. We shall return to this question at the end of this chapter; for the moment we may conclude that conventional perturbation methods give divergent results for strongly repulsive potentials of short range. We shall need more powerful tools.

6.3 PERTURBATION THEORY FOR THE ELECTRON GAS

6.3.1 Hartree–Fock Approximation for the Ground State

In the presence of a uniform positive charge distribution, we might expect that the electron gas should also be uniformly distributed in space. This is suggested by considering the electrons as a classical charged fluid; the Coulomb energy is minimized for a uniform charge distribution equal in magnitude to, but opposite in sign from, the positive charge distribution. The result is similar in the Hartree approximation, which we see as follows. Assume the positive charge density to be $+e\rho_0$, and the electron charge density to be

$$-e\rho(\mathbf{r}) = -e \sum_{\mu \leq F} |\phi_\mu(\mathbf{r})|^2 , \tag{23}$$

where ϕ_μ are Hartree–Fock orbitals, as in Chapter 4, whose form is to be determined by variation. Then the Hartree energy is given by e.g. Eq. 4.21 (omitting, for the moment, the exchange energy) and may be written in the form

$$E_H = E_0 + E_D \,,$$

$$E_0 = \sum_{\mu \leq F} \int dr \, \phi_\mu^*(\mathbf{r}) \left(-\frac{\nabla^2}{2m} \right) \phi_\mu(\mathbf{r}) \,, \tag{24a}$$

$$E_D = \frac{e^2}{2} \int d\mathbf{r}_1 \frac{\rho d\mathbf{r}_2(\mathbf{r}_1) \, \rho(\mathbf{r}_2)}{r_{12}} \,, \mathbf{r}_{12} = |\, \mathbf{r}_1 - \mathbf{r}_2 \,. \tag{24b}$$

To this we add the Coulomb self-energy of the positive charge distribution, and the interaction of the electrons with that distribution:

$$E' = \frac{e^2}{2} \rho_0^2 \int \frac{d\mathbf{r}}{r} - e^2 \rho_0 \int d\mathbf{r} \, \frac{\rho(\mathbf{r})}{r} \,. \tag{24c}$$

The total Coulomb energy $E_D + E'$ is clearly minimized by the uniform number density $\rho(\mathbf{r}) = \rho_0$, for which the three (direct) Coulomb energies exactly cancel.* In this case the Hartree–Fock orbitals are actually plane waves. The kinetic energy E_0 is also minimized (for a given *average* density ρ_0) by the uniform distribution. So far, so good.

We have, however, ignored the exchange energy in Eq. 24, which comes from the antisymmetry of the many-electron wave function. For sufficiently high density (which we define a little later), this energy will not change the result just obtained, of a uniform distribution of electrons in the ground state. For certain ranges of lower densities, two interesting effects arise from the exchange energy. First, the ground state may become ferromagnetic, with all electron spins aligned† (Exercise 6.4). At even lower densities, the ground state may form what is called a Wigner solid, which is a regular close-packed crystal-like lattice of electrons within the uniform positive-charged background.‡ In this arrangement the mutual repulsion of the electrons is reduced, at the price of a high kinetic energy due to localization of the electrons near crystal sites. However, the latter decreases at low densities fast enough to lead to the lattice as the lowest energy state. This effect did not appear in our argument based on Eq. 24, since we have improperly included in the direct Coulomb energy among electrons, E_D, the self-interaction of each electron. This self-energy is exactly removed by including the exchange energy (Exercise 6.5). Considering only those densities high enough for the validity of uniform density, the Hartree–Fock ground state takes the form of a Fermi gas of electrons of density ρ_0, as in Chapter 1, Eqs. 1.53–1.56, with spin degeneracy $g = 2$; with a Fermi momentum k_F given by Eq. 1.55,

---

*Here (as in Chapter 4) we take the electron charge $e^2/\hbar c = \alpha \cong \frac{1}{137}$, so that the Coulomb potential is e^2/r. In Chapter 3, $e^2/4\pi\hbar c = \alpha$.
†F. Bloch, Z. Phys. **57**, 545 (1929).
‡E. P. Wigner, Trans. Faraday Soc. **34**, 678 (1938).

$$\rho_0 = k_F^3/3\pi^2 \; ; \tag{25}$$

and the kinetic energy per electron given by Eq. 1.57,

$$\frac{E_0}{N} = \frac{3}{5} \frac{k_F^2}{2m} \; . \tag{26}$$

The first-order interaction energy is given entirely by the exchange (Fock) energy (per electron)

$$\frac{\Delta E^{(1)}}{N} = -\frac{1}{2N} \sum_{\mu_1, \mu_2 \le F} \langle \mu_1, \mu_2 | V | \mu_2, \mu_1 \rangle \; , \tag{27}$$

where $\mu = (\mathbf{k}, m_s)$ labels plane-wave orbits with spin projection m_s, and V is the Coulomb potential. The direct (Hartree) term of Eq. 24b has been canceled by the rest of the Coulomb interaction of Eq. 24c for $\rho(r) = \rho_0$. The expression for Eq. 27 may be evaluated by summing spin indices and integrating:

$$\frac{\Delta E^{(1)}}{N} = -\frac{1}{(2\pi)^6 \rho_0} \int d\mathbf{k}_1 \, d\mathbf{k}_2 \, \theta(k_F - k_1)\theta(k_F - k_2) \int d\mathbf{r} \, \frac{e^2}{r} \, e^{i(\mathbf{k}_2 - \mathbf{k}_1)\cdot \mathbf{r}} \; , \tag{28a}$$

finally obtaining the value (see also Eq. 4.71)

$$\frac{\Delta E^{(1)}}{N} = -\frac{3}{4\pi} \, k_F e^2 \; . \tag{28b}$$

Note that the exchange energy is *attractive*, representing an effect of the anticorrelation of electrons due to the Pauli principle (keeping electrons apart), which reduces the Coulomb repulsion relative to that of a uniform charge distribution.

It has become conventional to express the energy in terms of the rydberg (Ry) and lengths in terms of the Bohr radius a_0,

$$\mathrm{Ry} = e^2/2a_0 \; , \qquad a_0 = (me^2)^{-1} \; , \tag{29a}$$

introducing a dimensionless ratio of lengths

$$r_s = r/a_0 \; , \tag{29b}$$

where r is an effective radius of a sphere containing one electron on the average, i.e.

$$r = \left(\frac{3}{4\pi\rho_0} \right)^{1/3} = \left(\frac{9\pi}{4} \right)^{1/3} k_F^{-1} = 1.919 \, k_F^{-1} \; . \tag{30}$$

In this notation, the total Hartree–Fock energy per electron, given by Eq. 26 plus 28b, becomes

$$\frac{E_{HF}}{N} \cong \left(\frac{2.210}{r_s^2} - \frac{0.916}{r_s} \right) Ry \,. \tag{31}$$

(For real metals, $2 \le r_s \le 5$.) The kinetic-energy term (going as k_F^2 or r_s^{-2}) dominates at high density ($r_s < 1$), while the exchange potential energy ($\propto k_F$ or r_s^{-1}) dominates at low density ($r_s \gg 1$). However, in the latter limit, the energy can be lowered from the uniform Hartree–Fock value by formation of a Wigner solid lattice of electrons, as noted earlier.

We shall now consider only the high-density domain, for which the kinetic term of Eq. 31 dominates over the exchange potential term. It might appear that the Coulomb interaction can therefore be treated as a weak perturbation, but we shall find that this is not so; in fact we shall encounter divergences in the perturbation corrections to Eq. 31. We shall later find a method of evaluating the leading corrections to the energy, using a technique which converges well at high density.

The corrections beyond the Hartree–Fock approximation involve correlations of the electrons beyond those given by the antisymmetry of the Fermi gas. These correlations tend to reduce further the average Coulomb repulsion between electrons, an effect called *screening*. (In an atom, the outer electrons are attracted to the nucleus by a potential *screened* by the inner electrons, and therefore weaker than the full Coulomb potential $-Ze^2/r$. The screening in the electron gas is analogous, but works in the opposite direction to weaken the repulsion.)

It is conventional to refer to the energy contributions *beyond* first order (Eq. 31) as the *correlation energy*

$$E = E_{HF} + \Delta E_{corr} \,, \tag{32}$$

in contrast, say, to the hard-sphere gas of Section 6.2, for which the first-order energy already diverges. We now examine the correlation energy of a high-density electron gas in perturbation theory.

6.3.2 Perturbation Theory: Divergences

We may obtain an expression for the correlation energy to second order in the strength (e^2) of the Coulomb interaction, starting with the standard Rayleigh–Schrödinger formula of Eq. 14. For the uniform electron gas, for which the unperturbed wave function is the Fermi gas, we may use the same arguments which lead to Eq. 16, with the replacement of the short-range potential v by the Coulomb potential e^2/r. The momentum-space matrix elements in the numerator are easily calculated in terms of the Fourier transform of the Coulomb potential,

$$V(q) = \int d\boldsymbol{\rho} \, e^{i\mathbf{q}\cdot\boldsymbol{\rho}} \frac{e^2}{\rho} = \frac{4\pi e^2}{q^2}, \tag{33}$$

as follows. We use the plane-wave states of Eq. 1.52, converting to relative and c.m. variables $\boldsymbol{\rho} = \mathbf{r}_1 - \mathbf{r}_2$, $\mathbf{R} = (\mathbf{r}_1 + \mathbf{r}_2)/2$, and write

$$\langle \mathbf{k}_1, \mathbf{k}_2 | V | \mathbf{p}_1, \mathbf{p}_2 \rangle = \left(\frac{1}{\Omega}\right)^2 \int d\mathbf{R} \exp i(\mathbf{p}_1 + \mathbf{p}_2 - \mathbf{k}_1 - \mathbf{k}_2) \cdot \mathbf{R}$$

$$\times \int d\boldsymbol{\rho} \exp \left[\frac{i}{2}(\mathbf{p}_1 - \mathbf{k}_1 + \mathbf{k}_2 - \mathbf{p}_2) \cdot \boldsymbol{\rho}\right] \left(\frac{e^2}{\rho}\right) \tag{34}$$

$$= \frac{1}{\Omega} \delta_{\mathbf{p}_1 + \mathbf{p}_2, \mathbf{k}_1 + \mathbf{k}_2} V(q)$$

where $\mathbf{q}$ is the momentum transfer, defined by

$$\mathbf{q} = \mathbf{p}_1 - \mathbf{k}_1 = \mathbf{k}_2 - \mathbf{p}_2, \tag{35}$$

and the Kronecker delta is equal to unity for momentum conservation ($\mathbf{p}_1 + \mathbf{p}_2 = \mathbf{k}_1 + \mathbf{k}_2$), and zero otherwise.

The numerator of Eq. 16 contains antisymmetrized matrix elements (see Eq. 1.50), so we shall also need the matrix element of Eq. 34 with $\mathbf{p}_1$ and $\mathbf{p}_2$ interchanged, which can easily be expressed as

$$\langle \mathbf{k}_1, \mathbf{k}_2 | V | \mathbf{p}_2, \mathbf{p}_1 \rangle = \frac{1}{\Omega} \delta_{\mathbf{p}_1 + \mathbf{p}_2, \mathbf{k}_1 + \mathbf{k}_2} V(q') \tag{36}$$

with

$$\mathbf{q}' = \mathbf{p}_2 - \mathbf{k}_1 = \mathbf{k}_2 - \mathbf{p}_1 = \mathbf{k}_2 - \mathbf{k}_1 - \mathbf{q}. \tag{37}$$

The result for the numerator of Eq. 16 can be separated into two terms:

$$V^2(q) + V^2(q') \tag{38a}$$

and

$$2V(q)V(q'), \tag{38b}$$

omitting factors of Ω^{-1} and Kronecker deltas. (We have also suppressed spin matrix elements.) The denominator of Eq. 16 can be reexpressed in terms of $\mathbf{k}_1$, $\mathbf{k}_2$, and either $\mathbf{q}$ or $\mathbf{q}'$, using Eqs. 35, 37, and momentum conservation. Then, for example, the sum corresponding to Eq. 38a can be written (combining the two terms with a change of variable) in the form

$$\left(\frac{\Delta E^{(2)}}{\Omega}\right)_a \propto \Omega^{-3} \sum_{\mathbf{k}_1,\mathbf{k}_2,\mathbf{q}} V^2(q)[\epsilon(\mathbf{k}_1) + \epsilon(\mathbf{k}_2) - \epsilon(\mathbf{k}_1 + \mathbf{q}) - \epsilon(\mathbf{k}_2 - \mathbf{q})]^{-1}$$

$$= -\frac{m}{\Omega^3} \sum_{\mathbf{k}_1,\mathbf{k}_2,\mathbf{q}} V^2(q)[\mathbf{q}\cdot(\mathbf{k}_1 - \mathbf{k}_2 + \mathbf{q})]^{-1},$$

(39a)

where we have omitted numerical factors, and the momentum restrictions in Eq. 16 now imply that

$$|\mathbf{k}_1|, |\mathbf{k}_2| < k_F,$$

$$|\mathbf{k}_1 + \mathbf{q}|, |\mathbf{k}_2 - \mathbf{q}| > k_F.$$

(39b)

Passing to the limit $\Omega \rightarrow \infty$, Eq. 39a becomes an integral,

$$\left(\frac{\Delta E^{(2)}}{\Omega}\right)_a \propto -m \int d\mathbf{q}\, V^2(q) \int d\mathbf{k}_1\, d\mathbf{k}_2\, [\mathbf{q}\cdot(\mathbf{k}_1 - \mathbf{k}_2 + \mathbf{q})]^{-1} \qquad (40)$$

with the constraints of Eq. 39b. The integral appears to be singular, both because of the q^{-4} behavior of $V^2(q)$, and because of the q^{-1} behavior of the denominator in brackets. The restrictions on the $\mathbf{k}$-integrals following from Eq. 39b bring in a factor of q^2, so the integrand goes as q^{-3}, giving a singular integral. We can express the degree of singularity by introducing a lower limit q_0 of integration, so that

$$\left(\frac{\Delta E^{(2)}}{\Omega}\right)_a \propto \int d\mathbf{q}\, q^{-3} \propto \int q^2\, dq\, q^{-3} \propto \ln q_0. \qquad (41)$$

The divergence goes as $\ln q_0$ as $q_0 \rightarrow \infty$.

The singular behavior of the integral can be traced to the long range of the Coulomb potential, which is characterized by the q^{-2} behavior of $V(q)$. If the interaction had a long but finite range, e.g. so that $V(q) \propto (q^2 + a^2)^{-1}$, with a the inverse range $(a \ll k_F)$, then the q-integral in Eq. 40 would behave as if the lower limit were $q_0 \sim a$, giving a finite result proportional to $\ln(a/k_F)$ or to $\ln r_s$. We shall indeed find later that the effect of correlations of the electrons in higher orders is to modify the effect of the Coulomb potential by *screening* the charge distribution, so that the correlation energy is finite and proportional to $\ln r_s$, *as if* the interaction had a finite range. However, this physical result does not appear directly in our perturbation calculation, which gives an (unphysical) infinite answer for one second-order term, as we have seen.

We have not evaluated the term corresponding to the interference of direct and exchange matrix elements in the numerator of Eq. 16, given by Eq. 38b. Here the integration is well behaved, since $V(q)$ and $V(q')$ are singular at different points in $\mathbf{k}_1$, $\mathbf{k}_2$, $\mathbf{q}$, as can be seen from Eq. 37. The result is integrable, and gives a finite contribution to the correlation energy, which will remain after we have learned how to remove the unphysical divergences from our perturbation treatment. Therefore, we give the numerical result for this perturbation contribution, *per electron*, as

$$\left(\frac{\Delta E^{(2)}}{N}\right)_b = 0.046 \, \text{Ry} \qquad (42)$$

which is independent of r_s, or density (compare Eq. 31).

We should note that the singular behavior of the term in Eq. 41 persists in every order of perturbation theory. We shall see this more clearly when we have introduced diagrammatic perturbation theory, in Chapters 8 and 9. Then, in Chapter 10, we shall classify diagrams for the electron gas, and shall find that the "ring" diagrams are the most singular in each order. But, by that stage we shall also have tools for extracting finite and meaningful results from a more useful form of perturbation expansion, and shall be able to evaluate the correlation energy of the electron gas at high density.

6.4 SUMMARY

We have carried our discussions of the hard-sphere and electron gas problems far enough to see that ordinary perturbation theory does not give us a meaningful technique for calculating physical properties of the systems, such as the ground-state energy per particle. Yet in both cases, we have physical reasons to believe that the ground-state properties are not only finite, but in the appropriate domain of particle density may be expressible as an expansion in a small parameter. For the hard-sphere gas, the quantity which could play the role of the small parameter is $k_F d$, and as we have discussed in Section 2, the domain of validity of an expansion would be at low density. For the electron gas, it is at high density that we might expect a valid expansion, and the small parameter could be r_s, the ratio of lengths defined in Eq. 29b, which measures density relative to the Bohr radius, the latter being the natural measure of quantum effects of the Coulomb potential. The Hartree–Fock (or first-order perturbation) energy, Eq. 31, gives the leading orders in inverse powers: r_s^{-2} and r_s^{-1}. The first corrections, as we shall see in Chapter 10, will be of order $\ln r_s$, due to modifications of the wave function of the electron gas at relative distances $r \sim a_0$, giving the screening of the Coulomb potential mentioned in connection with Eq. 41.

To acquire theoretical tools, we shall first need to develop a systematic perturbation theory for large systems of particles, which can be used for all types of interactions. This we shall do in the next three chapters, with the introduction of Goldstone and Feynman diagrams as a method of generating and calculating perturbation expansions. Then we shall have to learn how to combine an infinite series of diagram expressions into a single, well-defined quantity, in order to eliminate the kinds of singularities we encountered in this chapter, for the hard-sphere and electron gas problems. The method of summation of partial series will be developed, and applied to the calculation of ground-state properties of these systems, in Chapter 10.

One could imagine developing special methods for each problem, without all the machinery of diagrammatic perturbation theory. This is not only possible, but in fact is how these two problems were first attacked. The method of Huang and Yang* for the hard-sphere gas used a geometric approach with a "pseudopotential" to find the ground-state properties, based on the notion that an expansion in the range of the interaction is physically meaningful, while an expansion in the strength of the interaction is not. Similarly, the first successful approach by Bohm and Pines† to the electron gas problem was based on use of the random-phase approximation (RPA—see Chapter 4), where the excitation so represented is the plasma oscillation of the electrons. This collective mode is called a *plasmon*, in analogy with the *phonon* for sound waves. The coupling of the electron-gas ground state to this *plasmon* mode provides the mechanism for the screening of the Coulomb field, mentioned earlier.

Rather than pursue a different method for each problem, we prefer to take a unified approach, through diagrammatic perturbation theory, which will be applicable to a large variety of problems, including our two classical examples. The theoretical ideas we shall encounter have come to form the foundation for much of the thinking about interacting systems with many degrees of freedom in modern physics. In condensed-matter physics, dia-grammatic methods have been applied to all sorts of problems of Fermi fluids, including the electron gas, magnetic interactions, and even supercon-ductivity (thus incorporating into the general theory the particular mecha-nisms of the BCS theory discussed in Chapter 5). A variety of problems in solid-state physics involving the interaction of electrons and phonons are naturally treated by such methods. In atomic and nuclear physics, such approaches have been applied to binding energies (as for nuclear matter; Chapter 10.4) and spectroscopy of excited states, giving a theory of the shell model. They also provide a systematic language for the treatment of scattering from a complex target, as we shall see in Chapter 11.

EXERCISES

6.1 Consider a Fermi gas of fixed density interacting through a potential of the form (in three dimensions) $V(r) = \lambda \delta(\mathbf{r})$, with $\lambda > 0$.

   (a) Calculate the first-order energy, Eq. 6; show that Eq. 12 is then exact.

---

*Kerson Huang and C.N. Yang, Phys. Rev. **105**, 767 (1957).
†D. Bohm and D. Pines, Phys. Rev. **92**, 609 (1953).

(b) Find the exact value of the scattering length a for this potential, and compare with Eq. 8. (You might consider the limiting case of Eq. 5.) What does that imply about the ground-state energy of the system?

6.2 Consider two equal-mass particles moving in one space dimension in a domain $0 \le x \le L$, with periodic boundary conditions. The particles interact through a delta-function potential $V(x_1 - x_2) = g\delta(x_1 - x_2)$, with $g > 0$. Find the exact ground-state energy of the system, and compare this with the result of first-order perturbation theory (in the potential). (Hint: First show that the exact wave function can be written in the form

$$\psi(x_1, x_2) = a \exp i(k_1 x_1 + k_2 x_2) + b \exp i(k_2 x_1 + k_1 x_2)$$

for $x_1 < x_2$, with a similar form with two new constants a', b', for $x_1 > x_2$; $k_1 \ne k_2$. This is a special case of the *Bethe Ansatz* form, which holds for N identical particles.*)

6.3 (a) Show explicitly that for a two-body interaction V, only unperturbed states Φ_m with one or two particles excited (relative to the unperturbed ground state Φ_0) can contribute to the second-order energy, Eq. 14. (The operator methods of Chapter 1 are the most direct.)

(b) Verify that for a uniform system with plane-wave single-particle orbits (unperturbed), the states Φ_m with one particle excited do not contribute to Eq. 14.

(c) Obtain Eq. 16 for the uniform case.

6.4 In deriving the ground-state energy of the electron gas in the Hartree–Fock approximation (Eqs. 26, 28b, 31) we assumed equal population of spin states. This is only valid for a particular range of densities.

(a) Find the appropriate range of densities (in terms of r_s) by comparing Eq. 31 with the Hartree–Fock energy for all spins aligned (i.e. the magnetic state).

(b) Find the range of r_s over which the nonmagnetic state is meta-stable, although not the ground state.

6.5 (a) Write an expression for the Hartree–Fock energy (Eq. 4.21) for the electron gas (Section 6.3.1) in terms of the orbitals ϕ_μ of Eq. 23. Show that neglect of the exchange interaction leads to Eq. 24, the Hartree energy.

(b) Verify that the Hartree energy is minimized by constant $\rho(r) = \rho_0$.

(c) Show in what sense the Hartree energy includes the self-interaction of each electron, and that the self-energy is exactly removed in the full Hartree–Fock energy.

*See e.g. E. Lieb and W. Liniger, Phys. Rev. **130**, 1605 (1963).

CHAPTER **7**

Time Development

In this chapter we introduce the formulation for time evolution of an interacting quantum system, which will serve as the basis for most of what we do in the rest of this book. In particular we shall use the time-evolution formulation of this chapter to develop perturbation theory in terms of Goldstone and Feynman diagrams in Chapters 8 and 9. We shall apply this theory in Chapter 10 to the calculation of the ground-state energy of the Fermi-gas problems of Chapter 6, and in Chapter 11 to problems of propagation and scattering in many-body systems.

The general methodology we are about to encounter has proved to be incredibly useful in a large variety of problems involving many degrees of freedom. Aside from the applications to many-particle systems which we shall actually study in this book, the same theoretical techniques underlie many aspects of quantum statistical mechanics and relativistic quantum field theory.

When we discuss the ground state of a quantum system, or indeed any *stationary* state, time does not normally enter the discussion as an interesting variable, since only the *phase* of the state changes with time:

$$|\psi(t)\rangle = |\psi_E\rangle e^{-iEt}, \tag{1}$$

where the time-independent state vector $|\psi_E\rangle$ satisfies the Schrödinger eigenvalue equation

$$H|\psi_E\rangle = E|\psi_E\rangle \tag{2}$$

159

with H the Hamiltonian for the system. Thus we introduced *number representation* (Chapter 1), the Hartree–Fock theory (Chapter 4), and pairing theory (Chapter 5) for the ground state of many-fermion systems, without discussing time. On the other hand, physical *processes*, such as those in which a system interacts with an external probe, evolve in time in essential ways. For example, one may consider the interaction of a time-dependent electromagnetic field with matter, or the scattering of an electron from an atom or nucleus, as dynamical processes in which time development is a natural representation of the change induced in the systems involved.

It is not surprising that the development of a theoretical formalism for evolution in time is of considerable use for treating dynamical processes such as scattering or excitation of systems by varying external fields. However, we shall also find that the same formalism provides the most efficient method of developing perturbation theory for *stationary states* of many-body systems. We shall even find that it is possible to extend the time-development theory to treat a quantum system in thermodynamic equilibrium, through the formal device of transforming the time variable to a temperature parameter.

7.1 HEISENBERG PICTURE: FIELD OPERATORS

Let us review the Heisenberg picture (or representation) of states and operators in quantum mechanics, and its relation to the Schrödinger picture, with which we begin.

The time-dependent Schrödinger equation has as solutions time-dependent state vectors $|\psi_S(t)\rangle$:

$$ i \frac{d}{dt}|\psi_S(t)\rangle = H|\psi_S(t)\rangle , \qquad (3) $$

where we use subscript S to denote the Schrödinger picture. Stationary states are of the form given in Eq. 1. Measurable quantities (position, momentum, spin, etc.) are represented by time-independent linear operators, A_S, B_S, C_S, etc.

In the Heisenberg picture (denoted by subscript H) quantum state vectors are independent of time, and related to the Schrödinger vectors by the transformation

$$ |\psi_H\rangle = e^{iHt}|\psi_S(t)\rangle . \qquad (4) $$

Conversely, operators for measurable quantities become time dependent in this picture, and are given by

$$ A_H(t) = e^{iHt}A_S e^{-iHt} \qquad (5) $$

in terms of the Schrödinger operators. The equation of motion for *states* (Eq. 3) is replaced by an equation of motion for *operators*:

$$i \frac{d}{dt} A_H(t) = [A_H(t), H],$$ (6)

which follows directly from Eq. 5. From Eq. 6 we see that the Hamiltonian operator is the same in both pictures, $H_H(t) = H_S = H$. The same equality holds for any operator that commutes with H.

The field operators $\psi(r, t)$ and $\pi(r, t)$ introduced in Chapter 2 (Eqs. 2.1, 2.13, 2.29) are actually Heisenberg operators, as are the time-dependent creation and annihilation operators $a_\alpha^\dagger(t)$, $a_\alpha(t)$ which appear in the expansions of Eq. 2.39, i.e.

$$\psi(\mathbf{r}, t) = \sum_\alpha a_\alpha(t) u_\alpha(\mathbf{r}),$$ (7)

where we find from Eq. 6 that

$$a_\alpha(t) = e^{iHt} a_\alpha e^{-iHt}$$ (8a)

relates the Heisenberg annihilation operator to the Schrödinger operator a_α introduced in Eq. 1.5. Similarly,

$$a_\alpha^\dagger(t) = e^{iHt} a_\alpha^\dagger e^{-iHt}.$$ (8b)

For the special case of a system of noninteracting particles, the Hamiltonian can be written as a one-body operator in the form

$$H_0 = \sum_\alpha \epsilon_\alpha a_\alpha^\dagger a_\alpha$$ (9)

with α labeling a set of single-particle states of energy ϵ_α (see. Eqs. 1.38 and 1.40). Let us calculate the product

$$e^{iH_0 t} a_\alpha e^{-iH_0 t} = e^{i\epsilon_\alpha a_\alpha^\dagger a_\alpha t} a_\alpha e^{-i\epsilon_\alpha a_\alpha^\dagger a_\alpha t} = a_\alpha e^{-i\epsilon_\alpha t},$$ (10)

where the last result is easily obtained by considering the operation on an arbitrary state, expanded in the number representation (Chapter 1). The a_α changes the number of particles occupying α from n_α to $n_\alpha - 1$. The exponential operator is diagonal, finding one fewer particle in state α when it operates after (to the left of) a_α than when it operates before. The resulting Eq. 10 (and a similar one for $a_\alpha^\dagger$) allows us to give the Heisenberg operators of Eq. 8 in the explicit forms

$$a_\alpha(t) = a_\alpha e^{-i\epsilon_\alpha t},$$ (11a)

$$a_\alpha^\dagger(t) = a_\alpha^\dagger e^{i\epsilon_\alpha t},$$ (11b)

in which the time dependence appears only in the phase factor.

Since we are interested in dealing with interacting systems of identical nonrelativistic particles, either bosons or fermions, we shall treat the field operator as discussed in Section 2.2. For bosons, the operators 8a, b obey the equal-time commutation relations

$$[a_\alpha(t), a_\beta^\dagger(t)] = \delta_{\alpha\beta} , \qquad (12a)$$

$$[a_\alpha(t), a_\beta(t)] = [a_\alpha^\dagger(t), a_\beta^\dagger(t)] = 0 . \qquad (12b)$$

Eq. 12a was derived in Eq. 2.41 from the field commutation relations

$$[\psi(\mathbf{r}, t), \psi^\dagger(\mathbf{r}', t)] = \delta(\mathbf{r} - \mathbf{r}') , \qquad (13a)$$

and 12b can be similarly derived from

$$[\psi(\mathbf{r}, t), \psi(\mathbf{r}', t)] = [\psi^\dagger(\mathbf{r}, t), \psi^\dagger(\mathbf{r}', t)] = 0 . \qquad (13b)$$

One can equally well obtain Eq. 12 from the time-independent commutation relations given in Eq. 1.19.

For fermions, the creation and annihilation operators (8a, b) obey anticommutation relations at equal times,

$$\{a_\alpha(t), a_\beta^\dagger(t)\} = \delta_{\alpha\beta} , \qquad (14a)$$

$$\{a_\alpha(t), a_\beta(t)\} = \{a_\alpha^\dagger(t), a_\beta^\dagger(t)\} = 0 , \qquad (14b)$$

which follow directly from the time-independent relations, Eqs. 1.31a–c. For fermions, the single-particle state labels α, β include a spin label, and the expansion function $u_\alpha(\mathbf{r})$ is actually a spinor, represented by a column vector of dimension $2S + 1$ for spin S. One can make this explicit by labeling the spinor components by m, e.g. $m = \pm\frac{1}{2}$ for $S = \frac{1}{2}$, and replacing α with a, m. Then Eq. 7 reads, for each component,

$$\psi_m(\mathbf{r}, t) = \sum_a a_{a,m}(t) u_{a,m}(\mathbf{r}) . \qquad (7')$$

The field operators in Eq. 7′ and their adjoints can be seen to obey the equal-time anticommutation relations

$$\{\psi_m(\mathbf{r}, t), \psi_{m'}^\dagger(\mathbf{r}', t)\} = \delta_{mm'}\delta(\mathbf{r} - \mathbf{r}') , \qquad (15a)$$

$$\{\psi_m(\mathbf{r}, t), \psi_{m'}(\mathbf{r}', t)\} = \{\psi_m^\dagger(\mathbf{r}, t), \psi_{m'}^\dagger(\mathbf{r}', t)\} = 0 , \qquad (15b)$$

which can be seen to be $(2S + 1) \times (2S + 1)$ *matrix* equations. Other internal degrees of freedom can be similarly included in the matrices; similarly for the *boson* fields of Eq. 13.

The interpretation of the creation and annihilation operators is similar to that of Chapter 1: $a_\alpha^\dagger(t)$ adds one particle to the system into a particular single particle state α, at a particular *time* t; $a_\beta(t')$ removes a particle from the state β at time t'. Similarly, $\psi^\dagger(\mathbf{r}, t)$ or $\psi(\mathbf{r}, t)$ adds or removes a particle, respectively, at the position $\mathbf{r}$ at time t.

For fermions, it may be convenient to redefine the creation and annihilation operators for single-particle states with a given Fermi sea ($\lambda \leq F$), as we did in Chapter 4, Eq. 4.17, so that

$$b_\lambda^\dagger(t) = -a_\lambda(t), \qquad b_\lambda(t) = -a_\lambda^\dagger(t), \qquad \lambda \leq F. \tag{16}$$

The operator $b_\lambda^\dagger(t)$ creates or adds a *hole* to the state λ, at time t, and $b_\lambda(t)$ removes it. The $b_\lambda(t)$, $b_\mu^\dagger(t)$ obey the same anticommutation relations 14 as the $a_\lambda(t)$, $a_\lambda^\dagger(t)$. The Fermi sea is usually defined with respect to a one-body Hamiltonian such as that of Eq. 9, which can also be rewritten in terms of particle ($\rho > F$) and hole ($\lambda \leq F$) states as in Eq. 4.22c:

$$H_0 = \sum_{\rho > F} \epsilon_\rho a_\rho^\dagger a_\rho - \sum_{\lambda \leq F} \epsilon_\lambda b_\lambda^\dagger b_\lambda + \sum_{\lambda \leq F} \epsilon_\lambda, \tag{17}$$

where the summed energy of the Fermi sea is an additive constant and may be omitted.

For noninteracting fermions it is often useful to measure the particle or hole energy relative to the Fermi energy ϵ_F of the last filled orbit of the Fermi sea. Then, for particle states Eq. 11a becomes

$$a_\rho(t) = a_\rho e^{-i(\epsilon_\rho - \epsilon_F)t}, \qquad \rho > F \tag{18a}$$

while for holes

$$b_\lambda(t) = -a_\lambda^\dagger e^{-i(\epsilon_F - \epsilon_\lambda)t}, \qquad \lambda \leq F, \tag{18b}$$

with similar equations for the adjoints. In this form, since both $(\epsilon_\rho - \epsilon_F) > 0$ and $(\epsilon_F - \epsilon_\lambda) \geq 0$, it becomes clear that both particles and holes are positive-energy excitations of the system, relative to the Fermi sea, which may be considered a vacuum for particles ($\rho > F$) and holes ($\lambda \leq F$).

The fermion field operator of Eq. 7 may be written in the particle–hole form

$$\psi(\mathbf{r}, t) = \sum_{\rho > F} a_\rho(t) u_\rho(\mathbf{r}) - \sum_{\lambda \leq F} b_\lambda^\dagger(t) u_\lambda(\mathbf{r}). \tag{19}$$

For bosons the concept of holes is not useful, since single-particle states can never be fully occupied for bosons, unlike the fermion case. For noninteracting bosons in the ground state, only the lowest-energy single-particle state is occupied. For some purposes it is useful to remove that lowest-energy state from the field operator, to give excitations relative to the noninteracting ground state. We shall not make use of this technique further.

A further relabeling of the hole states is often convenient to help keep track of symmetries carried by the single-particle quantum numbers. For example, if momentum is one of these quantum numbers, i.e. $\alpha = (\mathbf{k}, m)$, then $a_\alpha(t)$ removes momentum $\mathbf{k}$ from the system, along with the particle, However, $b_\alpha(t)$ as defined by Eq. 16 *adds* momentum $\mathbf{k}$ to the system, while removing a hole. It would be better to let the hole also carry momentum $\mathbf{k}$, rather than $-\mathbf{k}$, as in Eq. 16. Similarly, we would like a hole labeled with spin projection m to carry $+m$ units of angular momentum in the z-direction. We also want to be able to couple the spin and angular momenta of holes by the same rules (i.e. Clebsch–Gordan coupling) as for particles: this requires an m-dependent phase in the definition of holes.

These ends can be accomplished by introducing a label $\bar{\alpha}$ corresponding to each α. This is usually taken to be the quantum numbers of the time-reversed single-particle state; for the present example, $\bar{\alpha} = (-\mathbf{k}, -m)$. Then we modify the labeling of the hole operators of Eq. 16

$$b_{\bar{\alpha}}^\dagger(t) = (-1)^{\phi(\alpha)} a_\alpha(t) , \quad b_{\bar{\alpha}}(t) = (-1)^{\phi(\alpha)} a_\alpha^\dagger(t) , \qquad \alpha \leq F , \qquad (20)$$

where $\phi(\alpha)$ is a phase depending on the representation $\{\alpha\}$; e.g. for $\alpha = (\mathbf{k}, m)$, we may take

$$\phi(\alpha) = S - m . \qquad (21)$$

The fermion operator (19) becomes

$$\psi(\mathbf{r}, t) = \sum_{\rho > F} a_\rho(t) u_\rho(\mathbf{r}) + \sum_{\lambda \leq F} (-1)^{\phi(\lambda)} b_{\bar{\lambda}}^\dagger(t) u_\lambda(\mathbf{r}) . \qquad (22)$$

Now, both $b_{\bar{\alpha}}^\dagger$ and $a_\alpha^\dagger$ *add* the quantum numbers α to the system. If $\epsilon_\alpha = \epsilon_{\bar{\alpha}}$, Eq. 17 is unchanged for $\lambda \to \bar{\lambda}$.

7.2 SINGLE-PARTICLE GREEN FUNCTION

For the simplest quantum systems, say a particle moving in one dimension in a potential, one may treat all the energy states of the system on essentially the same footing: ground state, first excited state, etc. The ground state does have some distinguishing properties, of course, such as the absolute minimal energy expectation in the variational theorem; the lowest number of nodes in the wave function, for a local potential; etc. But otherwise, the ground state is but "one among equals". Continuum states are, of course, different from discrete states. For a system with a few degrees of freedom, such as a few-electron atom, the same is more or less true. Different states may have different symmetries, denoted by quantum numbers like angular momentum, parity, and so on. If the symmetry is conserved exactly, even the variational property of the ground state is compromised, since the lowest-energy state of each distinct quantum number for a given symmetry has the variational property separately, as states of different quantum numbers do not mix.

However, in nature, not all energy states of even a rather simple system are equally accessible. Isolated atoms, molecules, and such are found in their ground states; even in ensembles in equilibrium, ground states are more probable than any other. Excited states are *seen* only if they can be formed, either by excitation of the system from the ground state, or by changing the system by addition of particles (e.g. electron capture by an ion) or by subtraction (e.g. ionization). The modes of excitation readily available have their own symmetries (selection rules) which limit the states which can be reached, e.g. from the ground state. For example, in the Hartree–Fock theory of an atom (Chapter 4) particle–hole states may be easily excited (e.g. by absorption of single photons) while 6p–6h states may be much harder to populate (requiring multiphoton absorption from intense radiation).

With large many-body systems it becomes clear that the ground state really is different from (almost) any particular excited state, which may or may not be accessible. For that reason one concentrates on those excited states which can be reached by a particular excitation mechanism from the ground state. (In the temperature formalism—see Section 7.6—the equilibrium state replaces the ground state.) This requires information about the ground state *and* about the excitation mechanism. If we wish to know how a particular type of excitation evolves in time, as a packet of excited states, we introduce an appropriate Green function.

Suppose the ground state Ψ_0 (Heisenberg picture) is excited by a time-dependent operator $A(t)$. The system is subsequently deexcited by an operator $B(t')$. The probability amplitude for the excitation at time t and deexcitation at t' to return to the ground state is given by

$$\langle \Psi_0 | B(t') A(t) | \Psi_0 \rangle .\tag{23a}$$

For this amplitude to correspond to a causal process, with excitation by A preceding deexcitation by B, one restricts the time order to $t' > t$. On the other hand, the amplitude

$$\langle \Psi | A(t) B(t') | \Psi_0 \rangle \tag{23b}$$

also stands for a causal process where B excites and A deexcites the system, with $t > t'$. Both processes will be of interest, and may be combined by the formal device of the time-ordered product of Heisenberg operators, which we define for the pair $A(t)$, $B(t')$ as follows:

$$\{B(t')A(t)\}_T \equiv B(t')A(t)\theta(t' - t) \pm A(t)B(t')\theta(t - t') ; \qquad (24)$$

the sign ($\pm$) depends on the commutation properties of A and B:

+ if A, B separately are boson operators, *or* contain any *even* power of fermion operators $\psi(t)$, $\psi^\dagger(t)$;

− if A, B separately contain *odd* powers of fermion operators $\psi(t)$, $\psi^\dagger(t)$.

We do not define the time-ordered product for a choice of A and B which combined have an odd number of fermion operators, such as A = boson, B = fermion, which will necessarily give vanishing amplitudes in Eq. 23.

The probability amplitude of Eq. 24,

$$\langle \Psi_0 | \{B(t')A(t)\}_T | \Psi_0 \rangle , \qquad (25)$$

includes both causal processes of Eq. 23a, b. Such amplitudes are called *propagators*, time-correlation functions, or Green functions, depending on the operators A, B which are chosen. Particular choices of interest include $A = B$, which must be of the even ($+$) type, since then A and B must separately conserve particle number to give nonvanishing amplitudes, and therefore must contain an equal number of $\psi^\dagger(t)$ and $\psi(t)$ operators, satisfying the *even* fermion condition for ($+$) given above. On the other hand, the choice $A = B^\dagger \neq A^\dagger$ can be either odd or even.

The causal single-particle Green function is defined conventionally by

$$iG(x', x) \equiv \langle \Psi_0 | \{\psi(x')\psi^\dagger(x)\}_T | \Psi_0 \rangle . \qquad (26)$$

Following the sign rules just discussed, we must take the $+$ sign in Eq. 24 if ψ, $\psi^\dagger$ are boson field operators, and $-$ if ψ, $\psi^\dagger$ are fermion field operators. We here introduce a space–time variable $x = (\mathbf{r}, t)$. If there are internal degrees of freedom (e.g. spin), then Eq. 26 is a matrix in the (spin) labels. The Green function expresses the two causal excitations of the many-body system which can be propagated by adding a particle at position $\mathbf{r}$, at time t, and removing a particle at position $\mathbf{r}'$, at time t', for either order of t, t'. Since these are simple excitations, the corresponding Green function is a basic tool in many-body theory.

As an example, we discuss the single-particle Green function for a system of noninteracting fermions, which will also be useful to us in later chapters of this book. The Hamiltonian $H = H_0$ is a one-body operator, which we write in the particle–hole form of Eq. 17. The field operator may be written, using Eqs. 18, 20, and 22,

$$\psi(x) = \sum_\alpha \{\theta(\epsilon_\alpha - \epsilon_F)a_\alpha u_\alpha(\mathbf{r})e^{-i(\epsilon_\alpha - \epsilon_F)t}$$

$$+ (-1)^{\phi(\alpha)}\theta(\epsilon_F - \epsilon_\alpha)b_{\bar\alpha}^\dagger u_\alpha(\mathbf{r})e^{-i(\epsilon_\alpha - \epsilon_F)t}\} . \qquad (27)$$

(Note that the sign in Eq. 18b is superseded by the phase $(-1)^\phi$ of Eqs. 20, 22.) The ground state for the noninteracting system is Φ_0, the Fermi-"gas" ground state, which is the "vacuum" state for particles and holes, as shown in Eqs. 4.19 and 4.20. Then we calculate the Green function

$$
\begin{aligned}
iG^0(x', x) &= \langle \Phi_0 | \psi(x') \psi^\dagger(x) | \Phi_0 \rangle \theta(t' - t) \\
&\quad - \langle \Phi_0 | \psi^\dagger(x) \psi(x') | \Phi_0 \rangle \theta(t - t') \\
&= i \sum_\alpha u_\alpha(\mathbf{r}') G_\alpha^0(t' - t) u_\alpha^\dagger(\mathbf{r})
\end{aligned}
\tag{28}
$$

by direct substitution of Eq. 27 and some elementary manipulations. Here $G_\alpha^0(t' - t)$ is a Green function in a *diagonal* single-particle representation, given by

$$
iG_\alpha^0(t' - t) = \{ \theta(\epsilon_\alpha - \epsilon_F)\theta(t' - t) - \theta(\epsilon_F - \epsilon_\alpha)\theta(t - t') \} e^{-i(\epsilon_\alpha - \epsilon_F)(t' - t)} .
\tag{29}
$$

The superscript zero on G^0 identifies it as a "free" Green function for a noninteracting system. The negative sign between terms in Eqs. 28 and 29 follows from the rule for odd-fermion operators, given after Eq. 24.

One may interpret Eqs. 28 and 29 to say that particle states propagate forward in time $(t' > t)$ with positive energy $\epsilon_\alpha - \epsilon_F > 0$, and that hole states may be thought of as propagating *backward* in time $(t' < t)$ with *negative* energy: $\epsilon_\alpha - \epsilon_F < 0$, in both cases time running from t to t'. But one may just as well consider only increasing time, in which case the interpretation of particle propagation remains unchanged, but now we say that hole states propagate forward in time (from t' to t) with *positive* energy $\epsilon_F - \epsilon_\alpha > 0$, by changing the signs of both $\epsilon_\alpha - \epsilon_F$ and $t' - t$ in the exponential factor in Eq. 29. Further, this coincides with the introduction of creation and annihilation operators for *holes*, given in Eq. 16 or 20: the $t > t'$ terms in Eqs. 28 and 29 had a hole created at t' and annihilated at the later time t (as can be easily verified by filling in the steps from Eq. 28 to 29). The use of the modified relabeling of Eq. 20 has the consequence that the forward-going hole state carries the quantum numbers α if the particle state emptied to make the hole was labeled with the time-reversed quantum numbers $\bar{\alpha}$. When hole states are considered to propagate forward in time, with positive energy and particle-like quantum numbers (α, not $\bar{\alpha}$), they are called *antiparticles*. (See Figure 7.1.)

Returning to Eq. 29, the time behavior may be transformed to *frequency* behavior by defining the Fourier transform

$$
G_\alpha^0(t' - t) = \int_{-\infty}^{\infty} \frac{d\omega}{2\pi} e^{-i\omega(t' - t)} G_\alpha^0(\omega) .
\tag{30}
$$

This can be easily calculated from Eq. 29 to be

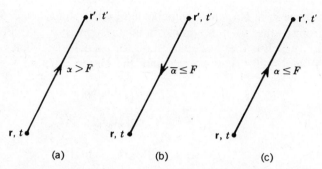

FIGURE 7.1 (a) Particle propagation forward; (b) hole propagating backward; (c) antiparticle propagating forward.

$$G_\alpha^0(\omega) = \frac{\theta(\epsilon_\alpha - \epsilon_F)}{\omega - (\epsilon_\alpha - \epsilon_F) + i\eta} + \frac{\theta(\epsilon_F - \epsilon_\alpha)}{\omega - (\epsilon_\alpha - \epsilon_F) - i\eta} \tag{31}$$

using the integral representation

$$\theta(t) = i \int_{-\infty}^{\infty} \frac{d\omega}{2\pi} \frac{e^{-i\omega t}}{\omega + i\eta} \tag{32}$$

with positive $\eta \to 0$. The first term represents particles, with positive ω, and the second, holes, with negative ω.

If the single-particle representation is $\alpha = (\mathbf{k}, m)$ (plane waves with spin projection), then $\epsilon_\alpha = \epsilon(k) = k^2/2M$, and the "free" Green function may be written as a multiple Fourier transform:

$$G_{m',m}^0 x', x) = (2\pi)^{-4} \int d\mathbf{k} \, e^{i\mathbf{k}\cdot(\mathbf{r}'-\mathbf{r})} \int d\omega \, e^{-i\omega(t'-t)} G_{m'm}^0(k, \omega) \tag{33a}$$

with

$$G_{m',m}^0(k, \omega) = \delta_{m',m} \left\{ \frac{\theta(\epsilon(k) - \epsilon_F)}{\omega - (\epsilon(k) - \epsilon_F) + i\eta} + \frac{\theta(\epsilon_F - \epsilon(k))}{\omega - (\epsilon(k) - \epsilon_F) - i\eta} \right\}. \tag{33b}$$

7.3 INTERACTION PICTURE

We now introduce yet another dynamic representation, called the *interaction picture*, which allows us to describe time development of an interacting system *relative* to a noninteracting system. This will be of great use in formulating perturbation theory in this chapter and again in Chapters 8 and

9. It is equally useful in scattering theory, as we shall see in Section 7.5 and again in Chapter 11. The basic idea of treating separately the time dependence of the noninteracting system is based on Eq. 11, which expresses the fact that each particle (or hole) in a noninteracting system carries its own time-dependent phase, so that the overall phase is simply the sum of these phases. This part of the time dependence is easily handled, as we shall see.

Assume that the full dynamics for the interacting system is given by a Hamiltonian which can be partitioned:

$$H = H_0 + V , \tag{34}$$

where H_0 is a one-body operator of the form of Eq. 9 and V is the interaction among particles.* The state vectors in the interaction picture (denoted by I) are defined relative to the Schrödinger vectors by

$$|\psi_I(t)\rangle = e^{iH_0 t}|\psi_s(t)\rangle , \tag{35a}$$

and obey the equation of motion

$$i \frac{d}{dt} |\psi_I(t)\rangle = V_I(t)|\psi_I(t)\rangle . \tag{35b}$$

This may be compared to Eq. 4, which defines the Heisenberg picture: note that H_0 replaces H in the phase, and that ψ_I is time dependent (unless $V \equiv 0$), unlike ψ_H. Now operators transform as

$$A_I(t) = e^{iH_0 t}A_s e^{-iH_0 t} . \tag{36}$$

We find that

$$H_0(t) = e^{iH_0 t}H_0 e^{-iH_0 t} = H_0$$

is time independent, since $[H_0, e^{-iH_0 t}] = 0$, but that

$$H_I(t) = H_0 + V_I(t) \tag{37}$$

is not, since $[V, H_0] \neq 0$, or $[V, e^{-iH_0 t}] \neq 0$.

The time dependence of the interaction picture states may be incorporated into a time-evolution operator

---

*This is a well-defined partition for a Schrödinger many-body system, where both the individual particles and the interacting system have independent meaning. Problems arise in relativistic quantum field theory in which the "bare" individual particle is not well defined, because of divergences in self-interactions.

$$U(t, t_0) \equiv e^{iH_0 t} e^{-iH(t-t_0)} e^{-iH_0 t_0} \tag{38}$$

which transforms a state from time t_0 to t,

$$|\psi_I(t)\rangle = U(t, t_0)|\psi_I(t_0)\rangle . \tag{39}$$

Clearly $U(t, t_0)$ is unitary. Using the Schrödinger equation (3) and Eqs. 35, 38, and 39, one may easily obtain an equation of motion for U:

$$i \frac{\partial}{\partial t} U(t, t_0) = V_I(t) U(t, t_0) \tag{40}$$

with the initial condition that $U(t_0, t_0) = 1$. Perturbation methods will be obtained (Chapters 8 and 9) by solving Eq. 40 for $U(t, t_0)$ in powers of V_I.

7.4   ADIABATIC GROUND STATE AND ENERGY

Time development can be used, under certain circumstances, to obtain the ground state of an interacting system with a Hamiltonian partitioned as in Eq. 34 from the ground state of the noninteracting Hamiltonian H_0. The method involves treating the interaction as depending on time, by introducing a function $\lambda(t)$ that governs the strength of the interaction, and writing a time-dependent Hamiltonian

$$H(t) = H_0 + \lambda(t)V . \tag{41}$$

For $\lambda(t) = 0$, we have $H(t) = H_0$, and the system is noninteracting, with a ground state we shall denote by Φ_0. For $\lambda(t) = 1$, we have $H(t) = H$, the full Hamiltonian, in whose ground state we are interested. If $\lambda(t)$ changes *slowly*, increasing monotonically from zero to unity, we would expect that the system initially in the state Φ_0 would adiabatically evolve into the interacting ground state of H. The time variation of $\lambda(t)$ must be sufficiently slow (compared to all physical frequencies) that Eq. 41 have almost stationary solutions, and the time dependence not give transitions to excited states; this is called *adiabatic switching*. The system may be thought of as passing through a sequence of states, each an eigenstate of Eq. 41 at that instant t.

Will this process necessarily lead to the ground state of the interacting system? The answer is "not always" and is intimately connected to the existence of a convergent perturbation expansion for the ground state. It seems reasonable, and we shall show explicitly later, that if there is a convergent perturbation expansion for the ground state based on the unperturbed ground state Φ_0, then one can construct a continuous sequence of ground states of $H = H_0 + \lambda V$ for all $0 \le \lambda \le 1$, and that the time sequence of these is related to adiabatic switching. Even more strongly, if a convergent perturbation expansion exists in the neighborhood of every value of $H(t)$, so that the ground state may be continued analytically in λ from Φ_0, this state will be obtained also by adiabatic switching. Such a ground state is called *normal*. As we shall see, for a normal ground state Ψ, the overlap

$$\langle \Phi_0 | \Psi \rangle \neq 0. \tag{42}$$

The true ground state may not be normal, for various different reasons. One case is that the ground state differs from the unperturbed ground state Φ_0 by an exact symmetry (such as angular momentum) which is conserved by *both* H and H_0. In this case, it may still be possible to expand the ground state convergently, based on a different state of H_0, say Φ_1, which is the lowest-energy state of H_0 with the *same* symmetry as the ground state. In the adiabatic switching process, the two states evolving from Φ_0 and Φ_1, respectively, cross (in energy) without mixing, because of the conserved symmetry. The state evolved from Φ_0 is still normal, but not the ground state.

Another circumstance is that of a *phase transition* between the noninteracting and interacting systems. This happens in superconductivity theory (see Chapter 5), in which an attractive interaction between electrons at the Fermi surface alters the ground state nonperturbatively, that is, no matter how weak the attraction. In both this case and the previous case of crossing of states of different symmetries, the overlap of the ground state* with Φ_0 is

$$\langle \Phi_0 | \Psi \rangle = 0. \tag{43}$$

It should be noted that for a phase transition, one needs an infinite number of particles. The ground state for a *finite* system with *pairing* can be analytically continued from Φ_0, such that Eq. 42 is valid. But the extension to an infinite system (thermodynamic limit) changes the analytic properties of the ground state, leading to the singular behavior of phase transitions.

We now discuss a basic theorem on adiabatic switching, proved by Gell-Mann and Low.† The assumption is that there is a convergent perturbation expansion based on Φ_0, giving a normal state, but not necessarily the ground state. Let the function $\lambda(t)$ be of the form $\lambda(t) = e^{-\alpha|t|}$, which "switches on" as t runs from $-\infty$ to 0. The parameter α will later be taken in the limit $\alpha \to 0^+$. Working in the interaction picture, we find from Eq. 36

$$V_I(t) = e^{iH_0 t} V e^{-iH_0 t} e^{-\alpha|t|}. \tag{44}$$

Consider the evolved state vector at $t = 0$, where we denote the evolution from $t = t_0$ to $t = 0$, writing

*The same result occurs in relativistic quantum field theory and is called Haag's theorem.
†M. Gell-Mann and F. Low, Phys. Rev. **84**, 350 (1951).

$$|\Psi(0)\rangle_\alpha = U_\alpha(0, t_0)|\Phi_0\rangle , \tag{45}$$

where U_α is given as a solution of Eq. 40, with $H(t)$ of Eq. 41, and $\lambda(t)$ given above. (Note that the noninteracting ground state Φ_0 is time indepen-dent in the interaction picture.) We want to take the limit of Eq. 45 as $t_0 \to -\infty$ to make the "switching on" completely smooth, but this does not lead to a well-defined state vector, as we discuss after Eq. 50. However, the limiting *ratio*

$$|\Psi_0\rangle = \lim_{\substack{\alpha\to0^+\\t_0\to-\infty}} \frac{|\Psi(0)\rangle_\alpha}{\langle\Phi_0|\Psi(0)\rangle_\alpha} \tag{46}$$

does exist, defining a state vector $|\Psi_0\rangle$, which is the adiabatically switched state. The theorem of Gell-Mann and Low states further that $|\Psi_0\rangle$ is an eigenvalue of H, with energy E_0

$$H|\Psi_0\rangle = E_0|\Psi_0\rangle , \tag{47}$$

which is the normal state generated in perturbation theory from Φ_0, although not necessarily the ground state. (Note that Ψ_0, having been defined at $t = 0$, is identical in the interaction and Schrödinger pictures.)

If the unperturbed ground state Φ_0 has energy $\mathscr{E}_0$ given by

$$H_0|\Phi_0\rangle = \mathscr{E}_0|\Phi_0\rangle , \tag{48}$$

then the shift of energy is easily calculated from Eq. 47:

$$\Delta E \equiv E_0 - \mathscr{E}_0 = \langle\Phi_0|V|\Psi_0\rangle , \tag{49}$$

where we note from Eq. 46 that the overlap

$$\langle\Phi_0|\Psi_0\rangle = 1 . \tag{50}$$

In Chapter 9 we shall develop explicit perturbation expressions for the state vector (Eq. 46) and the energy shifts (Eq. 49), which show that they are both well defined in a convergent theory (see Eqs. 9.14 and 9.15). The construction also shows that in the limits $t_0 \to -\infty$, $\alpha \to 0^+$ required in Eq. 46, both numerator and denominator develop a phase which diverges as $\alpha^{-1}\Delta E$ with ΔE the energy shift of Eq. 49, the divergence canceling in the ratio (see Eq. 9.4).

7.5 S-MATRIX

The method of time development in the interaction picture can also be used to formulate scattering theory. Here we shall briefly discuss the time evolution of a scattering state, and introduce the S-matrix. A more thorough treatment of scattering appears in Chapter 11.

We again assume that the full Hamiltonian can be partitioned as in Eq. 34. Now V represents the interaction between a projectile and a scattering target; H_0 represents not only the kinetic energy of the projectile and target (say, in the c.m. frame), but also the Hamiltonians for whatever internal structure the projectile and target may have. This allows us to include the scattering of complex systems with internal degrees of freedom (atoms, nuclei, etc.) as well as of elementary particles.

The scattering process takes place in three successive stages: preparation of the projectile beam, scattering of the beam by the target, and detection of the scattered projectile. It is an essential part of the description that the preparation and detection must take place far enough from the target region that there can be no interaction between projectile and target during these first and last stages. This can be accomplished by preparing the projectile initially in a wave packet of finite spatial size, far from the target, with an average momentum collimated toward the target. The scattering takes place some time later when the projectile packet reaches the target region (which is also localized in space, at rest in the lab frame). The detection takes place later still, when the scattered packet moves away from the target region.

Let us denote our wave packet by a state vector $\Psi(t)$ in the interaction picture (but suppressing the index I); it satisfies the equation of motion (40)

$$i\,\frac{d}{dt}\,|\Psi(t)\rangle = V_I(t)|\Psi(t)\rangle \ . \tag{51}$$

We set the time scale so that scattering takes place at finite times, centered at $t \cong 0$, whereas preparation is in the distant past ($t \to -\infty$), and detection in the distant future ($t \to +\infty$). Since we have assumed that the packet does not overlap the interaction region of V for $t \to \pm\infty$, the right-hand side of Eq. 51 vanishes in these limits, and the vector $\Psi(t)$ approaches constant vectors:

$$|\Psi(t)\rangle \xrightarrow[t \to -\infty]{} |\Phi_i\rangle \ , \tag{52a}$$

$$|\Psi(t)\rangle \xrightarrow[t \to +\infty]{} |\Psi_f\rangle \ , \tag{52b}$$

which represent the noninteracting packet states before and after interaction, respectively (i: initial; f: final).

Although both packets are superpositions of plane waves (and internal states of projectile and target) with the same average kinetic energy, they involve different superpositions, because of the interaction which has taken place during the time evolution, as can be seen as follows. The interacting state evolves from the initial packet state by the transformation of Eq. 39, in the limit $t_0 \to -\infty$:

$$|\Psi(t)\rangle = U(t, -\infty)|\Phi_i\rangle \ . \tag{53}$$

The final packet state is obtained from the $t \to +\infty$ limit

$$|\Psi_f\rangle = U(+\infty, -\infty)|\Phi_i\rangle$$

$$\equiv S|\Phi_i\rangle, \tag{54}$$

where the *scattering operator* S is defined by the limiting transformation

$$S \equiv \lim_{\substack{t\to\infty \\ t_0\to-\infty}} U(t, t_0). \tag{55}$$

As defined here, S is understood to operate on noninteracting packet states, of the type introduced in Eq. 52. This is not convenient for calculation: it is easier to work with noninteracting energy eigenstates like plane or spherical waves than with wave packets. However, such waves fill all of space, and therefore always overlap the interacting region, contrary to the assumptions leading to Eq. 55. Therefore we resort to the mathematical device of *adiabatic switching*, which was introduced in the previous section. The interaction is given a weak time dependence, as in Eq. 44, such that $V_I(t) \to 0$ for $t \to \pm\infty$, thus preserving the noninteraction of the preparation and detection stages. The scattering process is unaffected at finite t, in the limit $\alpha \to 0^+$ to be taken after the calculation. With the use of adiabatic switching,* Eqs. 54 and 55 can be applied to stationary states (e.g. plane waves) as well as to packets.

If we detect the scattered projectile in terms of given stationary final states Φ_j (e.g. plane waves in various directions relative to the beam), then the probability amplitude for each such component of the scattered wave is given by

$$\langle\Phi_j|\Psi_f\rangle = \langle\Phi_j|S|\Phi_i\rangle \equiv S_{ji}, \tag{56}$$

where S_{ji} are elements of the S-matrix. Cross sections may be calculated from Eq. 56, as will be shown in Chapter 11.

Lastly, we note that in the scattering problem we are able to define the state vector Eq. 53, while in the adiabatic switching of the ground state of the previous section, we found that the analogous vector Eq. 45 was ill defined because of a phase diverging as $\alpha^{-1}\Delta E$ (see discussion following Eq. 50). The difference in the present case is that there is no change of energy during scattering, even with adiabatic switching; therefore $\Delta E \equiv 0$, and there is no divergence. For scattering in a single channel (e.g. partial wave), the S-matrix is given by a *phase shift* (finite)

$$S_{ii} \equiv e^{2i\delta_i}. \tag{57}$$

---

*Or equivalent devices: see, e.g., Goldberger and Watson, *Collision Theory* (Wiley, New York, 1964) Chapter 5.1.

7.6 TEMPERATURE FORMALISM

The method of time development can also be applied to deal with complex systems in thermodynamic equilibrium, exploiting a formal relation between time and temperature. In this section we introduce some of the elementary notions; applications depend on the development of perturbation techniques which we undertake in Chapters 8 and 9.

We consider a quantum system with stationary states Ψ_m and energies E_m (in the Schrödinger picture). The system in equilibrium at absolute temperature T is described by a *mixture* of states, weighted by the Boltzmann factor $e^{-E_m \beta}$, where $\beta = 1/\kappa T$ is the inverse temperature in energy units, and κ is Boltzmann's constant. The normalized probability of each state may be written

$$P_m = e^{-E_m \beta}/Z \tag{58}$$

with the partition function given as a sum over all states,

$$Z = \sum_m e^{-E_m \beta} . \tag{59}$$

(Note that $\Sigma_m P_m = 1$.) One may introduce a distribution operator (or density matrix) $\rho(\beta)$ in terms of the Hamiltonian H:

$$\rho(\beta) = e^{-H\beta}/Z , \tag{60}$$

whose diagonal matrix elements give Eq. 58, that is

$$P_m = \langle \Psi_m | \rho(\beta) | \Psi_m \rangle , \tag{61}$$

and therefore

$$Z = \sum_m \langle \Psi_m | e^{-H\beta} | \Psi_m \rangle , \tag{62a}$$

by Eq. 59. This last expression will be recognized as the trace (tr) of the matrix of $e^{-H\beta}$ (i.e., sum over diagonal matrix elements),

$$Z = \text{tr } e^{-H\beta} . \tag{62b}$$

The equilibrium expectation value of a physical property described by an operator A is given by the weighted average

$$\langle A \rangle_\beta = \sum_m \langle \Psi_m | A | \Psi_m \rangle P_m \tag{63a}$$

$$= \text{tr } A\rho(\beta) . \tag{63b}$$

Note that for zero temperature ($\beta \to \infty$) we have $P_m \to \delta_{m,0}$ for a nondegenerate ground state ψ_0, and Eq. 63 become the usual ground-state expectation value of A. (For a ground state with d-fold degeneracy $v = 1, \ldots d$, we have $P_m \to d^{-1}\delta_{m,v}$, and Eq. 63 gives the *average* expectation value.)

If one wants to allow the system to have *indefinite* particle number, one uses the grand canonical ensemble of Gibbs. Then the set of states $\{\Psi_m\}$ is extended to include the complete energy eigenstates for all numbers of particles; the trace (tr) operation is extended accordingly. One modifies the Hamiltonian by a term proportional to the number operator, $H \to H - \mu N$, with μ the chemical potential. (Note that this was also done in the BCS approximation, with a Lagrange multiplier, in Eq. 5.30.) The distribution operator now becomes

$$\rho_G(\beta, \mu) = e^{-(H-\mu N)\beta}/Z_G \tag{64}$$

with the grand partition function

$$Z_G = \text{tr } e^{-(H-\mu N)\beta} = \sum_{m,N} e^{-(E_{m,N}-\mu N)\beta} . \tag{65}$$

The expectation of A is now written

$$\langle A \rangle_{\beta,\mu} = \text{tr } A\rho_G(\beta,\mu) . \tag{66}$$

The dynamical information about the system in equilibrium is carried by the distribution operator $\rho(\beta)$ [or $\rho_G(\beta, \mu)$]. This is analogous to saying that the dynamical information in a Schrödinger state vector is carried by the time-development operator e^{-iHt} (i.e. the inverse transformation to that of Eq. 5). In fact, comparison of the form of $\rho(\beta)$ given in Eq. 60 with e^{-iHt} shows a formal relation of the two which can be exploited to develop a general method of calculating properties of many-body systems at fixed temperature by techniques closely related to those which apply to ground-state properties of the same systems. The first observation* is that $e^{-H\beta}$ [$= Z\rho(\beta)$] can be considered as a time-development operator e^{-iHt} with an imaginary time parameter, $t = -i\beta$. One can further write an "equation of motion" for $Z\rho(\beta)$ in the temperature variable β (Bloch equation)

$$\frac{d}{d\beta} Z\rho(\beta) = -HZ\rho(\beta) . \tag{67}$$

Clearly this equation is no easier to solve than the time-dependent Schrödinger equation to which it is equivalent. The point is that it is also no harder to solve, at least in those situations in which perturbation (or related) methods can be used for time development. Then it is possible to use the

*F. Bloch, Z. Phys. **74**, 295 (1932).

time–temperature connection to obtain thermodynamic quantities for interacting systems, by treating the interactions as perturbations, and using the
time techniques we shall develop in the following chapters (suitably modified).

It is convenient to work in a representation based on the interaction
picture of Section 7.3. We define a time-development operator from $U(t, t_0)$
of Eq. 38, but at two imaginary times, $t = -i\tau$, $t_0 = -i\tau_0$, and write a new
function

$$u(\tau, \tau_0) \equiv U(-i\tau, -i\tau_0)$$
$$= e^{H_0\tau}e^{-H(\tau-\tau_0)}e^{-H_0\tau_0} . \qquad (68)$$

(For the grand canonical ensemble, replace H with $H - \mu N$.) This operator
obeys an "equation of motion" analogous to Eq. 40,

$$\frac{\partial}{\partial\tau} u(\tau, \tau_0) = -V(\tau)u(\tau, \tau_0) \qquad (69)$$

with $V(\tau)$ in the "temperature interaction picture" (compare Eq. 36),

$$V(\tau) = e^{H_0\tau}V_S e^{-H_0\tau} . \qquad (70)$$

Assume that one can solve for $u(\tau, \tau_0)$ by perturbation methods to some
approximation. Then one has the distribution operator in the form

$$\rho(\beta) = \frac{e^{-H_0\beta}u(\beta, 0)}{Z} \qquad (71)$$

with

$$Z = \mathrm{tr}\, e^{-H_0\beta}u(\beta, 0) , \qquad (72)$$

from which equilibrium properties may be calculated to the given order of
approximation.

The formal relation of the temperature theory to real time evolution can
be further exploited by introduction of temperature Green functions to
describe time-independent equilibrium properties or time and temperature
Green functions which allow consideration of excitations of equilibrium
systems.

EXERCISES

7.1 (a) Show that the field operator $\psi(\mathbf{r}, t)$ of Eq. 7, for the case of a
noninteracting system of particles $(H = H_0)$, obeys the equation
of motion

$$\left[i\frac{\partial}{\partial t} + \frac{\nabla^2}{2m} - U(\mathbf{r}) \right] \psi(\mathbf{r}, t) = 0 ,$$

where

$$H_0 = \sum_i \left(\frac{p_i^2}{2m} + U(\mathbf{r}_i) \right) .$$

(b) For the case of an interacting system $(H = H_0 + V)$ with a two-body potential $V = \sum_{i<j} v(\mathbf{r}_i - \mathbf{r}_j)$, the field operator $\psi(\mathbf{r}, t)$, obeys an equation of motion in which the zero on the right-hand side of (a) is replaced by an expression in terms of v, ψ, and $\psi^\dagger$. Find the expression.

7.2 (a) Show that the Green function $G^0(x, x')$ (Eq. 28) for a noninteracting system obeys the equation of motion [see Exercise 7.1(a)]

$$\left[i\frac{\partial}{\partial t} + \frac{\nabla^2}{2m} - U(r) \right] G^0(x, x') = \delta^4(x - x') = \delta(t - t')\delta(\mathbf{r} - \mathbf{r}') .$$

(b) Verify that the Green function of Eq. 33 satisfies this equation with $U(r) = $ constant.

(c) Find the equation of motion for the free-particle Green function in (diagonal) momentum representation: $G^0(\mathbf{k}, t - t')$ (see Eqs. 29, 33).

7.3 Consider a two-state system with unperturbed states Φ_0, Φ_1, and energies $\epsilon_0 < \epsilon_1$. The interaction V is a 2×2 matrix in the basis of these states. Using this basis:

(a) Express $V_I(t)$ (Eq. 37) as a time-dependent 2×2 matrix.

(b) Express $U(t, t_0)$ (Eq. 38) as a time-dependent 2×2 matrix. (Hint: The Pauli matrices are useful here, including the relation $e^{i\beta\boldsymbol{\sigma}\cdot\hat{\mathbf{n}}} = \cos\beta + i\boldsymbol{\sigma}\cdot\hat{\mathbf{n}}\sin\beta$.)

(c) Verify Eq. 40 as a matrix equation is this space.

CHAPTER 8

Diagrammatic
Perturbation Theory

We begin the development of perturbation theory for many-body systems with an approach in which we introduce perturbation diagrams as the basis for analysis. Theoretical methods based on diagrams have become standard in dealing with systems with many degrees of freedom for a number of reasons. The most immediate motivation is to have a method of classifying contributions to a perturbing process. In systems with few degrees of freedom, one begins by ordering perturbations by the power of the strength of the perturbing interaction: first order, second order, etc. The difficulty of calculating in each order increases roughly arithmetically; that is, the number of summations, integrations, and so on increases linearly with the order. For systems with many degrees of freedom, one also orders by powers of the interaction strength, but the number of distinct types of contribution in each order grows extremely rapidly with order. Diagrams provide a natural classification scheme for perturbation contributions.

Each diagram stands for a mathematical expression for a particular term in a perturbation expansion. A major advantage of diagram methods is that they provide explicit rules for the calculation of the term represented by each diagram. Particularly for high orders, this provides what is often the most efficient method of calculation. Further, the rules for calculating from diagrams exhibit explicitly the fact that there are simple relations among certain diagrams of different orders, but with some similarity of form. Often, these related forms, when combined, represent some simple description of a physical process contributing in all orders, e.g. the excitation of a particle–hole state, with repeated interactions in a Fermi gas. It is then possible to group perturbation contributions by types representing repeated processes, to all orders, and calculate them together. This is an extremely powerful method, called *partial summation of diagrams*, of which we shall make use several times in the following chapters.

The diagram expansion methods also help to exhibit some of the formal properties of the perturbation expansion, which may be useful, or in some cases necessary, for meaningful calculation. The most striking example of a necessary property whose discovery and utilization *requires* diagram methods is given by the *linked-cluster theorem*, which we discuss in Chapter 9. Here the topology, or degree of connectedness of the diagrams, is an essential element in this development. Without the use of this property, perturbation expansions will not in general converge in the limit of large numbers of particles (or degrees of freedom). Last, diagrams often serve as compact symbols for entire theoretical descriptions, serving as part of the "language" of many-body theory.

We start our discussion by expanding the time-development operator of Section 7.3 in orders of the interaction. There are two basic methods of expressing this expansion which lead to the introduction of two different but related forms of diagrams, commonly called Goldstone diagrams and Feynman diagrams, respectively. Although formally equivalent, each expansion has some advantages for certain applications, as we shall see explicitly in this and the following chapters. First we shall introduce the two methods, taking as our example an interacting system of fermions. In Section 8.5 we shall compare the methods, and discuss briefly the different treatment of energy variables in Section 8.6.

8.1 EXPANSION OF THE TIME-EVOLUTION OPERATOR

We work in the interaction picture, introduced in Section 7.3. The time-evolution operator, defined in Eq. 7.38, was shown to obey an equation of motion (Eq. 7.40)

$$i \frac{\partial}{\partial t} U(t, t_0) = V(t)U(t, t_0) \tag{1}$$

with $U(t_0, t_0) = 1$, and $V(t)$ is the perturbing interaction. [The label I indicating interaction picture is henceforth omitted, e.g. on $V(t)$.] This differential equation may be integrated to obtain an equivalent integral form

$$U(t, t_0) = 1 - i \int_{t_0}^{t} dt_1 V(t_1)U(t_1, t_0). \tag{2}$$

This equation provides a direct method of expanding $U(t, t_0)$ in orders of $V(t)$, by successive approximation:

$$U(t, t_0) = U^{(0)}(t, t_0) = 1 \,,$$

$$U(t, t_0) = U^{(0)}(t, t_0) + U^{(1)}(t, t_0)$$

$$= 1 - i \int_{t_0}^{t} dt_1 \, V(t_1) \,, \tag{3}$$

$$\vdots$$

Iteration of Eq. 2 n times gives the nth-order approximation

$$U(t, t_0) = 1 - i \int_{t_0}^{t} dt_1 \, V(t_1) + (-i)^2 \int_{t_0}^{t} dt_1 \, V(t_1) \int_{t_0}^{t_1} V(t_2)$$

$$+ \cdots + (-i)^n \int_{t_0}^{t} dt_1 \, V(t_1) \cdots \int_{t_0}^{t_{n-1}} dt_n \, V(t_n) \,. \tag{4}$$

The formal solution to Eqs. 1 or 2 may be given as an infinite series

$$U(t, t_0) = \sum_{n=0}^{\infty} (-i)^n \int_{t_0}^{t} dt_1 \, V(t_1) \cdots \int_{t_0}^{t_{n-1}} dt_n \, V(t_n) \,. \tag{5}$$

The interaction-picture operators $V(t)$ are related to the Schrödinger operators V by Eq. 7.36:

$$V(t) = e^{iH_0 t} V e^{-iH_0 t} \,. \tag{6}$$

Each term in Eq. 5 involves n interactions $V(t_i)$ at different times, the system propagating in the unperturbed Hamiltonian H_0 between interactions. The interactions at different times, $V(t_i)$, $V(t_j)$, do not commute, since $[H_0, V] \neq 0$. The time variables are ordered by the limits of integration

$$t \geq t_1 \geq t_2 \geq \cdots \geq t_n \geq t_0 \,, \tag{7}$$

so that the order of the operators (right to left) corresponds to the time order of interactions, with $V(t_n)$ earliest and rightmost. With this time-ordered form, Eq. 5, we shall obtain the Goldstone expansion.

An alternative form of the series expansion can be obtained in which the limits of integration are not ordered, by making use of a generalization of the time-ordered product $\{B(t')A(t)\}_T$ that was introduced in Eq. 7.24. We write

$$U(t, t_0) = \sum_{n=0}^{\infty} \frac{(-i)^n}{n!} \int_{t_0}^{t} dt_1 \cdots \int_{t_0}^{t} dt_n \, \{V(t_1) \cdots V(t_n)\}_T \,, \tag{8}$$

where the product enclosed in braces, $\{\cdots\}_T$, is to be ordered by time, with the earliest time to the right. Following the sign convention of Eq.

7.24, the overall sign is positive for boson operators, or even powers of fermion operators, as in the present case of the $V(t)$. (For fermion operators, or odd powers of them, the sign is given by the sign of the permutation which takes the times $t_1, t_2, \ldots t_n$ into the *ordered* set of times. We shall need this sign rule in Section 8.3.) Although the operators are ordered in Eq. 8, the times $t_1, \ldots, t_n$ may take any order, since each t_i is only constrained by the outer limits of (7), $t \geq t_i \geq t_0$. Then we can see that Eq. 8 is equivalent to Eq. 5, by comparing the nth terms of the two expressions. The time-ordered product integrated over all time variables t_i, with $t \geq t_i \geq t_0$, can be seen to give the same integral as in the nth term of Eq. 5, repeated $n!$ times for the number of possible permutations of $t_1, \ldots, t_n$: thus the additional factor $(1/n!)$. We shall use Eq. 8 (which is due to Dyson) to obtain the Feynman diagram expansion in Section 8.4.

Comparison of the two series expansions of $U(t, t_0)$ shows some possible advantages to be drawn from either. The Goldstone expansion proceeds by introducing intermediate states for the entire system, between the interactions in each term of Eq. 5. This leads directly to a perturbation series for the ground-state energy and wave function, as we see in Chapter 9. The Dyson form in Eq. 8 leads more naturally to an expansion in terms of the single-particle Green functions or propagators, which we introduced in Section 7.2. The fact that the limits of integration are not ordered means that time integrations may be done independently, which is a great simplification. Some calculations and proofs of formal properties are more easily carried out in the Feynman expansion. However, some properties, such as ground-state energies, are more cumbersome in the Feynman approach. Therefore there is an advantage to keeping both equivalent methods at hand for nonrelativistic many-body perturbation theory.*

8.2 GOLDSTONE DIAGRAMS

The Goldstone expansion begins with Eq. 5, in which the time-development operator is expressed as a series of integrals of products of the perturbing interaction,

$$V(t) = e^{iH_0 t} V e^{-iH_0 t} \tag{9}$$

(see Eqs. 7.36, 7.37). The time-dependent operators may be converted to simple phases by taking matrix elements of each factor of the form of Eq. 9 between eigenstates of the unperturbed Hamiltonian H_0. Let us introduce a complete set of eigenstates of H_0,

---

*For relativistic quantum system, considerations of covariance strongly favor the Feynman method, which treats particles and holes symmetrically.

$$H_0|m\rangle = \mathscr{E}_m|m\rangle \tag{10}$$

where the unperturbed state $|\Phi_0\rangle$ of Eq. 7.48 is now denoted by $|0\rangle$. (Remember that these states are time independent in the interaction picture.) To be specific, we assume a system of fermions, and with H_0 a one-body operator as in Eq. 7.9, with $|0\rangle$ the normal Fermi-gas ground state of orbits $\{\lambda\}$ filled to the Fermi level $\lambda = F$. Assume the state to be nondegenerate. The excited states $(m \neq 0)$ may be specified in a number representation by the quantum numbers of particles or holes excited, e.g., as in Eqs. 4.23 and 4.24. Here we use the generic notation $\{\lambda\}$ for all hole states and $\{\rho\}$ for particle states; then a state $|m\rangle$ is fully specified by

$$|m\rangle = |\{\rho\}\{\lambda\}\rangle = |\Phi_{\lambda_1\lambda_2\cdots}^{\rho_1\rho_2\cdots}\rangle \tag{11a}$$

with

$$\mathscr{E}_m = \sum_\rho \epsilon_\rho - \sum_\lambda \epsilon_\lambda \, . \tag{11b}$$

Let us calculate the matrix element of the nth-order term $U^{(n)}$ of Eq. 5 between two unperturbed states, $\langle m|$ and $|a\rangle$. Let us also introduce complete sets of intermediate states between the interactions $V(t_i)$: $\sum_b |b\rangle\langle b|$, $\sum_c |c\rangle\langle c|$, etc., to obtain

$$\langle m|U^{(n)}|a\rangle = (-i)^n \sum_{b,\ldots,l} \int_{t_0}^{t} dt_1 \, \langle m|V(t_1)|l\rangle$$

$$\times \langle l|\cdots|b\rangle \int_{t_0}^{t_{n-1}} dt_n \, \langle b|V(t_n)|a\rangle \, . \tag{12}$$

The matrix elements of the $V(t_i)$ each factor into a time-independent matrix element of V and a time factor:

$$\langle l|V(t_i)|k\rangle = \langle l|V|k\rangle e^{i(\mathscr{E}_l - \mathscr{E}_k)t_i} \, , \tag{13}$$

as can be seen from Eq. 9, where the coefficient of t_i depends on the difference of energies (Eq. 11b) between the two unperturbed many-body states $|l\rangle$ and $|k\rangle$.

For example, suppose $|k\rangle = |0\rangle$, the ground state, and $|l\rangle$ is a 2p–2h state (see Eq. 4.24)

$$|k\rangle = |0\rangle \, , \tag{14a}$$

$$|l\rangle = |\{\rho_1\rho_2\}\{\lambda_2\lambda_1\}\rangle$$
$$= a_{\rho_1}^\dagger a_{\rho_2}^\dagger b_{\lambda_2}^\dagger b_{\lambda_1}^\dagger |0\rangle \, . \tag{14b}$$

Then the matrix element

$$\langle l|V|k\rangle = \langle\{\rho_1\rho_2\}\{\lambda_2\lambda_1\}|V|0\rangle \tag{15a}$$

may be easily calculated, using Eqs. 1.47 or 1.51, and 4.24, and carrying out the anticommutation of the a, $a^\dagger$, b, $b^\dagger$, to obtain

$$\langle\{\rho_1\rho_2\}\{\lambda_2\lambda_1\}|V|0\rangle = \langle\rho_1\rho_2|v|\lambda_1\lambda_2\rangle_A \tag{15b}$$

$$= \langle\rho_1\rho_2|v|\lambda_1\lambda_2\rangle - \langle\rho_1\rho_2|v|\lambda_2\lambda_1\rangle \tag{15c}$$

in terms of the two-body antisymmetrized matrix elements Eq. 15b or the direct and exchange matrix elements Eq. 15c. The energy difference is

$$\mathscr{E}_l - \mathscr{E}_k = \epsilon_{\rho_1} + \epsilon_{\rho_2} - \epsilon_{\lambda_1} - \epsilon_{\lambda_2}. \tag{16}$$

The time integral of the matrix element (13) for the states given in Eq. 14, over the interval $t_0 \le t_1 \le t$, yields the first-order contribution to $U(t, t_0)$:

$$\langle l|U^{(1)}(t, t_0)|0\rangle = -i\int_{t_0}^{t} dt_1 \langle\rho_1\rho_2|v|\lambda_1\lambda_2\rangle_A$$

$$\times \exp i(\epsilon_{\rho_1} + \epsilon_{\rho_2} - \epsilon_{\lambda_1} - \epsilon_{\lambda_2})t_1. \tag{17}$$

This matrix element can be represented by a time-evolution diagram, as shown in Figure 8.1a. The vertical axis represents time, with t_1 understood to be integrated between the limits indicated in Eq. 17. Particles propagating in time are represented by vertical lines with upward arrows, while holes have downward arrows, indicating propagation backward in time. The vacuum state $|0\rangle$ is indicated by the absence of vertical lines, e.g., in the interval from t_0 to t_1. The antisymmetrized interaction of Eq. 15b at t_1 is represented by a dot (or vertex), into which the hole lines (λ_1, λ_2) converge and from which the particle lines (ρ_1, ρ_2) emerge. The horizontal order of the lines has no significance, and the lines are drawn curved to facilitate reading. The diagram, which is a variety of Goldstone diagram, is read as follows: The initial state at t_0 is the vacuum (Fermi ground state) which propagates without change to time t_1 (no vertical lines). The interaction $V(t_1)$ (vertex) excites the specific two-particle, two-hole state given in Eq. 14b, which then propagates to time t without change, as represented by the particle lines ρ_1, ρ_2 from t_1 to t, and hole lines λ_1, λ_2, from t to t_1. Excitations of other possible states from the ground state are represented by other diagrams.

It is sometimes more convenient to treat separately the direct and exchange matrix elements 15c in the expression 17. Diagrammatically this is done by dividing the diagram of Figure 8.1a into two parts, indicated in Figure 8.1b, showing the direct *minus* the exchange terms of Eq. 15c. The vertex becomes a dashed line connecting two vertices. In the direct case, $\lambda_1 \rightarrow \rho_1$ through one vertex and $\lambda_2 \rightarrow \rho_2$ through the second; for the ex-

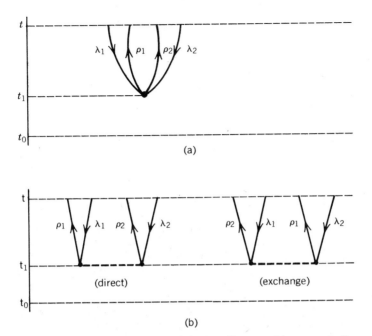

FIGURE 8.1 Representation of Eq. 17 by Goldstone diagrams: (a) represents the expression with an antisymmetrized interaction matrix element (vertex), while (b) represents the interaction divided into direct and exchange matrix elements.

change diagram, $\lambda_1 \to \rho_2$, $\lambda_2 \to \rho_1$ as in Eq. 15c. (The separation of direct and exchange matrix elements was illustrated earlier in Figure 1.1.) The more compact form of Fig. 8.1a is convenient for general discussion of the structure of diagrams. The separated form will prove useful in cases where the direct and exchange matrix elements have distinctly different behavior, e.g., for the Coulomb interaction (see Chapter 10).

We shall also be interested in matrix elements of the time-development operator which begin and end in the unperturbed ground state: $\langle 0|U(t, t_0)|0\rangle$. The first-order contribution in this case is easily found to be

$$\langle 0|U^{(1)}(t, t_0)|0\rangle = \frac{-i}{2} \int_{t_0}^{t} dt_1 \sum_{\lambda_1, \lambda_2 \leq F} \langle \lambda_1 \lambda_2|v|\lambda_1 \lambda_2\rangle_A , \qquad (18)$$

where there is no time dependence in the integrand, since the initial and final energies are equal (see Eq. 13). No particles or holes are created or annihilated in this operator; only filled orbits in the Fermi sea contribute. (The factor of $\frac{1}{2}$ assures that each pair of orbits is counted once.) The term given by Eq. 18 is represented by a Goldstone diagram, as shown in Figure 8.2a. There are no particles or holes propagating, only closed loops beginning and ending at the same time (t_1), representing interactions among the filled orbits. As in Figure 8.1, this diagram may also be decomposed into

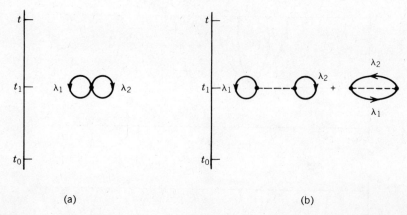

FIGURE 8.2 Representation of Eq. 18 by Goldstone diagrams: (a) dot vertex for antisymmetrized interaction matrix elements, (b) direct and exchange diagrams, with dashed line for interaction. The closed loops represent filled orbits in the Fermi sea; summation over λ_1, $\lambda_2 \leq F$ is implicit. Note that the *sum* of diagrams (b) gives the *difference* of matrix elements in Eq. 19.

two parts, as shown in Figure 8.2b, representing separately the direct and exchange contributions to the antisymmetrized matrix element

$$\langle \lambda_1\lambda_2|v|\lambda_1\lambda_2\rangle_A = \langle \lambda_1\lambda_2|v|\lambda_1\lambda_2\rangle - \langle \lambda_1\lambda_2|v|\lambda_2\lambda_1\rangle \ . \tag{19}$$

As a final example let us consider the second-order contributions to $\langle 0|U|0\rangle$. Since the system starts in the "vacuum" state $|0\rangle$, and V is a two-body interaction, the only intermediate states b which can contribute to the expansion of Eq. 12 are the "vacuum" state $|0\rangle$ itself; any one-particle, one-hole state; or any two-particle, two-hole state. The three separate kinds of terms may be calculated:

$$\langle 0|U^{(2)}(t, t_0)|0\rangle = -\int_{t_0}^{t} dt_1 \int_{t_0}^{t_1} dt_2 \, (I_a + I_b + I_c) \ , \tag{20}$$

where

$$I_a = \left\{ \frac{1}{2} \sum_{\lambda_1, \lambda_2 \leq F} \langle \lambda_1\lambda_2|v|\lambda_1\lambda_2\rangle_A \right\}^2 \ , \tag{21a}$$

$$I_b = \sum_{\substack{\lambda\mu_1\mu_2 \leq F \\ \rho > F}} \langle \lambda\mu_2|v|\rho\mu_2\rangle_A \langle \rho\mu_1|v|\lambda\mu_1\rangle_A$$

$$\times \exp i(\epsilon_\rho - \epsilon_\lambda)(t_2 - t_1) \ , \tag{21b}$$

$$I_c = \frac{1}{4} \sum_{\substack{\lambda_1\lambda_2 \leq F \\ \rho_1\rho_2 > F}} \langle \lambda_1\lambda_2|v|\rho_1\rho_2\rangle_A \langle \rho_1\rho_2|v|\lambda_1\lambda_2\rangle_A$$

$$\times \exp i(\epsilon_{\rho_1} + \epsilon_{\rho_2} - \epsilon_{\lambda_1} - \epsilon_{\lambda_2})(t_2 - t_1) \ , \tag{21c}$$

following steps similar to those leading to Eqs. 17 and 18. (The factor $\frac{1}{4}$ again assures that each pair of particles is counted only once.) Note that I_a, which represents no excitation of the system, consequently has no time dependence (as in Eq. 18). The Goldstone diagrams representing the three terms are given in Figure 8.3.

The general idea of associating Goldstone diagrams with each type of contribution to Eq. 12 should be clear. Each intermediate state (as well as the initial state and the final state) given as in Eq. 11a is represented by an appropriate number of particle and hole lines propagating between two times. At each such time, two particles interact, represented by a vertex with two lines entering and two leaving. The expression represented by each diagram has a rather simple connection to the elements of the diagram itself, as we see in Eq. 21: An antisymmetrized matrix element appears for each

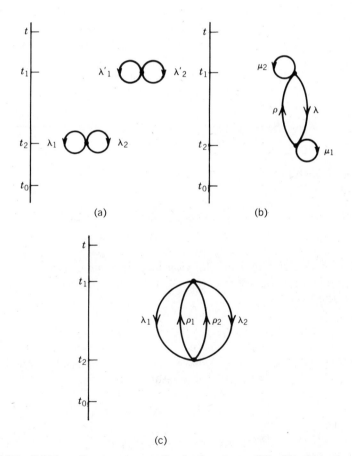

FIGURE 8.3 Goldstone diagrams representing the three terms of Eq. 21, with antisymmetrized matrix elements: (a) I_a, (b) I_b, (c) I_c.

vertex, labeled on the right by the two entering line labels, and on the left by the two emerging line labels (whether particles or holes). Each propagating particle gives a time factor $\exp[i\epsilon_\rho(t_{i+1} - t_i)]$, and each hole, a factor $\exp[-i\epsilon_\lambda(t_{i+1} - t_i)]$.

It is in fact possible to proceed without explicitly denoting the intermediate states, instead drawing the diagrams first and deriving expressions for the contribution of each diagram to $U(t, t_0)$ from simple rules of association, as just described. There are some rules needed for specifying distinct diagrams, so that all contributions are counted properly. Rather than derive these directly, we shall turn to an alternative form of diagrammatic expansion, which does a very similar thing.

8.3 TIME ORDERING AND WICK'S CONTRACTION THEOREM

For the present development we shall use the Dyson form of the expansion of $U(t, t_0)$, given in Eq. 8. The essential differences between this expansion and that of Eq. 5 from which we generated the Goldstone expansion are the following: First, in Eq. 8 the interaction times $t_1, \ldots, t_n$ may be taken in any order, and do not appear in the limits of integration, while in Eq. 5 the times are ordered by the limits of integration. Second, the integrands in Eq. 8 contain the time-ordered products of interaction operators $V(t_i)$:

$$\{V(t_1) \cdots V(t_n)\}_T , \tag{22}$$

while Eq. 5 contains simple operator products. It is the time-ordered property we exploit now.

The basic method of evaluating an expression with time-ordered products of operators is that given by Wick, in which one rewrites the original expression in terms of another kind of product of the same operators, called a *normal-ordered* product, which will be defined shortly. We first illustrate the method with a simple example. Consider the time-ordered product of a fermion annihilation operator $a_\alpha(t)$ and a creation operator $a_\beta^\dagger(t')$, which may be written out as

$$\{a_\alpha(t)a_\beta^\dagger(t')\}_T = a_\alpha(t)a_\beta^\dagger(t')\theta(t - t') - a_\beta^\dagger(t')a_\alpha(t)\theta(t' - t) \tag{23}$$

by definition (see Eq. 7.24). Now we define the *normal-ordered* product (denoted by $\{\cdots\}_N$) of two such operators as follows:

$$\{a_\alpha(t)a_\beta^\dagger(t')\}_N = -a_\beta^\dagger(t')a_\alpha(t) ,$$

$$\{a_\beta^\dagger(t')a_\alpha(t)\}_N = a_\beta^\dagger(t')a_\alpha(t) . \tag{24}$$

The rule is that the given product is to be rewritten with the creation operator to the left of the annihilation operator. The new product is assigned a positive sign if the new order can be reached from the given order by an even permutation of the fermion operators, and a negative sign if by an odd permutation. The times play no role in defining this product. For fermions we may also transform the hole operators for orbits below the Fermi sea, as in Eq. 7.18b. Then the normal-ordered product of b_λ and $b_\mu^\dagger (\lambda, \mu \leq F)$ will be defined as $\pm b_\mu^\dagger b_\lambda$, as in Eq. 24 for particle operators $(\alpha, \beta > F)$. A normal-ordered product gives no contribution when operating between vacuum states; this leads to great simplification in expressions, as we see later.

We may relate the two ordered products of Eqs 23 and 24 by simple recombination of terms, to give

$$\{a_\alpha(t)a_\beta^\dagger(t')\}_T = \{a_\alpha(t)a_\beta^\dagger(t')\}_N + \{a_\alpha(t), a_\beta^\dagger(t')\}\theta(t - t') . \quad (25)$$

The last term contains the anticommutator of the a, $a^\dagger$ operators, which is easily evaluated using Eq. 7.11 for the interaction-picture operators:

$$\{a_\alpha(t), a_\beta^\dagger(t')\} = \delta_{\alpha\beta}e^{-i\epsilon_\alpha(t-t')} . \quad (26)$$

The last term of Eq. 25 is called the *contraction* of $a_\alpha(t)$ and $a_\beta^\dagger(t')$, that is, the remainder on replacing the T-product by the N-product, and is seen in Eq. 26 to be a c-number. Further, the contraction can be identified with the causal one-particle Green function for the noninteracting Fermi system given in Eq. 7.29, i.e. for $\alpha, \beta = \rho, \sigma > F$:

$$\{a_\rho(t)a_\sigma^\dagger(t')\}_T - \{a_\rho(t)a_\sigma^\dagger(t')\}_N = i\delta_{\rho\sigma}G_\rho^0(t - t') . \quad (27)$$

Similarly, the *contraction* of $b_\mu^\dagger(t)$ and $b_\lambda(t')$ may be easily expressed by the same Green function

$$\{b_\mu^\dagger(t)b_\lambda(t')\}_T - \{b_\mu^\dagger(t)b_\lambda(t')\}_N = -\delta_{\mu\lambda}e^{-i\epsilon_\lambda(t-t')}\theta(t' - t)$$

$$= i\delta_{\mu\lambda}G_\lambda^0(t - t') , \qquad \lambda \leq F. \quad (28)$$

Note that ϵ_α and ϵ_λ are measured relative to the Fermi energy ϵ_F.

For completeness one may define the normal-ordered product for two fermion operators which anticommute as in Eq. 24, e.g.

$$\{a_\alpha(t)a_\beta(t')\}_N = a_\alpha(t)a_\beta(t') = -a_\beta(t')a_\alpha(t) , \quad (29)$$

where both orders on the right-hand side are normal-ordered, that is, annihilators are to the right of creators. The *contraction* for such anticommuting operators vanishes:

$$\{a_\alpha(t)a_\beta(t')\}_T - \{a_\alpha(t)a_\beta(t')\}_N = \{a_\alpha(t), a_\beta(t')\}\theta(t-t') = 0, \quad (30)$$

since the anticommutator vanishes.

For any two fermion operators, we see from Eqs. 25 and 30 that the time-ordered product can be replaced by the normal-ordered product plus a contraction, which is always a c-number (and which may vanish). The advantage of the reordering becomes evident for ground-state perturbation theory, for which we shall evaluate expectation values of these expressions in the unperturbed ground state Φ_0. Then the normal-ordered products give vanishing contributions, since annihilators on the right or creators on the left will give zero acting on Φ_0. Thus, only the contractions will contribute: these can always be expressed in terms of the unperturbed Green function, as in Eqs. 27 and 28.

The generalization of this simple example to other operators proceeds as follows. Suppose we want to evaluate a time-ordered product of operators $\{ABC\cdots\}_T$ (where we suppress the time labels). First we decompose each operator A, B, ... into products of its component creators $a^\dagger$, $b^\dagger$, and annihilators a, b. The entire product $(ABC\cdots)$ can then be written as a sum of products $(rstuv\cdots)$ of operators $r, s, t, \ldots$, each of which is one of the four: $a^\dagger$, $b^\dagger$, a, b. The time-ordered product $\{ABC\cdots\}_T$ can be expressed as a linear combination of time-ordered products of the form $\{rstuv\cdots\}_T$.

Now define the normal-ordered product $\{rstuv\cdots\}_N$ to be a reordering of $rstuv\cdots$ such that all creators stand to the left of all annihilators. The overall sign is given by the permutation required to obtain the new ordering from the original: $rstuv\cdots$. With this definition, one may now make use of a theorem due to Wick, which states that it is possible to rewrite the time-ordered products in the form

$$\{rstuv\cdots\}_T = \{rstuv\cdots\}_N$$
$$+ \{tuv\cdots\}_N\langle rs\rangle$$
$$+ \{rsv\cdots\}_N\langle tu\rangle \pm \text{other pairs}$$
$$+ \{v\cdots\}_N\langle rs\rangle\langle tu\rangle \pm \text{permutations}$$
$$+ \cdots$$
$$+ \langle rs\rangle\langle tu\rangle\langle vw\rangle\cdots \pm \text{permutations}, \quad (31)$$

where $\langle rs\rangle$ stands for the pair contraction

$$\langle rs\rangle = \{rs\}_T - \{rs\}_N \quad (32)$$

given in Eqs. 25, 27, 28, and 30, with $\pm$ for even/odd permutations. The proof is inductive, and is given in the Appendix. (We have assumed an even number of operators in Eq. 31.)

The power of the result given in Eq. 31 is again clear if we want ground-state expectation values of time-ordered products, for which the normal-ordered products will not contribute. The result is that only the last line of Eq. 31 survives,

$$\langle\Phi_0|\{rstuv\cdots\}_T|\Phi_0\rangle = \langle rs\rangle\langle tu\rangle\langle vw\rangle\cdots \pm \text{permutations}, \qquad (33)$$

giving an expression entirely in terms of pair contractions, which in turn can be written in terms of the unperturbed Green functions, as in Eqs. 27 and 28. We shall require just such expressions as Eq. 33 in the evaluation of the perturbation series of $U(t, t_0)$ (Eq. 8) for the ground state, with which we began this section. It is important for this result that we work in the interaction picture, in which Wick's contractions are c-numbers. We shall see how this works out in the next section.

8.4 FEYNMAN DIAGRAMS

The Wick method of handling time-ordering, which we have just studied, leads quite directly to the Feynman diagram representation of time evolution, and therefore also of perturbation theory. Let us see how this works by considering the expectation value of the time-development operator in the unperturbed ground state Φ_0, using the expansion given in Eq. 8,

$$\langle\Phi_0|U(t, t_0)|\Phi_0\rangle = \sum_{n=0}^{\infty} \frac{(-i)^n}{n!} \int_{t_0}^{t} dt_1 \cdots \int_{t_0}^{t} dt_n \langle\Phi_0|\{V(t_1)\cdots V(t_n)\}_T|\Phi_0\rangle . \qquad (34)$$

For the present we shall study only the integrand of the nth term in Eq. 34, which is the expectation value of Eq. 22 in the state Φ_0.

To make use of the Wick analysis, we first write each of the $V(t_i)$ in terms of creators and annihilators at the same time t_i; i.e., for a two-body interaction,

$$V(t_i) = \frac{1}{2} \sum_{\alpha\beta\gamma\delta} \langle\alpha\beta|v|\gamma\delta\rangle a_\alpha^\dagger(t_i)a_\beta^\dagger(t_i)a_\delta(t_i)a_\gamma(t_i) . \qquad (35)$$

This operator is normal-ordered with respect to the physical vacuum, but not with respect to the Fermi ground state Φ_0. When we rewrite Eq. 35 making the particle–hole transformation of Eq. 7.16, some creators $b_\lambda^\dagger$ will appear to the right of annihilators a_δ or b_μ, and so on. Each term in nth order in Eq. 34 contains the expectation value of a time-ordered product of $4n$ operators $(a^\dagger b^\dagger ab)$ at n different times, to be integrated over the times as indicated. For each product, we use the Wick result of Eq. 33 to reduce the expression to a sum of products of Green functions of the various times, and two-body matrix elements. Let us see how this works in some simple examples.

For $n = 1$, we may evaluate the expectation value directly, since there is no time ordering:

$$\langle \Phi_0 | V(t_1) | \Phi_0 \rangle = \frac{1}{2} \sum_{\lambda, \mu \leq F} \{ \langle \lambda\mu | v | \lambda\mu \rangle - \langle \lambda\mu | v | \mu\lambda \rangle \} . \tag{36}$$

The two terms on the right-hand side of Eq. 36 (direct and exchange) correspond to two ways of making contractions in the Wick method. The only contribution of Eq. 35 to Eq. 36 comes from occupied states, $\lambda \leq F$, for which

$$V(t_1) = \frac{1}{2} \sum_{\lambda\mu\lambda'\mu' \leq F} \langle \lambda\mu | v | \lambda'\mu' \rangle b_\lambda(t_1) b_\mu(t_1) b_{\mu'}^\dagger(t_1) b_{\lambda'}^\dagger(t_1) , \tag{37}$$

making the hole transformation of Eq. 7.16. The Wick normal ordering of Eq. 37 will give contractions (or anticommutators), all at equal times ($= t_1$),

$$+ \{ b_\lambda, b_{\lambda'}^\dagger \} \{ b_\mu, b_{\mu'}^\dagger \} = + \delta_{\lambda\lambda'} \delta_{\mu\mu'} , \tag{38a}$$

$$- \{ b_\lambda, b_{\mu'}^\dagger \} \{ b_\mu, b_{\lambda'}^\dagger \} = - \delta_{\lambda\mu'} \delta_{\mu\lambda'} , \tag{38b}$$

which are independent of time. The signs correspond to the permutations required to reach the orders of Eq. 38a, b.

In Section 8.2, we represented the time integral of Eq. 36, given in Eq. 18, by a pair of diagrams (Figure 8.2b) standing for the contribution of the direct and exchange parts of Eq. 36. These diagrams are now shown in Figure 8.4a and b, where again the dashed horizontal line represents the matrix element of $V(t_1)$ acting at t_1, and the occupied orbits λ, μ are represented by the solid lines beginning and ending at the same time, and

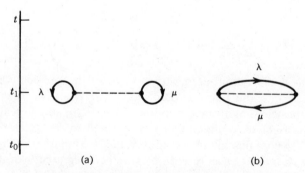

(a)   (b)

FIGURE 8.4   Diagrammatic representation of Eq. 36, separated into (a) direct and (b) exchange contributions, corresponding to the first and second terms of that equation.

are to be summed $\lambda, \mu \leq F$. We may think of Figure 8.4a as representing the equal-time contraction of Eq. 38a, and Figure 8.4b, that of Eq. 38b.

For $n = 2$ we must consider the time-ordered product

$$\langle \Phi_0 | \{V(t_1)V(t_2)\}_T | \Phi_0 \rangle$$

$$= \frac{1}{4} \sum_{\alpha\beta\gamma\delta} \sum_{\alpha'\beta'\gamma'\delta'} \langle \alpha\beta | v | \gamma\delta \rangle \langle \gamma'\delta' | v | \alpha'\beta' \rangle$$

$$\times \langle \Phi_0 | \{ a_\alpha^\dagger(t_1) a_\beta^\dagger(t_1) a_\delta(t_1) a_\gamma(t_1) a_{\delta'}^\dagger(t_2) a_{\gamma'}^\dagger(t_2) a_{\alpha'}(t_2) a_{\beta'}(t_2) \}_T | \Phi_0 \rangle .$$

$$(39)$$

First make the transformation to holes, for $\mu \leq F$: $a_\mu(t) \rightarrow (-1)^\phi b_\mu^\dagger(t)$. Then using the Wick theorem, Eq. 39 can be reduced to a sum of contractions on all possible pairs of the eight operators, as follows.

(i) For pairs of operators at the same time (t_1 or t_2), contractions produce time-independent numbers, as in Eq. 38. Combining all such equal-time contractions in Eq. 39 yields the time-independent expression

$$D(\text{i}) = \frac{1}{4} \left[\sum_{\lambda\mu \leq F} \{ \langle \lambda\mu | v | \lambda\mu \rangle - \langle \lambda\mu | v | \mu\lambda \rangle \} \right]^2 \qquad (40)$$

which is clearly just the square of Eq. 36. The four terms in Eq. 40 may be represented by the four diagrams shown in Figure 8.5, for $t_1 > t_2$.

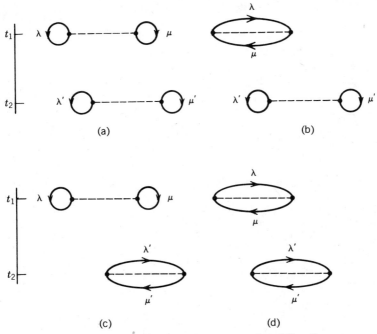

FIGURE 8.5 Diagrammatic representation of Eq. 40.

(ii) Contract one pair at t_1, one pair at t_2, and two pairs, each of which has one operator at t_1 and one at t_2. The resulting contribution is found to be (for $t_1 > t_2$)

$$D(\text{ii}) = 4 \times \frac{1}{4} \sum_{\substack{\rho > F \\ \lambda\mu\nu \le F}} \{\langle \rho\lambda|v|\nu\lambda\rangle - \langle\lambda\rho|v|\nu\lambda\rangle\}$$

$$\times \{\langle\nu\mu|v|\rho\mu\rangle - \langle\nu\mu|v|\mu\rho\rangle\}$$

$$\times i^2 G^0_\rho(t_1 - t_2)G^0_\nu(t_2 - t_1). \qquad (41)$$

The four terms in Eq. 41 may be represented by four time-evolution diagrams as shown in Figure 8.6. The lines λ and μ stand for the contractions at t_1 and t_2 separately. The lines ρ and ν connect t_1 and t_2 (in opposite directions), representing contractions of one operator at each time, and contributing the Green functions in the expression Eq. 41. There are 16 contractions of this type, giving each diagram four times. For $t_1 < t_2$, the particle–hole lines $\rho \leftrightarrow \nu$ interchange, repeating the form of the diagrams.

(iii) Contract four pairs, each with one operator at t_1 and one at t_2 (take $t_1 > t_2$); this yields an expression with four Green functions

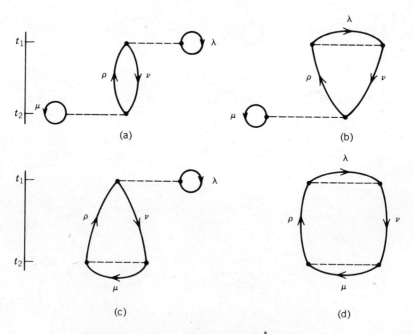

FIGURE 8.6 Diagrammatic representation of Eq. 41.

$$D(\text{iii}) = 2 \times \frac{1}{4} \sum_{\substack{\lambda\mu \leq F \\ \rho, \sigma > F}} \{\langle \lambda\mu|v|\rho\sigma\rangle\langle\rho\sigma|v|\lambda\mu\rangle - \langle\lambda\mu|v|\rho\sigma\rangle\langle\rho\sigma|v|\mu\lambda\}$$

$$\times (i)^4 G_\lambda^0(t_2 - t_1)G_\mu^0(t_2 - t_1)G_\rho^0(t_1 - t_2)G_\sigma^0(t_1 - t_2), \tag{42}$$

the two parts of which may be represented by the diagrams of Figure 8.7. The lines λ, μ represent hole-operator contractions, and the lines ρ, σ, particle-operator contractions. There are four contractions of this type, giving two distinct diagrams, each twice. For $t_1 < t_2$, exchange λ, $\mu \leftrightarrow \rho$, σ, repeating the diagrams.

The diagrams we have introduced are examples of *Feynman diagrams* for the expectation values

$$\langle \Phi_0|\{V(t_1)\cdots V(t_n)\}_T|\Phi_0\rangle \tag{43}$$

(so far, with $n = 1, 2$), each diagram representing a distinct way of carrying out the contraction of all pairs of creation and annihilation operators appearing in the expansion of Eq. 43 into products, using Eq. 35. Each interaction $V(t_i)$ is represented by a horizontal dashed line at time t_i; each end of the interaction line is connected to another interaction line (or itself) by a solid (particle or hole) line, representing a pair contraction in the Wick reduction. Particle lines run forward in time, holes backward, and equal-time contractions give interactions with unexcited occupied states. Since each diagram stands for a distinct way of contracting pairs in the Wick reduction of Eq. 43, one may directly associate with each diagram a unique expression in terms of the two-body matrix elements of V and the unpertur-bed Green functions $G_\alpha^0(t_i - t_j)$, for contractions at $t_i \neq t_j$. The time integra-tions are still to be done, as specified in Eq. 34.

The Feynman diagrams of Figure 8.4 look exactly like the Goldstone diagrams of Figure 8.2. Similarly, the Feynman diagrams of Figures 8.5, 8.6, and 8.7 correspond to the Goldstone diagrams of Figure 8.3a, b, c, although

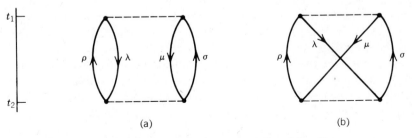

(a)   (b)

FIGURE 8.7 Diagrammatic representation of Eq. 42.

in this case the latter are given in the antisymmetrized interaction (dot vertex) form. The expressions which correspond to the two kinds of diagrams clearly have the same two-body matrix elements, as we can see by comparing Eq. 36 with Eq. 18, Eq. 40 with Eq. 21a, Eq. 41 with Eq. 21b, and Eq. 42 with Eq. 21c. Only the time dependences look different (and only for the last two cases). Also, the forms of the time integrations are different for the Goldstone and Feynman expressions, which can be traced back to the different forms on which they are based: Eq. 5 and 8. Yet these two expansion methods are calculating the same function, $\langle \Phi_0 | U(t, t_0) | \Phi_0 \rangle$, to the same order (second), and must give the same answer. We shall see how this comes about in the following section.

8.5   COMPARISON OF FEYNMAN AND GOLDSTONE DIAGRAMS

Both Goldstone and Feynman diagrams resulted from expansion of $U(t, t_0)$ in orders of the interaction, using either Eq. 5 or Eq. 8 respectively, and reducing the expression for each order to the sum of those corresponding to a finite set of diagrams. For the Goldstone expansion, the diagrams were generated by inserting intermediate unperturbed states of Eq. 11a between every two interactions in the terms of Eq. 5 as shown in Eq. 12. The Feynman diagrams were generated by the process of contraction, using the Wick theorem on the terms of Eq. 8. Both methods must (and do) lead to the same expression for each order n of the perturbation expansion. In some cases, a particular diagram may have the same expression whether it is meant as a Goldstone or as a Feynman diagram; in other cases, certain sums of the former are equivalent to the latter. Let us see how this works out in some examples.

We have already remarked on the similarity of the diagrams of Figures 8.4–8.7 to those of Figures 8.2 and 8.3; let us look more closely at the relation between the expressions represented by those diagrams, now including the time integrations implied by Eqs. 5 or 8. We found that the $n = 1$ term of the Dyson expansion (Eq. 34), with the matrix element given in Eq. 36, gives the same result as Eq. 18, which is the $n = 1$ contribution to the Goldstone expansion of the same matrix element of $U(t, t_0)$. In a similar way, the second-order term denoted by $D(\text{i})$ in Eq. 40 leads to the same expression as the term I_a in Eqs. 20, 21a. Let us pass on to the comparison of Eq. 42 with Eq. 21c, and their associated diagrams. (A similar analysis will serve for Eqs. 41 and 21b; we leave this as an exercise.)

First, consider the Feynman diagram of Figure 8.7a; this represents the first term in the contribution $D(\text{iii})$ of Eq. 42. With the time integrations and factors of Eq. 34 restored, this diagram contributes

$$ -\left(\frac{1}{2}\right)^2 \sum_{\alpha\beta\gamma\delta} |\langle \alpha\beta | v | \gamma\delta \rangle|^2 I(t, t_0) , \qquad (44) $$

where

$$I(t, t_0) = \int_{t_0}^{t} dt_1 \int_{t_0}^{t} dt_2 \, G_\alpha^0(t_2 - t_1) G_\beta^0(t_2 - t_1) G_\gamma^0(t_1 - t_2) G_\delta^0(t_1 - t_2) .$$

(45)

The Feynman diagram of Figure 8.7b represents the second term in Eq. 42, and contributes

$$+ \left(\frac{1}{2}\right)^2 \sum_{\alpha\beta\gamma\delta} \langle \alpha\beta|v|\gamma\delta \rangle \langle \gamma\delta|v|\beta\alpha \rangle I(t, t_0) .$$

(46)

The labels α, β, γ, δ are automatically restricted to be particle ($>F$) or hole ($\leq F$) states by the time order $t_2 - t_1$, through the form of the Green functions, as can be seen from Eq. 7.29. Using this last equation, the time integral of Eq. 45 becomes

$$\int_{t_0}^{t} dt_1 \int_{t_0}^{t} dt_2 \, [\exp\{-i(\epsilon_\gamma + \epsilon_\delta - \epsilon_\alpha - \epsilon_\beta)(t_1 - t_2)\}$$

$$\times \theta(t_1 - t_2)\theta(\epsilon_\gamma - \epsilon_F)\theta(\epsilon_\delta - \epsilon_F)\theta(\epsilon_F - \epsilon_\alpha)\theta(\epsilon_F - \epsilon_\beta)$$

$$+ \exp\{-i(\epsilon_\alpha + \epsilon_\beta - \epsilon_\gamma - \epsilon_\delta)(t_2 - t_1)\}$$

$$\times \theta(t_2 - t_1)\theta(\epsilon_\alpha - \epsilon_F)\theta(\epsilon_\beta - \epsilon_F)\theta(\epsilon_F - \epsilon_\gamma)\theta(\epsilon_F - \epsilon_\delta)] .$$

(47)

For comparison, consider the Goldstone diagram of Figure 8.3c, which represents the I_c term of Eqs. 20, 21c, which we now write in the form

$$-\frac{1}{2} \sum_{\substack{\alpha,\beta \leq F \\ \gamma,\delta > F}} \{|\langle \alpha\beta|v|\gamma\delta \rangle|^2 - \langle \alpha\beta|v|\gamma\delta \rangle \langle \gamma\delta|v|\beta\alpha \rangle\} I'(t, t_0) ,$$

(48)

where

$$I'(t, t_0) = \int_{t_0}^{t} dt_1 \int_{t_0}^{t} dt_2 \, \exp\{-i(\epsilon_\gamma + \epsilon_\delta - \epsilon_\alpha - \epsilon_\beta)(t_1 - t_2)\} .$$

(49)

(Note that reexpressing the antisymmetrized matrix elements of Eq. 21c in the ordinary product matrix elements of Eq. 48 brings in a factor of 2.)

Now the expression of Eq. 48 equals the sum of Eqs. 44 and 46, as can be easily seen by comparing Eqs. 47 and 49. So the diagrams of Figure 8.7 have identical expressions whether calculated by the Goldstone or by the Feynman rules.

This is not always the case: sometimes one Feynman diagram is equivalent to the sum of a number of Goldstone diagrams. An example in third order is shown in Figure 8.8, where the Feynman diagram in (a) includes all three Goldstone diagrams in (b). The expression for the Feynman diagram is unchanged by changing the time order of the interactions (in this case the interaction with occupied orbit μ) without changing the topology of the

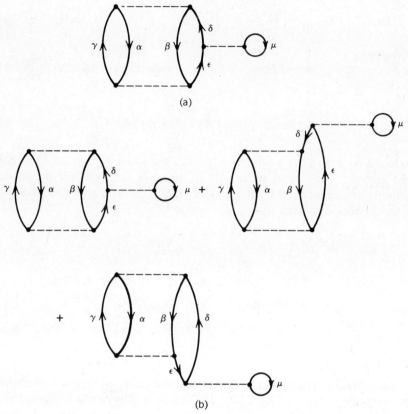

FIGURE 8.8 The Feynman diagram (a) is equivalent to the sum of three Goldstone diagrams in (b), with three different time orderings.

connecting lines. Of course, lines may change from particle to hole and back with the change of order, as can be seen in the figure. The Green functions obtained from the Wick contractions keep track of these changes, as in Eq. 47. On the other hand, the Goldstone expressions for each of the three diagrams of Figure 8.8b differ, since they involve different intermediate states. Only the sum over all time orders will produce the Feynman result.

In some nomenclatures, the diagrams we have introduced are referred to jointly as Feynman–Goldstone diagrams. However, even though the graphical forms may be similar, we shall continue to distinguish the two types, depending on which expansion method we are using.

Why do we introduce the two kinds of diagrams, if they are completely equivalent? The answer is that each type is more convenient for certain purposes. Many general properties of the perturbation expansion, such as the linked-cluster theorem (see Section 9.1) are easier to demonstrate with Feynman diagrams. As we shall also see later, it is often possible to calculate parts of a Feynman diagram separately from the whole diagram, as in the Dyson equation for the self-energy of a propagating particle, discussed in Section 11.2, or the polarization part for the electron gas, in Section 10.5. This separation often leads to a natural description of some approximation scheme, as we shall see with the quasiparticle approximation, in Section 11.2, and in the ring approximation in Section 10.6.

Goldstone diagrams, on the other hand, and more closely related to the perturbation expansion of the energy, which is why we have developed that expansion first. The Feynman expansion, although formally equivalent to that of Goldstone, is generally less directly calculated, as we shall see explicitly in Section 9.3. Also, some approximations are simpler to develop in the Goldstone scheme, for example, any scheme which keeps track of the number of particle or hole excitations at any time. The ladder approximation of Chapter 10 (Sections 10.2–10.4) is of this type.

The major difference between the two expansion methods is that the Goldstone expansion is based, like the Rayleigh–Schrödinger method to which it is closely related, on the labeling of many-body intermediate states between all interactions. The Feynman expansion does not refer to many-body states, but to excitations of particle and hole orbits from the unperturbed (vacuum) Fermi ground state. These appear in the one-body Green functions which are the result of Wick contractions. The Goldstone expansion is sometimes called "global", and the Feynman expansion "local", referring to this comparison.

8.6 ENERGY VARIABLES; ENERGY CONSERVATION

Although the Feynman and Goldstone expansions are closely related, some properties are treated differently, following from the two distinct treatments of time variables in Eqs. 5 and 8. For a given Feynman diagram the time dependence of the term it represents is given by a combination (sum of products) of unperturbed single-particle Green functions $G_\alpha^0(\tau)$, where $\tau = (t_i - t_j)$ is a relative time referring to two interactions $V(t_i)$, $V(t_j)$ which have been connected by a Wick contraction, as we have seen in the examples of Section 8.4. We have also noted that the integration over each time t_i may be done independently, e.g. in Eq. 34. It is often useful to perform the time integrations explicitly, by employing the Fourier transforms of the Green functions,

$$G_\alpha^0(\omega) = \int_{-\infty}^{\infty} d\tau \, e^{i\omega\tau} G_\alpha^0(\tau) , \qquad (50)$$

where we have inverted the defining equation 7.30. The frequency or energy variable ω may be thought of as the energy required to excite the particle (or hole) orbit α in the intermediate state between t_j and t_i. The transformed functions of Eq. 50 are simply expressed in terms of the excitation energy, as in Eq. 7.31, which can also be written

$$G_\alpha^0(\omega) = [\omega - (\epsilon_\alpha - \epsilon_F) \pm i\eta]^{-1} \qquad (51)$$

with $\pm i\eta$ for particles or holes, with $\epsilon_\alpha \gtrless \epsilon_F$.

The energy ω is a conserved quantity, in the following sense. Every Feynman diagram may be thought of as an integral of the form

$$D = \int_{-\infty}^{\infty} d\tau \sum_\alpha G_\alpha^0(\tau) X_\alpha(\tau) , \qquad (52)$$

where we have singled out one single-particle (-hole) line, representing $G_\alpha^0(\tau)$, and $X_\alpha(\tau)$ is the rest of the diagram (which may or may not depend on the orbit label α) as illustrated in Figure 8.9. Then, using the Fourier transform $X(\omega)$ of $X(\tau)$, Eq. 52 may be transformed to an integral over the single-particle energy ω:

$$D = \int_{-\infty}^{\infty} \frac{d\omega}{2\pi} \sum_\alpha G_\alpha^0(\omega) X_\alpha(-\omega) . \qquad (53)$$

If the particle in orbit α is understood to be carrying energy ω (relative to the unperturbed ground state), then the rest of the system, given by $X(-\omega)$, carries excitation energy $-\omega$, so that the overall energy of the system is unchanged.

The argument may be extended to diagrams which can be represented by two pieces X and Y which interact twice, as in Figure 8.10. The diagram may be written as a τ-integral

$$I = \int_{-\infty}^{\infty} d\tau \, X(\tau) Y(\tau) , \qquad (54)$$

or, using Fourier transforms, as

$$I = \int_{-\infty}^{\infty} \frac{d\omega}{2\pi} \, X(\omega) Y(-\omega) . \qquad (55)$$

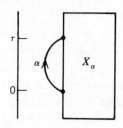

FIGURE 8.9   Feynman diagram D of Eq. 52, with one particle line α singled out; X_α represents the rest of the diagram. For a hole line, $\tau < 0$.

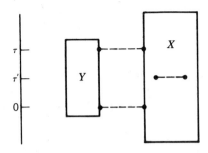

FIGURE 8.10 Feynman diagram which can be considered as two subsystems which interact twice: the internal energy of each subsystem is unchanged by interactions entirely within one subsystem, such as that illustrated within X at τ'.

Now the conservation of ω has the further consequence that ω for the X-system is unchanged by interactions within the X-system (e.g. at τ' in the figure) or by interactions within the Y-system. Only further interactions between the X- and Y-systems could change the energy balance between them. This property of the Feynman diagram method is a very attractive feature, which allows separate treatment of parts of a complicated system, independent of the state of the rest of the system. We shall use this in the discussion of ring diagrams in Chapter 10, and of the optical potential in Chapter 11.

This separation of noninteracting subsystems with energy conservation does not appear naturally in the Goldstone method. The problem is that for a single Goldstone diagram one keeps track of the excitation energy of the entire system in the various intermediate states, and not the separate energies of the subsystems. Of course, one can recover the description of separate local energies by combining Goldstone diagrams. But then one is back at the Feynman expansion. We shall actually do something like that in Section 9.1, in discussing the linked-cluster theorem.

Lastly, we remark that for a uniform, infinite many-body system, the single particle orbits conserve linear momentum $\mathbf{k}$, so the Fourier-transformed Green functions may be written in the form (see Eq. 7.33b)

$$G_\alpha^0(\omega) = G_\mu^0(\mathbf{k}, \omega) \tag{56}$$

with μ a spin projection, or any other intrinsic orbit label. Although the form of Eq. 56 suggests the utility of a four-vector notation, with $k_0 = \omega$, covariance of the theory is not implied. The material system does provide a preferred rest frame.

EXERCISES

8.1 Consider a two-fermion system with only four possible orbits: λ_1, λ_2, which are occupied, and ρ_1, ρ_2, which are unoccupied, in the unperturbed ground state. The only nonzero matrix elements of the interaction V are those which excite both particles, i.e.

$$\langle \lambda_1 \lambda_2 | v | \rho_1 \rho_2 \rangle \neq 0 \, ,$$

$$\langle \lambda_1 \lambda_2 | v | \rho_2 \rho_1 \rangle \neq 0 \, .$$

(a) Calculate the second-order matrix element $\langle 0 | U^{(2)}(t, t_0) | 0 \rangle$ of Eq. 20 explicitly for this system.

(b) Calculate the matrix element $\langle 0 | U(t, t_0) | 0 \rangle$ of the exact time-development operator from Eq. 7.38. (Note that this problem can be reduced to a two-dimensional matrix; see Exercise 7.3b.)

(c) Compare the result of part (b), to second order in V, with that of part (a).

8.2 Show explicitly that Eq. 39 can be reduced to the sum $D(\mathrm{i}) + D(\mathrm{ii}) + D(\mathrm{iii})$ given in Eqs. 40–42, using the Wick theorem of Eq. 33.

8.3 Show that the Goldstone diagram of Figure 8.3b, calculated using Eqs. 20 and 21b, gives the identical result to the Feynman diagrams of Figure 8.6, calculated from Eq. 41.

CHAPTER **9**

Perturbation Expansion
of the Ground State

We now have almost all the tools we need to calculate the ground-state wave function and energy of a normal many-body system, using diagrammatic perturbation theory. Let us review briefly the method so far. We use the time-evolution operator, with the trick of adiabatic switching, to generate the fully interacting ground state from the unperturbed ground state, as we discussed in Section 7.4. We have developed, in Chapter 8, diagrammatic expansions of the matrix elements of the time-evolution operator, which can now be used to evaluate the wave function and energy. In fact we have both the Goldstone and Feynman expansions available. We shall start with the former, which leads to the most convenient method for calculating the ground-state properties. Then we shall examine an alternative form using Feynman diagrams. To complete the method of adiabatic switching, we must evaluate our results in the limit $(\alpha \to 0^+)$ of infinitely slow "turning on" of the interaction. But before we go into the full development of the perturbation results, we shall need one more tool, the *linked-cluster theorem*, which we discuss next. This new result will allow us to ensure that the limiting processes involved in the calculation of ground-state properties are all well defined.

9.1 LINKED-CLUSTER THEOREM

If we start enumerating diagrams representing time evolution (either Goldstone or Feynman), we notice that some diagrams are linked, or connected,

203

in the sense that all the particle and hole lines communicate with each other through interactions and possibly through other particle or hole lines. These are called *linked* (or fully linked) diagrams. For example, the diagrams of Figure 8.8 are linked. In contrast, other diagrams may consist of *unlinked* or disconnected parts, in which not all particle and hole lines communicate. An example of this is shown in Figure 9.1, where three Goldstone diagrams are illustrated, representing different contributions to the matrix element of $U(t, t_0)$ to third order:

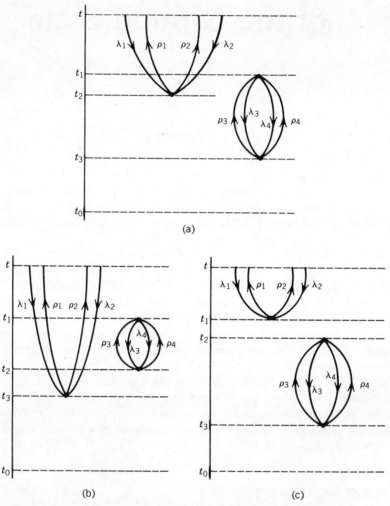

(a)

(b)   (c)

FIGURE 9.1   Three Goldstone diagrams representing different contributions to the matrix element of Eq. 1, coming from different time orderings of the interactions. The combination of all three is equivalent to one Feynman diagram. All the diagrams are unlinked.

$$\langle l|U^{(3)}(t, t_0)|0\rangle = \langle\{\rho_1\rho_2\}\{\lambda_2\lambda_1\}|U^{(3)}(t, t_0)|0\rangle \ . \qquad (1)$$

As in the first-order diagrams of Figure 8.1, the operator connects the vacuum state of Eq. 8.14a to the two-particle, two-hole state of Eq. 8.14b. In the Goldstone expansion of Eq. 8.12, the three diagrams correspond to different possible intermediate states between t_1 and t_2, and between t_2 and t_3. All three Goldstone diagrams have two separate linked parts, not connected to each other: one corresponding to the first-order diagram of Figure 8.1, and one corresponding to the second-order diagram of Figure 8.3c (with slightly different labeling). However, the time ordering of the interaction vertices differs among the three diagrams.

There is also a Feynman diagram related to the three diagrams of Figure 9.1, which represents the sum of all three, corresponding to the inclusion of all time orderings (as we have discussed in Sections 8.4 and 8.5). The present Feynman diagram, which could be drawn like any one of the three of Figure 9.1, is therefore also not fully linked; it consists of two separate linked parts. [The present example is a generalization of the sort we studied in Section 8.4, in that unlike Eq. 8.34, we are considering matrix element of $U(t, t_0)$ between states other than the vacuum, as in Eq. 1.]

The special property we wish to note about the diagrams of Figure 9.1 (considered either as a set of Goldstone diagrams or as one Feynman diagram) is that the relevant matrix element of Eq. 1 factors in the form

$$\langle l|U^{(3)}(t, t_0)|0\rangle = \langle l|U^{(1)}(t, t_0)|0\rangle\langle 0|U^{(2)}(t, t_0)|0\rangle \ , \qquad (2)$$

where the first factor is the first-order matrix element given in Eq. 8.17, with the diagram shown in Figure 8.1. The second factor is the second-order matrix element of Eqs. 8.20, 8.21c given by the vacuum–vacuum diagram of Figure 8.3c (with relabeling: $1, 2 \rightarrow 3, 4$). We notice that the diagrams corresponding to the factors of Eq. 2 are the separate linked parts of Figure 9.1. The factoring property of Eq. 2 can be shown directly by summing the Goldstone-diagram expressions for Eq. 1, or by examining the form of the Feynman-diagram expression. The first is lengthy (and will be given as an exercise); the second is much more direct, and goes as follows.

Considered as a Feynman diagram, the existence of two separate linked parts in Figure 9.1 is connected to the order of contracting pairs of creation and annihilation operators in the Wick method. In particular, the second-order linked part on the right corresponds to exactly the same operation which leads to Eq. 8.42 (and Figures 8.3c or 8.7), and which contributes to $\langle 0|U^{(2)}(t, t_0)|0\rangle$, when integrated over the times of the two interactions (t_1 and t_3 in Figure 9.1a). Because this part of the diagram is not linked to the first-order part, this integration can be performed independently of anything pertaining to the first-order part. Thus the two unlinked contributions factor, leading to Eq. 2.

The factoring property just discussed is an example of a general property associated with all unlinked diagrams, and may be stated in the form of a *linked-cluster* theorem. In the present context, the theorem says that any matrix element of $U(t, t_0)$ may be put in the factored form

$$\langle m|U(t, t_0)|n\rangle = \langle m|U(t, t_0)|n\rangle_l \langle 0|U(t, t_0)|0\rangle \qquad (3)$$

where the first factor represents a sum of linked diagrams (denoted by subscript l), while the second factor is a sum of vacuum–vacuum diagrams. The class of diagrams is illustrated in Figure 9.2a, where the left-hand part of the figure stands for a linked piece of the diagram, beginning in state $|n\rangle$ at t_0 and ending in state $|m\rangle$ at t, with any number of interactions within the rectangular box, all connected by particle or hole lines to at least one of

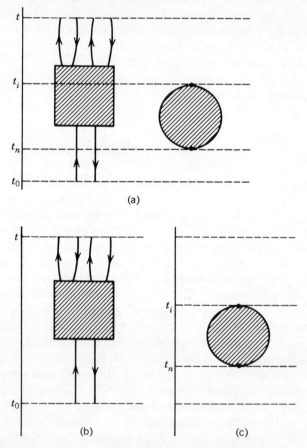

FIGURE 9.2 (a) Class of diagrams which lead to a factored matrix element of the form of Eq. 3. (b) Linked part of diagram (a). (c) Vacuum–vacuum part of diagram (a).

those lines shown going into or out of the box. The right-hand part of the figure stands for any vacuum–vacuum diagram which starts at t_n and ends at t_i, with any arrangement of lines and vertices in between, but *not* connected in any way to the left-hand piece. The right-hand part could itself have disconnected parts. The left-hand part could also have parts not connected to each other, but each part must be connected to some ingoing or outgoing particle line. In this sense, it is considered to be the linked part of the entire diagram; if it stands alone, as in Figure 9.2b, it is called a linked diagram. As for the example of Eq. 2 and Figure 9.1, we may understand the diagram of Figure 9.2a in either the Feynman or the Goldstone sense; in the latter case the figure stands for a set of diagrams to be summed, in which all time orderings of the interactions in the vacuum part relative to those in the linked part are included.

The linked-cluster theorem of Eq. 3 follows from the following considerations. The expansion of the matrix element $\langle m|U(t, t_0)|n \rangle$ in Feynman diagrams contains both linked and unlinked diagrams. The unlinked diagrams can be shown to lead to a product of contributions for the linked and unlinked parts, as in the example of Eq. 2. Each factor is calculated by the Wick contraction method, and is independent of the other. The complete expansion will have each possible linked diagram combined with each possible vacuum–vacuum diagram. Since the contributions factor, the whole series can be reordered as a series of linked diagrams, each of which is *multiplied* by the entire series of vacuum–vacuum diagrams. A more detailed proof is given in Section B of the Appendix. It is clear from this that the factoring property of Eq. 3 may not hold for an individual, time-ordered unlinked Goldstone diagram (e.g. Figure 9.1a), but will hold for the sum over all time orders (Figure 9.1a, b, c), which is *equivalent* to a simple unlinked Feynman diagram.

The theorem of Eq. 3 may be used in several different ways. First, a given unlinked Feynman diagram may be factored, as in Eq. 2. Second, one may consider a set of diagrams having a given specific linked part, with the vacuum–vacuum part summed over all possible diagrams. Then Eq. 3 says that the sum of matrix elements factors into the matrix element for the given linked part, multiplied by the exact (i.e. summed over all diagrams) vacuum expectation value of $U(t, t_0)$. Last, if one considers the complete expansion of $\langle m|U(t, t_0)|n \rangle$ in diagrams, it follows that the expression for that matrix element also factors into an expression including the linked parts and an expression for the vacuum part: $\langle 0|U(t, t_0)|0 \rangle$.

One consequence of Eq. 3 is based on the fact that the vacuum expectation value of $U(t, t_0)$ becomes the normalizing factor $\langle \Phi_0|\Psi(0) \rangle_\alpha$ for the adiabatic ground state (Eqs. 7.45, 7.46), in the limits $t \to 0$, $t_0 \to -\infty$. We shall use this property in the following to remove the normalizing (second) factor from expressions of the form of Eq. 3, so that ground-state properties will be given in terms of linked diagrams only. This is the main significance of the linked-cluster theorem.

A second consequence of the factoring of disconnected parts of diagrams is that the vacuum expectation of U can be reexpressed in terms of linked vacuum–vacuum diagrams, in the exponential form

$$\langle 0|U(t, t_0)|0\rangle = \exp\langle 0|U(t, t_0)|0\rangle_l \qquad (4)$$

(see Appendix). This compact result can be used to demonstrate the divergence of the phase of $\langle\Phi_0|\Psi(0)\rangle_\alpha$, which was discussed at the end of Section 7.4.

PAULI EXCLUSION PRINCIPLE

One peculiarity of the use of the linked-cluster theorem for fermion systems should be pointed out, namely, that diagrams which *seem* to violate the exclusion principle will often appear, and must be included in the perturbation expansion. The point is illustrated by returning to Figure 9.1, considering the three parts either as Goldstone diagrams or as one Feynman diagram drawn with different time orderings. Then, if we sum over the particle and hole labels, we encounter the following puzzle: in Figure 9.1c, it is possible for any of the particles or holes in the linked part to carry the same labels as particles or holes in the vacuum–vacuum part, e.g. $\rho_1 = \rho_3$, since those states are not occupied at the same time. However, in Figure 9.1a or b, we should not allow e.g. $\rho_1 = \rho_3$, since we have assumed antisymmetrized states for the system in our construction of the Goldstone expansion. This constraint would not allow us to sum all possible time orders in this case, to obtain Eq. 2. In the Feynman method, we know that the exclusion principle is also built into the Wick algebra, through the anticommutation relations on which it is based. However, if we apply it to diagrams, we find that certain time orders are excluded, as in Figure 9.1a, b, but not in Figure 9.1c. This would cause difficulty not only in applications of the linked-cluster theorem, but in time integrations of Feynman diagrams in general.

The resolution of these difficulties follows from recognizing that although individual diagrams appear to violate the Pauli exclusion priniple, no error in calculation follows from this if all diagrams of a given order n are included together. What happens is that any given "Pauli-violating" diagram is exactly canceled by another "Pauli-violating" diagram (or diagrams) of the same order. For the present example, the contribution of the $\rho_1 = \rho_3$ case of Figure 9.1a is exactly canceled by that of the diagram shown in Figure 9.3. Notice that the two diagrams have all the same particle and hole labels; they differ only in how the two particle lines labeled by ρ_1 are connected to the interactions at t_1 and t_2. In fact, Figure 9.3 is a fully linked diagram, while Figure 9.1a is unlinked, as we have noticed previously. The fact that the two diagrams differ only in their overall sign, and therefore give no contribution when added, follows from the two ways of connecting the ρ_1-lines, which in turn reflects the antisymmetry of the system of particles.

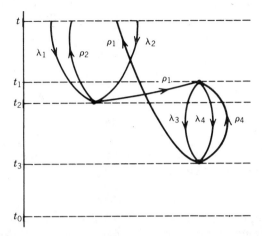

FIGURE 9.3 **Connected Pauli-violating diagram which cancels the diagram of Figure 9.1a, for**
$\rho_1 = \rho_3$.

The general result, which follows from the fact that antisymmetry has been built into the perturbation expansion, is that the sum of Pauli-violating diagrams in each order n must give no contribution. (This is proved in the Appendix.) However, as we noted in the example of Figures 9.1a and 9.3, the cancellation may involve both linked and unlinked diagrams. We see that we had to include some Pauli-violating unlinked diagrams to obtain the factored forms of Eqs. 2 and 3. Therefore for consistency we must also include linked Pauli-violating diagrams, like Figure 9.3, in the perturbation expansion. Later, when we make use of the linked-cluster expansion to obtain, e.g., the ground-state energy (next section), only linked diagrams will appear directly in the calculation. These therefore must include Pauli-violating diagrams in general, since the *unlinked* Pauli-violating diagrams have already implicitly been included.

9.2 LINKED-CLUSTER EXPANSION FOR THE GROUND STATE: GOLDSTONE EXPANSION

We are now ready to study the perturbation theory of the ground state of a normal Fermi system with the help of Goldstone diagrams and the linked-cluster theorem. We start with the expression given in Eq. 7.46 for the adiabatic ground-state vector, which we now express as a set of expansion amplitudes, using the unperturbed representation states $|m\rangle$ of Eq. 8.11a, with coefficients given by the ratio

$$\langle m|\Psi_0\rangle = \lim_{\alpha \to 0^+} \frac{\langle m|\Psi(0)\rangle_\alpha}{\langle 0|\Psi(0)\rangle_\alpha}. \tag{5a}$$

As we saw in Section 7.4, the numerator and denominator are given by the amplitudes

$$\langle m | U_\alpha (0, -\infty) | 0 \rangle \tag{5b}$$

and

$$\langle 0 | U_\alpha (0, -\infty) | 0 \rangle \tag{5c}$$

respectively, where U_α is the time-evolution operator of Eq. 7.38 with the interaction adiabatically switched, as in Eq. 7.44, and taken in the time limits $t = 0$, $t_0 \to -\infty$.

The numerator can be expanded in a series represented by a summation of Goldstone diagrams, starting from Eq. 8.5, following the discussion of Section 8.1. Summation over all time orders of unlinked diagrams leads to the factored form of Eq. 3 for Eq. 5b,

$$\langle m | U_\alpha (0, -\infty) | 0 \rangle = \langle m | U_\alpha (0, -\infty) | 0 \rangle_l \langle 0 | U_\alpha (0, -\infty) | 0 \rangle . \tag{6}$$

Clearly, the second factor in the numerator is identical to the denominator, Eq. 5c, so that Eq. 5a becomes

$$\langle m | \Psi_0 \rangle = \lim_{\alpha \to 0^+} \langle m | U_\alpha (0, -\infty) | 0 \rangle_l , \tag{7}$$

which gives an expansion of the ground-state vector in terms of linked diagrams only. Note that the unlinked parts of the diagrams have served to remove the normalizing denominator.

The ground-state energy shift is evaluated by a similar procedure, starting with Eq. 7.49 (and the normalization, Eq. 7.50):

$$\Delta E = E - \mathscr{E}_0 = \langle \Phi_0 | V | \Psi_0 \rangle = \lim_{\alpha \to 0^+} \frac{\langle 0 | V U_\alpha (0, -\infty) | 0 \rangle}{\langle 0 | U_\alpha (0, -\infty) | 0 \rangle} . \tag{8}$$

The numerator of this last expression may also be expanded as a series of Goldstone diagrams, in which the last interaction $V = V(0)$ is always at $t = 0$, after which the system is in the "vacuum" state $|0\rangle$ (which we recall is the unperturbed ground state). Again using the linked-cluster theorem (Eq. 3), the numerator may be factored, so that

$$\Delta E = \lim_{\alpha \to 0^+} \frac{\langle 0 | V U_\alpha (0, -\infty) | 0 \rangle_l \langle 0 | U_\alpha (0, -\infty) | 0 \rangle}{\langle 0 | U_\alpha (0, -\infty) | 0 \rangle}$$

$$= \lim_{\alpha \to 0^+} \langle 0 | V U_\alpha (0, -\infty) | 0 \rangle_l . \tag{9}$$

This last gives an expression for ΔE in terms of linked diagrams only. The time integrations for the linked diagram sums in Eqs. 7 and 9 can be performed explicitly, and the $\alpha \to 0^+$ limit obtained directly, as follows. First write the amplitude of Eq. 5b, using Eqs. 8.5 (with $t = 0$, $t_0 \to -\infty$) and 7.44, in the form

$$\langle m|U_\alpha(0, -\infty)|0\rangle = \langle m| \sum_{n=0}^{\infty} (-i)^n \int_{-\infty}^0 dt_1 \, e^{(\alpha + iH_0)t_1} V e^{-iH_0 t_1}$$

$$\times \cdots \times \int_{-\infty}^{t_{n-1}} dt_n \, e^{(\alpha + iH_0)t_n} V e^{-iH_0 t_n} |0\rangle .$$
(10)

Changing integration variables from t_i to $t_i' = t_i - t_{i-1}$ (i.e., $t_i = t_i' + t_{i-1}' + \cdots + t_1'$), one may regroup the factors so that

$$\langle m|U_\alpha(0, -\infty)|0\rangle = \langle m| \sum_{n=0}^{\infty} (-i)^n \int_{-\infty}^0 dt_1' \, e^{[n\alpha + i(H_0 - \mathscr{E}_0)]t_1'} V$$

$$\times \int_{-\infty}^0 dt_2' \, e^{[(n-1)\alpha + i(H_0 - \mathscr{E}_0)]t_2'} V \cdots \int_{-\infty}^0 dt_n' \, e^{[\alpha + i(H_0 - \mathscr{E}_0)]t_n'} V |0\rangle .$$
(11)

The integrals may now be evaluated separately to obtain the series

$$\langle m|U_\alpha(0, -\infty)|0\rangle = \sum_{n=0}^{\infty} \langle m|[\mathscr{E}_0 - H_0 + in\alpha]^{-1} V$$

$$\times [\mathscr{E}_0 - H_0 + i(n-1)\alpha]^{-1} V \cdots [\mathscr{E}_0 - H_0 + i\alpha]^{-1} V |0\rangle .$$
(12)

So far we have not made use of the diagram expansion. This can be accomplished by introducing intermediate states, as in Eq. 8.12, from which Goldstone diagrams are defined. (This will include modification to allow Pauli-violating states: see Appendix.) The time factors in Eq. 12 have been replaced by energy denominators of the form

$$\langle j|[\mathscr{E}_0 - H_0 + in_j\alpha]^{-1}|j\rangle = (\mathscr{E}_0 - \mathscr{E}_j + in_j\alpha)^{-1} .$$
(13)

Now, let us restrict the sum in Eq. 12 to those terms that correspond to linked-cluster diagrams only, as has been specified in Eq. 7. For linked-cluster diagrams, no intermediate state $|j\rangle$ can ever be the vacuum state $|0\rangle$, since $|j\rangle = |0\rangle$ means that there is an intermediate time with no lines in the diagram, as in Figure 9.1c. Therefore, the denominators in Eq. 12 are never singular: $\mathscr{E}_0 - \mathscr{E}_j < 0$. Thus, for linked diagrams, one may take the $\alpha \to 0^+$ limit directly in Eq. 12 to obtain a linked-cluster expression for the ground-state vector,

$$\langle m|\Psi_0\rangle = \langle m|U(0, -\infty)|0\rangle_l$$

$$= \sum_{n=0}^{\infty} \langle m|\{[\mathscr{E}_0 - H_0]^{-1}V\}^n|0\rangle_l. \qquad (14)$$

(This constitutes a proof of the existence of the limit defined in the Gell-Mann–Low theorem, expressed in Eq. 7.46. The divergent phase has been removed by the factoring of the unlinked diagrams.)

Similarly, the ground-state energy shift is also given by a linked-cluster expansion

$$\Delta E = \sum_{n=0}^{\infty} \langle 0|V\{[\mathscr{E}_0 - H_0]^{-1}V\}^n|0\rangle_l, \qquad (15)$$

which is obtained by taking the $\alpha \to 0^+$ limit of the linked-diagram part of

$$\langle 0|VU_\alpha(0, -\infty)|0\rangle = \sum_m \langle 0|V|m\rangle\langle m|U_\alpha(0, -\infty)|0\rangle. \qquad (16)$$

The linked-cluster expressions for the ground-state vector (Eq. 14) and the ground-state energy shift (Eq. 15) are the primary results of the Goldstone perturbation theory. We shall study explicitly the lowest orders in the Goldstone expansion in Section 9.4. Let us first summarize what has been accomplished so far in the development of Eqs. 14 and 15.

First, we have found compact expressions for ground-state quantities in terms of the unperturbed Hamiltonian H_0 and the perturbing interaction V. The expressions are given in the form of an expansion. Explicit expressions in each order of expansion may be developed in terms of the unperturbed energies $\mathscr{E}_m$ of Eqs. 8.10, 8.11b, and the two-body antisymmetrized matrix elements $\langle\alpha\beta|v|\gamma\delta\rangle_A$, as in Eqs. 8.15, 8.19. Although in principle these expressions may be obtained by first introducing complete sets of intermediate states $|m\rangle$ as in Eq. 8.11a, we have modified this treatment by introducing diagrams, eliminating Pauli restrictions in intermediate states, and factoring out unlinked parts of diagrams to obtain Eqs. 14 and 15. The result is that the most direct method of obtaining explicit expressions for contributions to the perturbation expansions is through *diagram rules*, which automatically incorporate the steps we have gone through. The rules are rather simple and straightforward to apply. They will be summarized in Section 9.5, after we have worked out some explicit examples.

The compactness of the linked cluster expansions of Eqs. 14 and 15 is to be contrasted with the equivalent forms for the Rayleigh–Schrödinger perturbation expansion, which may be found (for low orders) in any elementary textbook on quantum mechanics. Higher orders in Rayleigh–Schrödinger are rarely written down explicitly, since they are rather complicated. It is in fact the normalization of the ground state vector which causes considerable complication, a difficulty which is removed entirely from the linked cluster method by the steps from Eq. 4 to 7, and Eqs. 8 and 9. If we had not done the division in Eq. 8, but simply expanded the ratio in powers of V, we would, of course, reproduce the Rayleigh–Schrödinger series for ΔE, but not in the linked-cluster form of Eq. 15. (This is complicated, but can be carried out directly in low orders: see Exercise 9.2.)

It is also true that Eq. 15 must agree, order by order in V, with the Rayleigh–Schrödinger expansion, since they both represent the same power series in V. In low orders this is seen directly, as we show explicitly in Section 9.4. The correspondence in first order is immediate, as seen in Eq. 23. In second order the exclusion of $|0\rangle\langle0|$ in the intermediate state expansion (see Eqs. 12–14), which is the only effect of restriction to linked diagrams, leads directly to the usual Rayleigh–Schrödinger form in Eq. 24. The equivalence in higher order is more complicated. Here the diagrams which violate the Pauli exclusion principle can be seen to play an important role in the linked cluster expansion (see, e.g. Exercise 9.4).

The removal of the normalization problem is not just a question of ease of expression, however. It is crucial to guarantee that the Goldstone perturbation expansion will be well behaved in the limit of a very large system, where the number of particles (or degrees of freedom) $N \to \infty$. In such a limit, the state of the perturbed system as a whole looks rather different from the unperturbed ground state, since a small perturbation of the state of each particle, by an overlap factor of, say $1 - \lambda^2 \lesssim 1$, vanishes in the product over all particles:

$$\lim_{N \to \infty} (1 - \lambda^2)^N \to 0 . \tag{17}$$

This problem invalidates the use of Rayleigh–Schrödinger theory in its standard form for $N \to \infty$, since individual terms will diverge, due to the behavior of unlinked diagrams in the limit. The linked-cluster method avoids this limit problem by calculating properties of finite parts of the entire system only—the "linked clusters". The series has been rearranged so that the divergent terms exactly cancel.

The fact that the linked-cluster expansion remains well behaved for $N \to \infty$ does not guarantee that the expansion is convergent in general. That question is related to the physics of the problem. As we discussed in Chapter 7, we do *not* expect a convergent perturbation expansion for the ground state of a system that differs qualitatively from the unperturbed case, by symmetry or by phase change in general. In addition, even normal systems may have ill-behaved perturbation expansions because of some specific property of the perturbing interaction. We have encountered such problems for two interesting physical cases: the hard-sphere potential and the Coulomb potential in Chapter 6. In both cases we shall find systematic methods for removing the singular behavior, such that the linked-cluster expansion can be made into a convergent and useful series for the analysis of the relevant physical systems. We return to these questions in Chapter 10.

9.3 GROUND-STATE ENERGY IN FEYNMAN DIAGRAMS

In the previous section we used the Goldstone expansion of the time-evolution operator to obtain compact, time-independent expressions for the ground-state energy and wave function of a normal Fermi system. In this section we shall consider the alternative of the Feynman expansion. The result is not as compact or convenient in general, so we shall often prefer the Goldstone expansion for ground-state properties. However, for many purposes Feynman diagrams are easier to work with, even if the reduction needed for the ground state requires some special treatment. A particularly good example of this will be spelled out in Section 10.5, where we calculate the correlation energy for the ground state of the electron gas.

We discuss here only the energy shift, which is given in Eq. 8. Since the linked-cluster theorem of Eq. 3 applies to Feynman diagrams as well as to Goldstone, we may formally divide out the numerator, again obtaining Eq. 9, but understanding the expression to refer to linked Feynman diagrams:

$$\Delta E = \lim_{\alpha \to 0^+} \langle 0|VU_\alpha(0, -\infty)|0\rangle_l \,.$$

In Section 8.4 we discussed the generation of Feynman diagrams for expressions like the denominator of Eq. 8 [which is the unperturbed ground-state expectation value of $U(t, t_0)$ in the limit $t = 0$, $t_0 \to -\infty$], using the Wick reduction method. The same method may be used for the numerator as well by rewriting VU_α in terms of time-ordered products. This can be done in two ways as follows.

First, since the interaction V is evaluated at time $t_0 = 0$ [i.e., $V = V(0)$], and all the intermediate interaction times in $U_\alpha(0, -\infty)$ are earlier ($t_i < 0$), then $V(0)$ operating on the nth-order expansion of U_α, using Eqs. 8.8 or 8.34, is already time-ordered; that is,

$$V(0)\{V(t_1)\cdots V(t_n)\}_T = \{V(0)V(t_1)\cdots V(t_n)\}_T \,. \tag{18}$$

The Wick reduction of the expectation values of Eq. 18 proceeds as before, producing the same Feynman diagrams, now in the $(n + 1)$th order. However, the integration is over n time variables t_i; one time variable must be kept fixed at $t_0 = 0$, the last permissible time. This means that a Feynman diagram of a given topology may appear repeatedly, once with each of the interaction vertices stretched to be in the last position at $t_0 = 0$. This is somewhat awkward to carry out, but the counting problem can be handled somewhat more symmetrically by the second method, which follows.

The second approach makes use of the time-translational invariance of the theory, specifically, that expressions for Feynman diagrams depend only on relative times. Therefore, rather than requiring that the last interaction be fixed at time $t_0 = 0$, one may simply require that *any* one of the $n + 1$

interactions be fixed at time $t_i = 0$ $(i = 0, 1, \ldots, n)$. The same expression will be obtained from the integration over the other n time variables, independent of i. To make use of this result, first write the integral of Eq. 18 in a symmetrical form:

$$\int_{-\infty}^{0} dt_1 \cdots dt_n \, \{V(0)V(t_1) \cdots V(t_n)\}_T$$

$$= \int_{-\infty}^{\infty} dt_0 \delta(t_0) \int_{-\infty}^{0} dt_1 \cdots dt_n \, \{V(0)V(t_1) \cdots V(t_n)\}_T$$

$$= \frac{1}{n+1} \int_{-\infty}^{\infty} dt_0 dt_1 \cdots dt_n \, \{V(t_0) \cdots V(t_n)\}_T \delta(t_0) , \quad (19)$$

where in the last step we have let each $t_i = 0$ in turn, using the time-translational invariance, and relabeled $t_i = t_0$. Now all integrations run from $-\infty$ to $+\infty$. Then, the expansion of Eq. 9 is given to $(n+1)$th order by

$$\frac{(-i)^n}{(n+1)!} \int_{-\infty}^{\infty} \cdots \int dt_0 dt_1 \cdots dt_n \, \langle 0 | \{V(t_0) \cdots V(t_n)\}_T | 0 \rangle \delta(t_0) , \quad (20)$$

as we see by combining Eq. 8.34 and Eq. 19. (We continue to denote the unperturbed ground state Φ_0 by $|0\rangle$.) This is equivalent (up to a factor $-i$) to the sum of $(n+1)$th-order Feynman diagrams for $\langle 0 | U(\infty, -\infty) | 0 \rangle$, calculated with one intermediate time fixed.

Either method allows us to express Eq. 9 in terms of Feynman diagrams. We have already noted that linked diagrams have the same meaning for Feynman diagrams as for Goldstone diagrams, and that the factoring of the expressions for disconnected Feynman diagrams corresponds to the similar factoring of *sums* of Goldstone diagrams. Thus the linked-cluster theorem discussed in Section 9.1 obtains in both formalisms, and has the same meaning (only the counting of diagrams differs). The energy shift of Eq. 9 may then be rewritten in a linked-cluster expansion in Feynman diagrams, in the form

$$\Delta E = \sum_{n=1}^{\infty} \frac{(-i)^{n-1}}{n!} \int_{-\infty}^{\infty} \cdots \int dt_1 \cdots dt_n \, \langle 0 | \{V(t_1) \cdots V(t_n)\}_T | 0 \rangle_l \delta(t_1) , \quad (21)$$

where we have relabeled the symmetrical expression (20). The time integrations are still to be done; these were discussed in Section 8.6. Compare Eq. 21 with the Goldstone linked-cluster expansion, Eq. 15, in which the time integrations have already been performed. The different ordering of the operations produces quite different expressions at various stages of calculation, even though the final (exact) results must agree in both expansions. It is these differences which are exploited in devising approximation methods.

Finally we remark that the special devices needed in Eqs. 18 or 19 to evaluate Eq. 9 by singling out one interaction time in the Feynman expansion are just those which decompose each Feynman diagram into a sum of Goldstone diagrams of the same topology but different time orderings. In this sense, the Goldstone expansion is the more natural for the ground-state energy shift.

9.4 THE GOLDSTONE EXPANSION IN LOW ORDERS

Let us see how the diagrammatic method works for the first few orders in the linked-cluster expansion, Eq. 15, for the ground-state energy shift. In these orders it is just as easy to obtain explicit expressions for contributions to ΔE from the intermediate-state expansions as from the diagram rules, so these examples serve to help us understand the connection between the two approaches. We write ΔE explicitly as a series

$$\Delta E = \sum_{n=1}^{\infty} \Delta E^{(n)} \tag{22}$$

with $\Delta E^{(n)}$ given by the terms on the right-hand side of Eq. 15, the indices now shifted by one to indicate the power of V.

9.4.1 First Order

The first-order result is obtained immediately from the expectation value of V, calculating with the rules for the number representation given in Chapter 1:

$$\Delta E^{(1)} = \langle 0|V|0\rangle_l = \langle 0|V|0\rangle = \frac{1}{2} \sum_{\lambda,\mu \leq F} \langle \lambda\mu|v|\lambda\mu\rangle_A , \tag{23a}$$

or

$$\Delta E^{(1)} = \frac{1}{2} \sum_{\lambda,\mu \leq F} \{\langle \lambda\mu|v|\lambda\mu\rangle - \langle \lambda\mu|v|\mu\lambda\rangle\} . \tag{23b}$$

The appropriate diagrams are shown in Figure 9.4; (a) corresponds to the

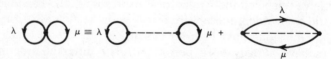

FIGURE 9.4 Goldstone diagrams for $\Delta E^{(1)}$: (a) corresponding to an antisymmetrized matrix element, and (b) decomposed into two diagrams corresponding to direct and exchange matrix elements in Eq. 23.

antisymmetrized form in Eq. 23a, and (b) to the separation into direct and exchange matrix elements in Eq. 23b. The interaction V acts at time $t = 0$. Since in first order there is no time integration, the interacting orbitals λ, μ do not propagate in time; they are occupied orbits in the unperturbed Fermi ground state which do not make transitions. Therefore they are represented by lines which close back on themselves. There are no disconnected parts in first order; all diagrams are linked. The result is identical with that in Rayleigh–Schrödinger theory.

9.4.2 Second Order

The second-order term may be written, using an intermediate-state expansion,

$$\Delta E^{(2)} = \langle 0|V(\mathcal{E}_0 - H_0)^{-1}V|0\rangle_l = \sum_{m \neq 0} \frac{|\langle 0|V|m\rangle|^2}{\mathcal{E}_0 - \mathcal{E}_m}, \qquad (24)$$

which is the conventional Rayleigh–Schrödinger result. The Goldstone diagrams for $\langle 0|U^{(2)}(t, t_0)|0\rangle$ were shown in Figure 8.3. To get to Eq. 24, we have integrated over times, and eliminated unlinked diagrams, as indicated in the steps from Eq. 9 to Eq. 15. The restriction $m \neq 0$ eliminates the only unlinked diagram in second order, shown in Figure 9.5, which does not contribute to $\Delta E^{(2)}$. The allowed intermediate states that couple to the vacuum state $|0\rangle$ through the two-body interaction V can have one particle and one hole (p–h) or two particles and two holes (2p–2h), represented by

$$|m\rangle = |\{\rho\}\{\nu\}\rangle, \qquad (25a)$$

$$|m\rangle = |\{\rho\sigma\}\{\mu\lambda\}\rangle \qquad (25b)$$

respectively (see Eq. 8.14b). The contributions to $\Delta E^{(2)}$ are easily evaluated to yield

$$\Delta E^{(2)}(a) = \sum_{\substack{\lambda,\mu,\nu \leq F \\ \rho > F}} \frac{\langle \mu\nu|v|\mu\rho\rangle_A \langle \lambda\rho|v|\lambda\nu\rangle_A}{\epsilon_\nu - \epsilon_\rho} \qquad (26a)$$

$$\Delta E^{(2)}(b) = \frac{1}{4} \sum_{\substack{\lambda,\mu \leq F \\ \rho,\sigma > F}} \frac{|\langle \rho\sigma|v|\lambda\mu\rangle_A|^2}{\epsilon_\lambda + \epsilon_\mu - \epsilon_\rho - \epsilon_\sigma}, \qquad (26b)$$

FIGURE 9.5 Unlinked diagram in second order: it does not contribute to $\Delta E^{(2)}$.

respectively, using the antisymmetrized matrix elements. Note the factor of $\frac{1}{4}$ in Eq. 26b, which corrects for the fact the sum on λ, μ, which can take identical values, counts states twice, as does the sum on ρ, σ. The linked diagrams corresponding to Eq. 26 are shown in Figure 9.6a, b. We no longer indicate times, which have been integrated in Eq. 26. The expansion of Eq. 26a in direct and exchange matrix elements gives four distinct terms, represented by the four diagrams on the right-hand side of Figure 9.6a. The expansion of Eq. 26b in direct and exchange matrix elements gives the two distinct diagrams of the right-hand side of Figure 9.6b. Finally, the linked-cluster result in second order is

$$\Delta E^{(2)} = \Delta E^{(2)}(a) + \Delta E^{(2)}(b) . \tag{27}$$

We can obtain the same results starting from the diagrammatic point of view. Start by drawing all linked second-order (antisymmetrized) diagrams for $\Delta E^{(2)}$. These must have two vertices at different "times": each vertex

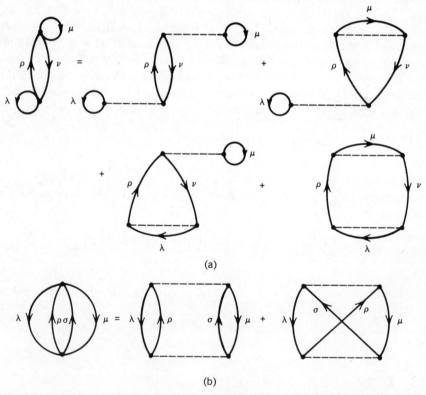

(a)

(b)

FIGURE 9.6 Linked Goldstone diagrams for $\Delta E^{(2)}$, shown in both antisymmetrized and direct-exchange form, with (a) p–h intermediate states, and (b) 2p–2h intermediate states; see Eq. 26.

must have two lines entering and two emerging. (Since the times have been integrated, the vertices indicate the order of interaction, i.e. right to left in Eq. 15.) All lines must end on vertices, since for $\Delta E^{(n)}$, the initial and final states are the vacuum state, which has no lines. The two possibilities are in fact those of Figure 9.6a, b (left side). For each diagram, the evaluation proceeds by assigning to each vertex an antisymmetrized two-body matrix element, according to the lines in and lines out of the vertex, as in the numerators of Eq. 26. The denominator is simply given by the excitation energy of Eq. 8.11b. Factors of $\frac{1}{2}$ should be included whenever there is a pair of lines whose labels can be interchanged without changing the diagram (equivalent lines): there are two such pairs in Figure 9.6b [(λ, μ) and (ρ, σ)], giving the $\frac{1}{4}$ in Eq. 26b, as noted earlier. The contributions are to be summed over all particle states ($\rho > F$) for upward lines, and hole states ($\lambda \leq F$) for downward lines. Closed loops representing occupied orbits, as those labeled (λ, μ) in Figure 9.6a, are summed ($\lambda \leq F$). Using these simple diagram rules, one obtains the contributions of Eq. 26 directly. There are also sign rules; for these cases the signs are positive. We return to the general rules in Section 9.5.

9.4.3 Hartree–Fock and the Removal of Single-Particle Excitations

We have expressed our perturbation expansion in terms of a given single-particle Hamiltonian H_0, which defines the basis of orbital states, as in Eq. 8.10. It is possible to vary the basis by introducing an auxiliary single-particle potential

$$U = \sum_{\alpha, \beta} \langle \alpha | U | \beta \rangle a_\alpha^\dagger a_\beta \,, \tag{28}$$

which then is added to the original H_0, defining a new unperturbed Hamiltonian

$$H_0' = H_0 + U \,. \tag{29}$$

The perturbation is now given by

$$H - H_0' = V' = V - U \,, \tag{30}$$

which has a one-body part $(-U)$ in addition to the two-body part (V). To do diagrammatic perturbation theory with this new separation of $H = H_0' + V'$, we need a new vertex which represents the one-body part $(-U)$ of Eq. 30. This is illustrated in Figure 9.7, where the interaction is denoted by a dot connected by a dashed (interaction) line to ×. One particle line enters and one leaves, denoting the matrix element

$$-\langle \sigma | U | \rho \rangle \,. \tag{31}$$

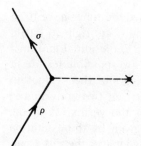

FIGURE 9.7 Diagram for the auxiliary potential $-U$, whose matrix element is given by Eq. 31. This is understood to be part of a larger diagram in which ρ, σ are particle lines.

The added interaction contributes a new diagram to the first-order energy shift $\Delta E^{(1)}$, shown in Figure 9.8, whose contribution to $\Delta E^{(1)}$ is

$$- \sum_{\mu \leq F} \langle \mu | U | \mu \rangle \,. \tag{32}$$

This new term cancels the $+U$ term in the new *unperturbed* ground-state energy, so that

$$\mathscr{E}_0' + \Delta E^{(1)'} = \langle 0' | H_0' + V' | 0' \rangle = \langle 0' | H_0 + U + V - U | 0' \rangle$$

$$= \langle 0' | H_0 + V | 0' \rangle \,, \tag{33}$$

in which U does not appear explicitly. However, the state $|0'\rangle$ has orbits defined by H_0', which does depend on the potential U (unless $[H_0, U] = 0$). The first-order energy in Eq. 33 has been changed by the introduction of U, that is, in general

$$\mathscr{E}_0' + \Delta E^{(1)'} \neq \langle 0 | H_0 + V | 0 \rangle = \mathscr{E}_0 + \Delta E^{(1)} \,. \tag{34}$$

The possibility of changing basis seems to add complications, but it can be used to simplify calculation, if U is chosen appropriately. This can be seen by making the *self-consistent* choice of the matrix elements of U and V,

$$\langle \alpha | U | \beta \rangle = \sum_{\mu \leq F} \langle \alpha \mu | v | \beta \mu \rangle_A \,, \tag{35}$$

which is identical with the Hartree–Fock choice of Eq. 4.14. The diagonal orbitals of $H_0' = H_0 + U$ are the solutions of the Hartree–Fock equation (Eq. 4.15). With this choice of U, we find that the contribution of the new diagrams for $-U$, such as are shown in Figure 9.7, now cancels the

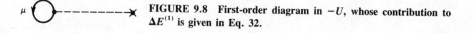

FIGURE 9.8 First-order diagram in $-U$, whose contribution to $\Delta E^{(1)}$ is given in Eq. 32.

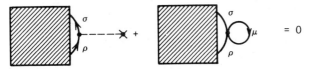

FIGURE 9.9 Illustrates the cancellation of $-U$ diagrams with certain V diagrams (single-particle insertions) when U obeys the self-consistent or Hartree–Fock relation, Eq. 35. It is assumed that the shaded parts of both diagrams are identical. (Sum over all $\mu \leq F$ understood.)

contribution of those diagrams which have a particle interacting through V with an occupied orbit, denoted by a closed loop, as illustrated in Figure 9.9, with $\mu \leq F$. These are called single-particle *insertion diagrams*; the entire class is removed by the addition of $-U$ to the perturbation, which by virtue of Eq. 35 means shifting to a Hartree–Fock orbital basis.

For the particular example of second order, the addition of the $-U$ diagrams to first and second order in $-U$ cancels the diagrams of Figure 9.6a, whose contribution were given in Eq. 26a. So, in the Hartree–Fock basis, the total second-order energy shift is given by

$$\Delta E^{(2)} = \Delta E^{(2)}(b) \quad \text{(HF)} . \tag{36}$$

(The removal of the p–h excitations is an example of Brillouin's theorem of Chapter 4.) In first order, Eq. 35 leads to the result

$$\Delta E^{(1)} = \frac{1}{2} \sum_{\lambda, \mu \leq F} \langle \lambda\mu | v | \lambda\mu \rangle_A - \sum_{\lambda \leq F} \langle \lambda | U | \lambda \rangle$$

$$= -\frac{1}{2} \sum_{\lambda \leq F} \langle \lambda | U | \lambda \rangle ; \tag{37}$$

the cancellation is incomplete here.

9.4.4 Third Order with Hartree–Fock Orbits

With the self-consistent choice of potential given in Eq. 35, the linked-cluster contribution

$$\Delta E^{(3)} = \langle 0 | V'(\mathscr{E}_0 - H_0)^{-1} V'(\mathscr{E}_0 - H_0)^{-1} V' | 0 \rangle_l \tag{38}$$

is represented by three types of diagrams shown in Figure 9.10, without explicit labeling of particle and hole lines. There are no other linked diagrams, since all single-particle insertion diagrams have been canceled by the $-U$ diagrams. Each of the three types shown in Figure 9.10 is typical of a class, two of which we shall deal with further for particular physical problems in Chapter 10. In Section 10.1, type (a) will be recognized as

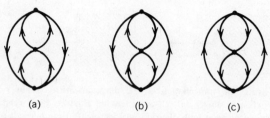

FIGURE 9.10 Third-order linked diagrams for $\Delta E^{(3)}$ (see Eq. 38): (a) p–p "ladder" or "chain"; (b) p–h "ring"; (c) h–h "ladder" or "chain".

important for short-range correlations, while in Section 10.5 (b) will be part of the ring-diagram contribution to the energy of a Coulomb gas. An explicit expression for each type illustrated is easily found; for (a) it may be written

$$\Delta E^{(3)}(a) = \frac{1}{8} \sum_{\substack{\rho\sigma\tau\omega > F \\ \lambda\mu \leq F}} \frac{\langle \lambda\mu|v|\rho\sigma \rangle_A \langle \rho\sigma|v|\tau\omega \rangle_A \langle \tau\omega|v|\lambda\mu \rangle_A}{(\epsilon_\lambda + \epsilon_\mu - \epsilon_\rho - \epsilon_\sigma)(\epsilon_\lambda + \epsilon_\mu - \epsilon_\tau - \epsilon_\omega)} . \tag{39}$$

Expressions for (b) and (c) are also easily derived.

9.4.5 Pauli-Principle-Violating Diagrams

The first diagrams that violate the Pauli principle for intermediate states enter in third and fourth order; examples are shown in Figure 9.11. The

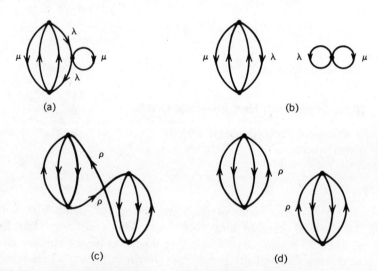

FIGURE 9.11 Pauli-violating diagrams, in which the linked examples (a), (c) exactly cancel the unlinked examples (b), (d) of the same order: (a), (b), third order; (c), (d), fourth order. Only repeated orbits are labeled here.

linked diagrams (a) and (c) are equal in magnitude but opposite in sign to the unlinked diagrams (b) and (d), respectively, so there is no contribution to the original series for the numerator in Eq. 8. However, following the discussion of Sections 9.1 and 9.2, the factoring which leads to the linked-cluster series of Eq. 15 removes all unlinked diagrams, *including* the Pauli-violating diagrams (b) and (d). Therefore, the linked diagrams (a) and (c) do contribute to the calculation of ΔE in the linked-cluster method.

9.5 SUMMARY: DIAGRAM RULES

The method we have used in the previous section to construct and evaluate all linked-cluster terms for $\Delta E^{(n)}$ in each order follows from the method of derivation, and can in principle be carried on to higher orders. One starts by inserting sums over intermediate states $|m\rangle$ in the general expression for $\Delta E^{(n)}$, from Eq. 15, reducing the expression by number-representation techniques to a sum of products of two-body matrix elements of v, divided by energy denominators which contain the single-particle or -hole energies ϵ_α. Each individual term may be associated with a distinct Goldstone diagram with n vertices, connected by particle and hole lines—two into and two out of each vertex—including occupied orbit lines which begin and end at the same vertex (single-particle insertions). Only the contributions corresponding to linked diagrams are to be retained. In low orders this separation was easy to carry out by inspection, but it becomes more involved in higher orders. As discussed above, the Pauli exclusion principle must be ignored in summing over particle and hole lines. This actually requires modification of the number-representation methods of evaluation, since states that are not antisymmetrized are included. If one works in the Hartree–Fock or self-consistent basis, the single-particle insertion diagrams do not contribute (except in first order).

A more direct method of evaluating $\Delta E^{(n)}$ is by diagram rules, as we indicated in the previous section. The logic of associating Goldstone diagrams with terms in the linked-cluster expansion, Eq. 15, is reversed. We may imagine first expanding Eq. 9 in terms of Feynman diagrams, as discussed in Section 9.3. The Wick reduction method specifies the counting of distinct Feynman diagrams (including Pauli-violating ones). Each Feynman diagram generates a set of distinct Goldstone diagrams, one for each order of the vertices. Then the contribution of each Goldstone diagram may be evaluated, using a modified form of the Wick algebra.

DIAGRAM RULES FOR $\Delta E^{(n)}$

The rules are most easily given for diagrams that have been decomposed into direct and exchange parts, so that interactions are represented by horizontal dashed lines, as in Figure 9.6 (right side), and stand for nonantisymmetrized matrix elements $\langle \alpha\beta|v|\gamma\delta \rangle$. Then, for each diagram, with each line labeled by a particle or hole orbit quantum number:

(1) Each interaction line contributes a matrix-element factor

$$\langle \alpha\gamma | v | \beta\delta \rangle \,, \tag{40}$$

where line β enters and α leaves one end of the interaction line, and δ enters and γ leaves the other end; e.g. see Figure 9.12a.

(2) With the (time) interval between two successive interactions, associate an energy-denominator factor given by the sum of hole energies minus particle energies,

$$\sum_\lambda \epsilon_\lambda - \sum_\rho \epsilon_\rho \,, \tag{41}$$

for the hole and particle lines which cross that interval; e.g. see Figure 9.12b, for which the denominator is $\epsilon_\lambda - \epsilon_\rho$.

(3) For each labeled diagram, the value is given by the product of the matrix elements (Eq. 40) divided by the product of the denominator factors (Eq. 41). There is an overall sign corresponding to the topology of the diagram, given by

$$(-1)^{h+l} \,, \tag{42}$$

where h is the number of hole lines ($\lambda \leq F$, including insertions), and l is the number of distinct closed loops made by continuous particle and hole lines, including insertions. For example, in Figure 9.6b, the first right-hand diagram has two loops, while the second has one loop. Since they both have two holes, the first gets $(-1)^4 = +1$ from Eq. 42, while the second gets $(-1)^3 = -1$. This sign comes from the number of anticommutations occurring in the Wick reduction leading to a given diagram.

(4) In summing over particle and hole labels to add contributions to $\Delta E^{(n)}$, one must ensure that no distinct diagrams are repeated simply by permutation of labels. For example, the diagram in Figure 9.13a, when

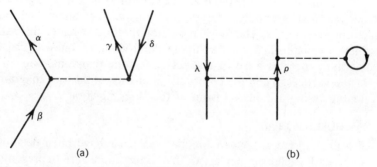

(a)   (b)

FIGURE 9.12 Examples of parts of diagrams contributing a factor of (a) Eq. 40 to the numerator, and (b) Eq. 41 to the denominator of the expression for the diagram.

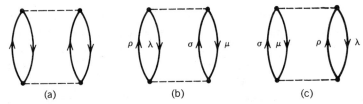

FIGURE 9.13 **Example of a diagram (a) which is repeated twice when summed over particle and hole labels: diagrams (b) and (c) are equivalent.**

summed over line labels, will produce both Figs. 9.13b and c, which are topologically equivalent. Therefore, the unrestricted summation over ρ, $\sigma >$ F; λ, $\mu \leq F$ must be modified by a factor of $\frac{1}{2}$ to correct for the double counting.

It is easy to show that these rules will reproduce the results quoted in Section 9.4. The rules may also be modified to allow calculation directly with the antisymmetrized matrix elements $\langle \alpha\beta|v|\gamma\delta \rangle_A$. This clearly must be possible, since that is the only form in which the interaction V can contribute to a Fermi system. The prescription is to choose any *one* of the dashed-line diagrams corresponding to a given dot-vertex diagram, e.g. any one of the four on the right side of Figure 9.6(a); calculate with rules (1)–(3) above; and then replace $\langle \alpha\beta|v|\gamma\delta \rangle$ with $\langle \alpha\beta|v|\gamma\delta \rangle_A$ throughout.

The economy of the use of diagram rules for the calculation of perturbation contributions comes from the elimination of the need to specify intermediate states, as in Eqs. 25 and 26. Rather than enumerating all possible intermediate states in each order, one must enumerate all possible distinct (Feynman or Goldstone) diagrams in each order. In higher orders than $n = 2$ or 3, this is not a trivial task, but still considerably easier than the route from the intermediate-state expansion.

In the following chapter, we shall see how the use of diagrams can be extended beyond any given order, and infinite series of particular diagrams may be evaluated together, that is, "summed." What generally happens when this is done is that a new perturbation expansion is defined, in an *effective interaction* defined by the process of summing. It is often possible then to evaluate the low orders in the new series, using the diagram rules for the effective interaction. We shall return to this matter at the end of Chapter 10.

EXERCISES

9.1 **(a)** Obtain explicit expressions for the contributions to Eq. 1 for each of the three Goldstone diagrams of Figure 9.1, in terms of the (antisymmetrized) two-body matrix elements of the form $\langle \lambda_i \lambda_j |v| \rho_k \rho_l \rangle_A$, single-particle energies ϵ_λ, ϵ_ρ, and integrals over time variables.

(b) Show that the sum of the time-integrals in (a), for a given set of orbits $\lambda_1-\lambda_4$, $\rho_1-\rho_4$, may be rewritten so that the expression factors (into a single and a double time integral).

(c) Show that the sum of the three Goldstone diagrams with different time orders (of Figure 9.1) gives the factored result of Eq. 2.

9.2 (a) Using Rayleigh–Schrödinger perturbation theory, find explicit expressions for the ground-state energy to third order, in terms of matrix elements of V and energy denominators.

(b) Evaluate Eq. 8 explicitly through third order, by direct expansion of numerator and denominator, and integration over times. (Take the limit $t_0 \to -\infty$ after expansion.) Show that the Rayleigh–Schrödinger expansion is obtained by this procedure.

9.3 Derive the Brillouin–Wigner perturbation series for the ground state energy shift:

$$\Delta E = \sum_{n=0}^{\infty} \langle 0|V\{[E_0 + \Delta E - H_0]^{-1}Q_0V\}^n|0\rangle$$

where $Q_0 = 1 - |0\rangle\langle 0|$ removes the unperturbed ground state from an intermediate-state expansion of ΔE. [Hint: Convert the Schrödinger equation, $(E - H_0)|\Phi\rangle = V|\Phi\rangle$, into integral equation form, with the normalization condition of Eq. 7.50, and iterate in powers of V. Note that ΔE appears in the denominators. If the denominators are expanded in powers of V, one obtains the Rayleigh–Schrödinger series.]

9.4 For the two-fermion system of Exercise 8.1:

(a) Draw and label all nonzero (linked) Goldstone diagrams through fourth order for the ground-state energy shift ΔE. Identify any Pauli-violating diagrams.

(b) Calculate the ground-state energy through fourth order.

(c) Since this problem is equivalent to a 2×2 matrix eigenvalue problem, calculate the exact ground-state energy, expand through fourth order in V, and compare with part (b).

9.5 Consider a system of two spin-$\frac{1}{2}$ fermions, each moving in a one-dimensional harmonic-oscillator potential and interacting through a potential V_{12}, so that the total Hamiltonian can be written

$$H = H_0 + V_{12}, \qquad H_0 = H_1 + H_2,$$

$$H_i = \frac{p_i^2}{2m} + \frac{m\omega^2 x_i^2}{2}, \quad i = 1, 2, \qquad V_{12} = \alpha\left(\frac{m\omega^2}{2}\right)x_1 x_2.$$

(a) This problem of coupled oscillators may be solved exactly. Find an expression for the ground-state energy in terms of $\hbar\omega$ and α. Expand the exact expression to order α^2.

(b) Draw and label all relevant Goldstone (linked) diagrams for the perturbation energy (shift) ΔE, to second order in V_{12} (or α). It is convenient to label lines by single-particle (H_0) oscillator states $(n = 0, 1, 2, \ldots)$ and spin projections $(\pm\frac{1}{2})$. Give an expression for the energy shift corresponding to each diagram, and show which diagrams will be zero for the particular V_{12} (and H_0) we are considering.

(c) Calculate ΔE through second order in V_{12} (or α) explicitly from the diagram expressions in (b), and compare the result with the expansion (to α^2) of the exact result in part (a).

(d) It is also possible in this case to calculate ΔE to second order, using closure in Rayleigh–Schrödinger theory:

$$\Delta E^{(2)} = \frac{\langle 0|V_{12}^2|0\rangle}{2\hbar\omega} \ .$$

Show this, and calculate the shift, comparing with (a) and (c).

9.6 For the system of Exercise 9.5:

(a) Draw and label all *nonzero* Goldstone (linked-cluster) diagrams for the energy shift, to fourth order in V_{12} (or α). Show typical cases only, and explain how to generate all diagrams by permutation of labels, etc.

(b) Find and identify: Pauli-violating diagrams; ladder diagrams (see Figures 9.10a, c and 10.3); ring diagrams (see Figures 9.10b and 10.7a).

(c) Carry out the energy calculation in Exercise 9.5 to fourth order, comparing the expansion of the exact result (a) to order α^4 with the result of perturbation theory. Here it is easier to use Rayleigh–Schrödinger theory.

CHAPTER **10**

Applications to
Fermi-Gas Problems

We are now ready to attack the problems of the interacting Fermi gas that were introduced in Chapter 6: the hard-sphere system and the electron gas. These provide two interesting examples of interacting systems that have normal Fermi ground states, so that we may hope to describe their behavior in terms of a perturbation expansion. At the same time, when we tried to calculate the first few terms of the expansion for the ground state energy using Rayleigh–Schrödinger theory, we found divergent behavior for both these systems. We should not expect any difference in the results on recalculating in the diagrammatic perturbation theory to low orders in the strength of the interaction, since the diagrammatic theories reduce to Rayleigh–Schrödinger form through second order.

What we shall now have to do is make use of the systematic structure of diagrammatic perturbation theory to extract a well-defined expansion from the apparently divergent series. The method is based on the idea of treating together sets of diagrams of all orders whose structure is related and which have similar divergent behavior. These sets are infinite subsets of the complete set of perturbation diagrams, defining partial (infinite) series. If we can "sum" these partial series of divergent terms to obtain finite and meaningful results, then we will have "rearranged" the original divergent series into a new series with finite terms. A new expansion parameter (or parameters) will replace the interaction strength. The new series may converge or not in the sense of an analytic expansion, but it will in general have some useful range of validity. We have used the words "sum" and "rearrange" in a symbolic sense: their operation in this method must be defined carefully, since we are dealing with apparent divergences.

We apply the method first to the hard-sphere problem, in the following two sections. Then we shall extend the approach to include the case of the Bose hard-sphere gas (Section 10.3), and also to the related problem of nuclear matter (Section 10.4). Finally we shall find how closely-related methods work out for the electron gas (Sections 10.5 and 10.6).

10.1 SUMMATION OF PARTIAL SERIES: LADDER DIAGRAMS

The basic notion we shall use in this section is the replacement of a selected infinite series of terms in the perturbation expansion by a function which is formally equivalent to the sum of the series and which can be evaluated by some other means. By formally equivalent we mean that for small values of the perturbation parameter the function is represented convergently by the series; for larger values of the parameter, the function is defined by a process continuous in the parameter. For the present case, the parameter is V_0, the strength of the repulsive potential (see Eq. 6.5). An example of this is, of course, analytic continuation. The function $f(\lambda) = (1 - \lambda)^{-1}$ may be defined by the series

$$f(\lambda) = \sum_{n=0}^{\infty} \lambda^n \tag{1}$$

within the (complex) circle of convergence $|\lambda| < 1$. For $|\lambda| > 1$, the function is given by analytic continuation. However, the process we shall generally use to define the function will be to specify a linear *integral equation* of which it is the solution. For small values of the perturbation parameter, the original perturbation series gives a convergent expansion. For larger values, the function is still well defined (and generally unique) although the perturbation series may diverge. If this process can be accomplished, the contribution represented by the original divergent series is shown to be finite, and is calculated in a convergent way. This will even give well-defined results in the hard-sphere limit. $V_0 \to \infty$, for which each *term* in the original series diverges as V_0^n. [In the example of Eq. 1, $f(\lambda) \to 0$, a finite limit, as $\lambda \to \infty$.]

The problem remains that the infinite series summed by this method is not the entire perturbation series; other divergence problems will occur as $V_0 \to \infty$. We shall find, however, that the method of partial summation may be repeated, so that the original expansion V_0 is replaced by a new expansion in a function of V_0 (well defined even as $V_0 \to \infty$), which is properly behaved.

For the hard-sphere problem, the partial series of interest is the series of terms represented by the diagrams of the type shown in Figure 10.1, which are commonly called *ladder diagrams*, referring to the structure of any number of "rungs" (horizontal interaction lines) connecting two particle lines at various times. It is the series of these diagrams which will be replaced by a function. In Eq. 6.22 we gave an estimate that this series diverges as an alternating series in powers of V_0. To see the origin of this estimate, and to see how the "summation" of diagrams is to work, we first return to Eq. 6.16 for the second-order energy shift. We perform the sum over the two particle states $\mathbf{p}_1$, $\mathbf{p}_2$, by first introducing a projection operator

$$Q = \sum_{\mathbf{p}_1\mathbf{p}_2 > F} |\mathbf{p}_1, \mathbf{p}_2\rangle \langle \mathbf{p}_1, \mathbf{p}_2| \qquad (2)$$

which operates on the space of two-particle states, selecting only those plane-wave states whose momenta are outside of the ground-state Fermi sphere: i.e. particle states. Now if H_0 is the kinetic-energy operator in this two-body space, we may write

$$H_0 Q = Q H_0 = \sum_{\mathbf{p}_1, \mathbf{p}_2 > F} |\mathbf{p}_1, \mathbf{p}_2\rangle [\epsilon(p_1) + \epsilon(p_2)] \langle \mathbf{p}_1, \mathbf{p}_2| . \qquad (3)$$

Similarly, we may express the operator

$$Q[\epsilon - H_0]^{-1} = \sum_{\mathbf{p}_1, \mathbf{p}_2 > F} |\mathbf{p}_1, \mathbf{p}_2\rangle [\epsilon - \epsilon(p_1) - \epsilon(p_2)]^{-1} \langle \mathbf{p}_1, \mathbf{p}_2| , \qquad (4)$$

where ϵ is an energy parameter. Now we set $\epsilon = \epsilon_0 = \epsilon(k_1) + \epsilon(k_2)$, and take the antisymmetrized two-body matrix elements in $|k_1, k_2\rangle$ of the operator (also two-body) $vQ[\epsilon_0 - H_0]^{-1}v$, where v is the two-body potential. Finally, summing over k_1, $k_2 \leq k_F$, we find that Eq. 6.16 can be reexpressed in the compact form

$$\frac{\Delta E^{(2)}}{\Omega} = \frac{1}{2\Omega} \sum_{k_1, k_2 \leq F} \langle k_1, k_2 | vQ(\epsilon_0 - H_0)^{-1}v | k_1, k_2 \rangle_A . \qquad (5)$$

(To get from Eq. 6.16 to Eq. 5 it is useful to write out the $\langle\ \rangle_A$ matrix elements in direct and exchange form: this accounts for the factor-of-2 change in the coefficient of the sum. Note also that internal quantum numbers have again been suppressed here.)

The same compact notation can be used to write the nth-order term in the ladder series as follows:

$$\left(\frac{\Delta E^{(n)}}{\Omega}\right)_{\text{ladder}} = \frac{1}{2\Omega} \sum_{k_1, k_2 \leq F} \langle k_1, k_2 | [vQ(\epsilon_0 - H_0)^{-1}]^{n-1}v | k_1, k_2 \rangle_A , \qquad (6)$$

where the matrix elements are again of a two-body operator, of nth order in v. This can easily be shown to express the contribution of the nth-order diagrams of the form of Figure 10.1, considered as Goldstone diagrams, to

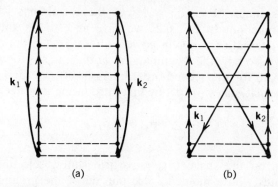

FIGURE 10.1 Examples of ladder diagrams (Eq. 6): (a) direct and (b) exchange. These examples are of sixth order in V; higher orders are constructed by adding interactions ("rungs") between the particle lines at arbitrary intermediate times.

the linked-cluster expression Eq. 9.15 (per unit volume). Referring to the nth-order term of Eq. 9.15 and to Figure 10.1, the first V annihilates the particles $\mathbf{k}_1, \mathbf{k}_2$ in the unperturbed ground state Φ_0, creating two particles and two holes. The particles interact $n-2$ times before the last interaction V destroys the 2p–2h state by returning the excited particles to the original orbits $\mathbf{k}_1, \mathbf{k}_2$ in Φ_0. Since the only interactions involve $\mathbf{k}_1, \mathbf{k}_2$, and $\mathbf{p}_1, \mathbf{p}_2$, the entire expression can be written as a two-body matrix element of a two-body operator, as in Eq. 6. The notation is valid for $n=1$ as well (see Eq. 1.58).

A closure estimate of Eq. 6 in the short-range limit may now be found, following the argument leading from Eq. 6.16 to Eq. 6.21, yielding

$$\frac{\Delta E_{\text{ladder}}^{(n)}}{\Delta E^{(1)}} \cong \frac{V_0^{n-1}}{D^{n-1}} \cong (-md^2 V_0)^{n-1} . \tag{7}$$

The last step, which makes use of Eq. 6.17, gives the result quoted in Eq. 6.22.

In order to "sum" the series of ladder diagrams, we introduce a two-body operator defined by the series

$$\tilde{t}(\epsilon) = \sum_{n=1}^{\infty} [vQ(\epsilon - H_0)^{-1}]^{n-1} v . \tag{8}$$

The sum over n of Eq. 6 can now be expressed as a sum of antisymmetrized matrix elements of $\tilde{t}(\epsilon_0)$:

$$\Omega^{-1} \sum_{n=1}^{\infty} \Delta E_{\text{ladder}}^{(n)} = (2\Omega)^{-1} \sum_{\mathbf{k}_1, \mathbf{k}_2 \leq F} \langle \mathbf{k}_1, \mathbf{k}_2 | \tilde{t}(\epsilon(k_1) + \epsilon(k_2)) | \mathbf{k}_1, \mathbf{k}_2 \rangle_A . \tag{9}$$

The operator series in Eq. 8 can be formally summed as a geometric series similar to Eq. 1. Let $x = vQ$, $a = \epsilon - H_0$, and $m = n - 1$; then

$$\tilde{t}(\epsilon) = \sum_{m=0}^{\infty} (xa^{-1})^m v = (1 - xa^{-1})^{-1} v$$

$$= a(a - x)^{-1} v = v + x(a - x)^{-1} v ,$$

or

$$\tilde{t}(\epsilon) = v + vQ[\epsilon - H_0 - vQ]^{-1} v . \tag{10}$$

This compact expression replaces the series expansion of Eq. 8, each term of which diverges with $V_0 \to \infty$, by an operator *function* of V_0, which we shall find is well behaved in this limit. The matrix element of Eq. 9 will be well defined if the operator of Eq. 10 is used on the right-hand side. However, we need a method to evaluate the operator $\tilde{t}(\epsilon)$ or its matrix element, which we shall do by writing a linear equation for $\tilde{t}(\epsilon)$ as follows. We use the operator identity

$$(a - x)^{-1} = a^{-1} + a^{-1}x(a - x)^{-1} \tag{11}$$

to construct the equation

$$v + x(a - x)^{-1}v = v + xa^{-1}[v + x(a - x)^{-1}v] . \tag{12}$$

Substituting Eq. 10 on both sides, we obtain

$$\tilde{t}(\epsilon) = v + vQ(\epsilon - H_0)^{-1}\tilde{t}(\epsilon) . \tag{13}$$

This linear equation for the operator $\tilde{t}(\epsilon)$ becomes a linear *integral* equation for the two-body matrix elements

$$\langle \mathbf{k}_1, \mathbf{k}_2 | \tilde{t}(\epsilon) | \mathbf{k}_1', \mathbf{k}_2' \rangle_A = \langle \mathbf{k}_1, \mathbf{k}_2 | v | \mathbf{k}_1', \mathbf{k}_2' \rangle_A$$

$$+ \frac{\Omega^2}{(2\pi)^6} \int d\mathbf{p}_1 d\mathbf{p}_2 \langle \mathbf{k}_1, \mathbf{k}_2 | v | \mathbf{p}_1, \mathbf{p}_2 \rangle \frac{\theta(p_1 - p_F)\theta(p_2 - p_F)}{\epsilon - \epsilon(p_1) - \epsilon(p_2)}$$

$$\times \langle \mathbf{p}_1, \mathbf{p}_2 | \tilde{t}(\epsilon) | \mathbf{k}_1', \mathbf{k}_2' \rangle_A , \tag{14}$$

where we have passed to the integral form of Eq. 13.

The kernel of Eq. 14 is nonsingular for $\epsilon < 2\epsilon_F$ and for bounded finite-range potentials, like the finite repulsive case of Eq. 6.5. The solutions $\tilde{t}(\epsilon)$ will also be well behaved, contributing finite matrix elements to Eq. 9 even in the limit $V_0 \to +\infty$. The behavior of the $\tilde{t}$-matrix elements can be better understood by turning to a very similar operator, $t(E)$, the t-operator or t-matrix for the two-body scattering through the potential v (see Chapter 11). The operator equation for scattering at energy E is

$$t(E) = v + v(E - H_0 + i\eta)^{-1}t(E) ,$$ (15)

where H_0 is the kinetic-energy operator for two particles. This differs from Eq. 13 for $\tilde{t}(\epsilon)$ by the presence of the projection operator Q in Eq. 13, prohibiting excitation of pairs into occupied states in the Fermi-gas ground state Φ_0, and also by the infinitesimal $i\eta$ in Eq. 15, which guarantees that solutions to that equation correspond to outward-going scattered waves. The solutions to Eqs. 13 or 14 will be real for a real potential, as would be expected for a contribution to an energy shift, as in Eq. 9. The matrix elements of the scattering t-matrix are in general complex. Other than these differences, the behavior of $\tilde{t}(\epsilon)$ and that of $t(E)$ are qualitatively similar; one may expect that when the two-body scattering problem has a finite scattering amplitude for a given potential v, then the integral equation Eq. 14 will also have a well-behaved solution for the same v. The first condition is true for a finite-range repulsive potential, and even for the hard-sphere potential, Eq. 6.1.

For two-body scattering of particles of equal mass (m), the t-matrix elements may be written

$$\langle \mathbf{k}_1, \mathbf{k}_2 | t(E) | \mathbf{k}_1', \mathbf{k}_2' \rangle = (2\pi)^3 \delta(\mathbf{k}_1 + \mathbf{k}_2 - \mathbf{k}_1' - \mathbf{k}_2') \langle \mathbf{k} | t(E) | \mathbf{k}' \rangle ,$$ (16a)

where the δ-function expresses conservation of momentum, $\mathbf{k} = \frac{1}{2}(\mathbf{k}_1 - \mathbf{k}_2)$ is the relative momentum, and $E = k^2/m$ in the c.m. frame, where the reduced mass is $m/2$. As is usual for scattering theory, the plane waves in Eq. 16a are normalized in a unit volume. For comparison with the matrix elements of $\tilde{t}$, we must change to normalization in a box of large volume Ω, with periodic boundary conditions, as in Eq. 1.52, which changes the δ-function of Eq. 16a [times $(2\pi)^3$] to a Kronecker delta (divided by Ω):

$$\langle \mathbf{k}_1, \mathbf{k}_2 | t(E) | \mathbf{k}_1', \mathbf{k}_2' \rangle = \Omega^{-1} \delta_{\mathbf{k}_1 + \mathbf{k}_2, \, \mathbf{k}_1' + \mathbf{k}_2'} \langle \mathbf{k} | t(E) | \mathbf{k}' \rangle .$$ (16b)

The forward scattering amplitude is related to the c.m. matrix element with $\mathbf{k}' = \mathbf{k}$ by

$$f(E, 0°) = -\frac{m}{4\pi} \langle \mathbf{k} | t(E) | \mathbf{k} \rangle .$$ (17)

The scattering length a was defined in Eq. 6.9 by the limit $k(E) \to 0$:

$$a = \frac{m}{4\pi} \lim_{k \to 0} \langle \mathbf{k} | t(E) | \mathbf{k} \rangle .$$ (18)

In this limit, the term $i\eta$ in Eq. 15 has no effect: a is real for real v.

We may obtain the same limit for matrix elements of $\tilde{t}(E)$ by taking $k_1, k_2 \to 0$, $E \to 0$ and letting $Q \to 1$. These limits combined correspond to taking the limit of a zero-density gas, for which $k_F \to 0$ and no states are blocked. Now Eq. 13 corresponds to the $k \to 0$ limit of Eq. 15, so their solutions must be identical. Therefore, in the low-density limit, we find that

$$\lim_{k_1, k_2 \to 0} \langle \mathbf{k}_1, \mathbf{k}_2 | \tilde{t}(0) | \mathbf{k}_1, \mathbf{k}_2 \rangle = \frac{4\pi a}{\Omega m} . \tag{19a}$$

This shows directly that in the low-density limit the energy shift is finite for finite scattering length a. Remember that for the hard-sphere potential, $a = d$, the sphere diameter. For a finite, but strong repulsion, like that of Eq. 6.5 with $V_0 \gg E$, then $a \cong d$. Equation 19a actually gives only the direct term of the ladder series, Figure 10.1a. We may include the exchange series of Figure 10.1b, obtaining

$$\lim_{\substack{k_1, k_2 \to 0 \\ \Omega \to \infty}} \Omega \langle \mathbf{k}_1 \mu_1, \mathbf{k}_2 \mu_2 | \tilde{t}(0) | \mathbf{k}_1 \mu_1, \mathbf{k}_2 \mu_2 \rangle_A = \frac{4\pi}{m} a(1 - \delta_{\mu_1 \mu_2}) , \tag{19b}$$

where we have now explicitly included the internal quantum numbers μ, and assumed the potential to be μ-independent.

10.2 LOW-DENSITY EXPANSION FOR A HARD-SPHERE FERMI GAS

The connection we have just established between the ladder series at low density and the two-body scattering length (at low energy) allows us to write the ground-state energy density of the Fermi gas with strong repulsion as a low-density expansion in the dimensionless parameter $k_F a$. To first order, the energy density is given by the sum of the noninteracting energy (Eq. 6.4), $E_0 / \Omega = \frac{3}{5} \epsilon_F \rho_0$, with the contribution of the ladder series (Eq. 9) in the limit given in Eq. 19b. Other diagrams will contribute terms of higher order, as we shall see shortly. The sums in Eq. 9 are performed as for $\Delta E^{(1)} / \Omega$, Eqs. 6.6–6.12, giving a similar result:

$$\frac{\Delta E_{\text{ladder}}}{\Omega} = \frac{2(g-1)}{3\pi} \epsilon_F \rho_0 (k_F a) , \qquad k_F a \ll 1 . \tag{20}$$

Comparing with the first-order result of Eq. 6.12, we see that the effect of summing the ladder series (at low density) is to replace the Born approximation a_B (Eq. 6.8) to the scattering length by its exact value a, through Eq. 18. Low density is defined by $k_F d = k_F a \ll 1$, for which the mean separation of particles is much larger than the interaction range. As we conjectured in Sections 6.1 and 6.2, the correction due to the hard-sphere repulsion is small relative to E_0 / Ω in this domain, as can be seen by comparing Eq. 20 with Eq. 6.4. The divergence of the perturbation series may be thought of as an expression of the very large changes required by the *wave function* of the system to exclude other particles from the repulsive sphere around each particle. These very large changes take place in a small fraction of configuration space, so that the average effect on the energy is in fact small.

So far we have only evaluated the ladder series, but have not investigated the effect of other diagrams in the perturbation expansion. Clearly it will do no good to evaluate them singly; we shall again have an ill-behaved series in powers of V_0 or a_B, as in Section 6.2. We can, however, arrange diagrams in partial series so that every interaction v becomes a rung in a "ladder," which forms part of a diagram. For example, the third-order Goldstone diagram of Figure 10.2 can be combined with an infinite series of diagrams of the same sort, so that each original interaction is replaced by a ladder series similar to that defined by the diagrams of Figure 10.2, representing a two-body operator like $\tilde{t}(\epsilon)$, defined in Eqs. 8 and 9. Then the triple-ladder diagrams represented in Figure 10.2 are of order $\langle \tilde{t} \rangle^3$, where $\langle \tilde{t} \rangle$ is an average matrix element of the ladder operator $\tilde{t}(\epsilon)$, evaluated at an appropriate energy. Again, the comparison with scattering amplitudes is useful to establish the order of magnitude of the term, which is $\sim a^3$, or $(k_F a)^3$. (We may use $\langle \tilde{t} \rangle \sim \langle t \rangle \sim 4\pi a/m$ for *magnitudes*.)

Since the ladder series includes all second-order Goldstone diagrams, there are no double-ladder series of order $(k_F a)^2$; the first diagram corrections to Eq. 20 are therefore from the triple-ladder diagrams, and are of order $(k_F a)^3$. However, there is a correction to the first-order term (Eq. 20) of order $(k_F a)^2$ which comes from the lowest-order *difference* between the first-order ladder $\tilde{t}(\epsilon)$ and the scattering t-matrix. This comes from the Pauli exclusion of occupied states in the gas, and was ignored to leading order in Eq. 19. We may recover this correction by using the following relation to compare $\tilde{t}(\epsilon)$ with $t(E)$ for $\epsilon = E$:

$$\tilde{t}(E) = t(E) + t(E)[Q(E - H_0)^{-1} - (E - H_0 + i\eta)^{-1}]\tilde{t}(E) . \qquad (21)$$

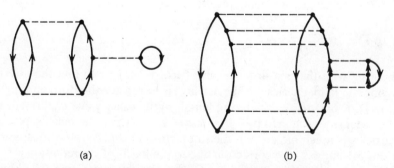

(a)   (b)

FIGURE 10.2   Third-order Goldstone diagram (a), and the triple-ladder diagram (b), of which it is the simplest example. Only the direct ladder terms have been shown.

This identity may be derived by solving Eq. 15 formally for v and substituting the result into Eq. 13; it can be seen to eliminate v explicitly. At low density, t is of order a, as we have seen; the leading correction to t of Eq. 19 is of order a^2, and may be calculated from Eq. 21 by iteration and integration. The result is a second-order correction to Eq. 20 due to the Pauli principle, which was first obtained by Huang and Yang.[*]

The expansion of the energy density may be written explicitly, incorporating this result:

$$\frac{E}{\Omega} = \epsilon_F \rho_0 \left[\frac{3}{5} + \frac{2(g-1)}{3\pi} (k_F a) + \frac{4(g-1)}{35\pi^2} (11 - 2\ln 2)(k_F a)^2 + O(k_F^3 a^3) \right],$$

(22)

where the third and higher orders come from the higher-order Goldstone diagrams, combined into multiple-ladder series, as described. For low density or short range ($k_F a \ll 1$), the corrections to the energy density are indeed small, the linear term giving the main effect of the potential. In fact, the result does not appear to depend on the form of the potential directly, only on the scattering length a. This is true only through second order in $k_F a$, to which only scattering amplitudes at vanishing energy contribute. The higher-order terms depend on the *shape* of the repulsive potential. Explicit results through third order have been obtained for the hard-sphere potential.[†]

Finally, although Eq. 22 gives an accurate expression for the ground-state energy density at low particle density, we cannot expect the expansion to give an analytic result; that is, there is no circle of convergence in $k_F a$. This follows from the fact that for $a < 0$ (attractive potential), the Fermi gas will have a superconducting ground state, with no analytic perturbation expansion.

10.3 BOSE GAS AT LOW DENSITY

The hard-sphere Bose gas has some similarities to the Fermi gas we have just discussed. We give a brief account of the low-density expansion for the ground-state energy, following the general methods we used for the Fermi case. This will easily lead us to the contribution to first order in the scattering length a, or equivalently, in the sphere diameter d. However, to continue the expansion, we shall need to include an important property of the repulsive Bose gas, namely, that the single-particle excitation spectrum is altered in an important way by the interaction. The change is analogous to that of a Fermi gas with a weak *attractive* interaction, which changes the free

*Kerson Huang and C. N. Yang, Phys. Rev. **105**, 767 (1957).
†C. DeDominicis and P. C. Martin, Phys. Rev. **105**, 1417 (1957).

Fermion excitation spectrum to that of *quasifermions*, with a gap, as in Eq. 5.73. The transformation to quasiparticles in the Bose case is actually the original application of the method by Bogolyubov.

Probably the best-studied Bose quantum fluid is liquid ^{4}He. The interatomic potential is of the type shown in Figure 6.2, with a strong repulsion at shorter range (from Pauli repulsion of the electrons) and a weaker attraction at larger distances. The short-range repulsion is responsible for the distortion of the excitation spectrum of the fluid in its ground state (or at low temperature: $T \lesssim 2.2$ K), which is connected with its *superfluid* properties. The Bose gas with a repulsive or hard-sphere interaction is an idealized model of such a system.

We start with the ground state of a noninteracting Bose gas of particle density ρ_0. Assuming a large box of length L, volume $\Omega = L^3$, with periodic boundary conditions, plane wave states of the lowest momenta $\mathbf{k}_0$ will be occupied, with components $(k_0)_i L = \pm 2\pi$, $i = 1, 2, 3$. The ground-state vector may be written

$$|\Phi_0\rangle = (N!)^{-1/2} [a^\dagger(k_0)]^N |0\rangle ,\qquad (23)$$

where we ignore the directions of the vectors $\mathbf{k}_0$, anticipating the limit $\Omega \to \infty$, $|\mathbf{k}_0| \to 0$. We also ignore internal degrees of freedom. The energy density of the system is given by

$$\frac{E_0}{\Omega} = \frac{k_0^2}{2m}\,\rho_0 ,\qquad (24)$$

which vanishes for $\Omega \to \infty$, at fixed ρ_0, unlike the Fermi case, Eq. 6.4. The first-order contribution of a two-body potential is easily calculated (Exercise 10.1):

$$\frac{\Delta E^{(1)}}{\Omega} = \frac{N(N-1)}{2\Omega} \langle \mathbf{k}_0, \mathbf{k}_0 | v | \mathbf{k}_0, \mathbf{k}_0 \rangle \qquad (25a)$$

and evaluated in the limit $\Omega \to \infty$:

$$\frac{\Delta E^{(1)}}{\Omega} \to \tfrac{1}{2}\rho_0^2 \int d\mathbf{r}\, v(r) = 2\pi\, \frac{\rho_0^2 a_B}{m} ,\qquad (25b)$$

where we have used Eq. 6.8 to connect the volume integral of the interaction to the scattering length in the Born approximation. The result in Eq. 25b differs from the Fermi-gas result of Eq. 6.12 only by a statistical factor $(g-1)/g$.

The problems we found in the use of perturbation theory for strong repulsive potentials (like that of Eq. 6.5) in the Fermi gas will also occur here: the lack of convergence of the series in V_0, and the divergence in each order, for $V_0 \to \infty$. However, the method of combining terms corresponding to the ladder diagrams into a series to be evaluated by an integral equation can also be used here. We may define a two-body operator $\tilde{t}(\epsilon)$ as in Eqs. 8, 10, and 13, with Q again projecting onto two-particle excitations of the ground state Φ_0. The contribution of the ladder series is then (note that $\epsilon = k_0^2/m$)

$$\frac{\Delta E_{\text{ladder}}}{\Omega} = \frac{N(N-1)}{2\Omega} \langle \mathbf{k}_0, \mathbf{k}_0 | \tilde{t}\left(\frac{k_0^2}{m}\right) | \mathbf{k}_0, \mathbf{k}_0 \rangle . \tag{26}$$

In the limits $\Omega \to \infty$, $k_0 \to 0$, the operator $\tilde{t}(\epsilon) \to \lim_{E \to 0} t(E)$, the t-matrix for scattering of two particles, as given in Eq. 16. Expressed in terms of the scattering length, Eq. 26 becomes

$$\frac{\Delta E_{\text{ladder}}}{\Omega} \xrightarrow[\Omega \to \infty]{} 2\pi\rho_0^2 \frac{a}{m} \qquad \text{(Bose gas)} , \tag{27}$$

as well-defined expression even for the hard-sphere potential ($V_0 \to \infty$, $a = d$). The contribution to the energy density to this order in a is almost the same as that for the Fermi gas, for which we have, rewriting Eq. 20 in comparable form,

$$\frac{\Delta E_{\text{ladder}}}{\Omega} \xrightarrow[\substack{\Omega \to \infty \\ k_F a \ll 1}]{} 2\pi\rho_0^2 \frac{a}{m} \cdot \frac{g-1}{g} \qquad \text{(Fermi gas)} . \tag{28}$$

The statistical factor $(g-1)/g$ expresses the reduction of the effect of repulsion due to the Pauli principle, which keeps fermions of the same spin (or other internal symmetry) state away from each other.

For the Fermi system, we found that $k_F a$ was the appropriate expansion parameter for low density. For bosons, we might also expect to expand in a dimensionless parameter proportional to a, say $\rho_0^{1/3} a$. However, here the special features of the repulsive Bose gas enter, giving a different result. One may summarize the effect on our present approach as follows. The single-particle excitation spectrum of the system is modified by the repulsion so that the *quasiparticle* energy (see Exercise 5.10)

$$E(k) = \sqrt{\left(\frac{k^2}{2m} + \frac{4\pi a\rho_0}{m}\right)^2 - \left(\frac{4\pi a\rho_0}{m}\right)^2} \tag{29}$$

replaces $k^2/2m$ in the unperturbed energy for excitations. This modifies H_0 to a new operator $\tilde{H}_0$ in the ladder operator $\tilde{t}_0(\epsilon)$. To lowest order in a, this still gives $\tilde{t}(\epsilon) \to t(E \to 0)$, leading directly to Eq. 27. The next correction can be obtained from Eq. 21 (with $\tilde{H}_0$ replacing H_0 in the first term in the brackets). The result of integration on k is a term of order $a^{5/2}$, rather than a^2 as expected, reflecting the nonanalytic distortion of the spectrum, given in Eq. 29. The first two terms in the expansion of the Bose gas can then be written

$$\frac{\Delta E}{\Omega} = 2\pi\rho_0^2 \frac{a}{m} \left[1 + \frac{128}{15} \left(\frac{\rho_0 a^3}{\pi} \right)^{1/2} \right], \tag{30}$$

a result first obtained by Lee and Yang.* The next corrections are of order $\rho_0 a^3 \ln \rho_0 a^3$ and $\rho_0 a^3$.

10.4   NUCLEAR MATTER AND BRUECKNER THEORY

We have seen that the ground-state energy of a hard-sphere gas at low density can be expressed in terms of the scattering length and the density of particles. Now we turn to the more physical problem of a system of particles—we shall discuss only fermions—whose interactions are strongly repulsive at short distances, like the hard-sphere gas, but which also have an attractive potential at somewhat larger distances. We shall see that the ground-state energy of this system can also be expressed as an expansion in terms of the ladder series discussed in Sections 10.1 and 10.2, but not in the zero-energy limit. The method we follow was introduced by Brueckner and collaborators, and has been developed and refined over more than twenty years. The original problem for which the method was developed is the theory of *nuclear matter*, as we discuss below.

As we noted in Section 6.1.1, many systems in nature have potentials of the general form of Figure 6.2, with both repulsion and attraction. The attractive part of the atom–atom interaction is responsible for the binding of two or more atoms into molecules; similarly, the nucleon–nucleon interaction binds atomic nuclei. The ground state of a system of very many particles interacting with such a potential may be bound, forming a liquid or a solid. With sufficiently strong repulsion at short distance, a large system will also *saturate*, that is, the ground state will be stable against compression at a particle density independent of the size of the system. Examples of such systems are the two forms of liquid helium: the Bose liquid of ^{4}He atoms and the Fermi liquid of ^{3}He atoms, at very low temperatures. Most other many-atom systems will form solids in their ground states.

Atomic nuclei are also a kind of Fermi liquid, which comes in small "droplets" limited in size by the repulsive Coulomb potential of the positive charge, which becomes stronger than the average attractive interaction at atomic number $A \cong 240$. If we consider a hypothetical system of protons and neutrons with nuclear interactions, but without any Coulomb repulsion, there would apparently be no limit to the size of the bound system formed. An infinite nuclear system of this sort is called *nuclear matter*. It is expected from the properties of real (finite) nuclei that nuclear matter is a saturated Fermi liquid in its ground state.

---

*T. D. Lee and C. N. Yang, Phys. Rev. **105**, 1119 (1957).

An infinite material consisting entirely of neutrons (neutron matter) is also of interest. On the one hand, it is a physical possibility since there is no Coulomb interaction for neutrons. On the other hand, it is not clear from known nuclear physics whether such a fluid has a bound state from nuclear interactions alone. With the additional attractive interaction of *gravity*, a very large mass of $(>10^{40})$ neutrons will form a bound system, constituting a *neutron star*.

For a saturated, bound fluid system in its ground state, without an external container (i.e., at zero pressure), the particle density takes a fixed value, the saturation density at which the total energy per particle is a minimum. In a container the material may be compressed to higher density, with a higher energy per particle. For the simplest potentials, e.g. spin independent, the ground state has a uniform distribution, except near the walls of the container. Under some circumstances, the ground state may exhibit nonuniform behavior, e.g. standing waves of spin polarization. Average density lower than the saturation value corresponds to negative pressure; the fluid is unstable against separation into drops or chunks or zero-pressure, saturated material.

Let us consider a uniform medium of fermions at fixed density ρ which we assume to be stable; therefore $\rho > \rho_0$, the saturation density. The potential between particles is assumed to be of the form of Figure 6.2. The Goldstone perturbation expansion for this system has all the same divergence problems we encountered in Section 6.2. Therefore we are again led to consider the method of summation of ladder diagrams, introduced in Section 10.1, to obtain a well-defined expansion for the ground-state energy. We again introduce the two-body operator $\tilde{t}(\epsilon)$, which replaces the ladder series, as given in Eqs. 9–14. The contribution to the energy density is then

$$\left(\frac{\Delta E}{\Omega}\right)_{\text{ladder}} = (2\Omega)^{-1} \sum_{\mathbf{k}_1, \mathbf{k}_2 \leq F} \langle \mathbf{k}_1, \mathbf{k}_2 | \tilde{t}(\epsilon(k_1) + \epsilon(k_2)) | \mathbf{k}_1, \mathbf{k}_2 \rangle_A , \quad (31)$$

where sums over spin states and other internal quantum numbers are implicit. We represent the $\tilde{t}$-matrix elements by a new diagram, in Figure 10.3.

In the case of the hard-sphere gas, we were able to go to the low-density limit, in which the right-hand side of Eq. 31 could be evaluated in terms of the scattering length, as in Eq. 20. In addition, we could show that Eq. 20 was the leading contribution to the interaction energy at low density, corrections going as higher order in $k_F a$. For the self-bound liquid, we are interested in densities at or above the saturation density ρ_0. Therefore we may not make direct use of the low-density limit and the related association of the operator $\tilde{t}$ with the free-scattering t-matrix of Eq. 15. However, the $\tilde{t}$-matrix may be obtained by solving the integral equation 14, as we discuss shortly. A second problem is to determine the conditions under which the reexpression of the original Goldstone expansion as an expansion in $\tilde{t}$ is

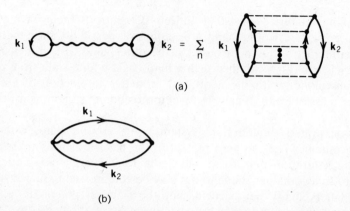

(a)

(b)

FIGURE 10.3   (a) The diagram on the left represents the direct matrix element of $\tilde{t}$ in Eq. 31, which is the formal sum of the ladder series represented on the right, and in Figure 10.1a. The exchange diagram is shown in (b).

convergent, so that accuracy of approximations to the ground-state energy are controlled. We return to this question later.

In order to evaluate the matrix elements of $\tilde{t}(\epsilon)$ for the system at finite density, $\rho > \rho_0$, one must solve an integral equation of the form of Eq. 14. This requires the explicit form of the two-body potential v, unlike the low-density limit, for which only the scattering length was required. The problem is analogous to finding the amplitude for two-body scattering at finite energy, given the scattering potential. For a central potential, this latter problem can be reduced to a separate one-variable integral equation in each partial wave. For the $\tilde{t}$-matrix problem, however, there are two additional complications: One is the projection operator Q (see Eqs. 2, 13) which eliminates two-particle intermediate states within the Fermi sea. The second is *self-consistency* of the one-particle energy spectrum, which we explain next. Both of these impose effects of the whole medium on the $\tilde{t}$-matrix, somewhat complicating the calculation. For example, the medium imposes a preferred frame, so that $\tilde{t}$ may depend on $\mathbf{k}_1 + \mathbf{k}_2$ as well as $\mathbf{k}_1 - \mathbf{k}_2$, unlike the free-space t-matrix. These technical modifications can be (and have been) incorporated into the integral equation for $\tilde{t}$, for practical calculations.

The idea of using a self-consistent single-particle energy spectrum is a generalization of the introduction of the Hartree–Fock basis into the perturbation expansion, which was discussed in Section 9.4.4. There one modified the unperturbed Hamiltonian H_0 by addition of an auxiliary potential U, Eqs. 9.28 and 9.29. With the self-consistent choice of U given by Eq. 9.35, we found that all single-particle insertion diagrams of the form shown in Figure 9.9 are exactly canceled. In the present case, we may choose U as follows:

$$U(k_1) = \langle \mathbf{k}_1 | U | \mathbf{k}_1 \rangle = \sum_{k_2 \leq F} \langle \mathbf{k}_1, \mathbf{k}_2 | \tilde{t}(\epsilon(k_1) + \epsilon(k_2)) | \mathbf{k}_1, \mathbf{k}_2 \rangle_A \qquad (32)$$

for the occupied plane wave orbits, $k_1 \leq k_F$, with

$$\epsilon(k) = \frac{k^2}{2m} + U(k) . \qquad (33)$$

These two relations are often called the Brueckner–Hartree–Fock self-consistency conditions: Eq. 32 simply defines U so that it will cancel the insertion of a ladder series, as illustrated in Figure 10.4. Thus the choice of the condition in Eq. 32 is equivalent to summing formally the infinite series of ladder or $\tilde{t}$-matrix insertions to all orders. Note that there is a *new* self-consistency here, over and above that of Hartree–Fock theory: the $\tilde{t}$-matrix element in Eq. 32 must be evaluated at an energy $\epsilon = \epsilon(k_1) + \epsilon(k_2)$, which depends on $U(k)$ through Eq. 33, and therefore on $\tilde{t}$ itself.

We have imposed Eqs. 32 and 33 only for the occupied orbits, $k \leq k_F$. There are a number of technical reasons for this, the simplest of which is that it turns out that the cancellation of insertions indicated in Figure 10.4 is exact for hole lines only; for particle lines, the ladder series is not evaluated at $\epsilon = \epsilon(k_1) + \epsilon(k_2)$, as in Eq. 32, but at a value of ϵ which depends on the rest of the diagram. An example is given in Figure 10.2b: the middle ladder insertion depends on the energies of the other particle and hole lines at the same times. It is possible to introduce an auxiliary potential $U(k)$ for $k > k_F$ which partially compensates for particle-line insertions. This has the advan-

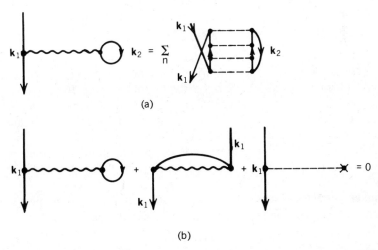

(a)

(b)

FIGURE 10.4   (a) Ladder insertion in hole line, defined by the direct term of the right-hand side of Eq. 32. The self-consistency condition given by Eq. 32 will result in cancellation of diagrams shown in (b), where the last diagram represents $-U$, as in figures 9.7, 9.9.

tage of giving a smooth single-particle potential, as a function of k, through the region of the Fermi momentum, which well represents physical single-particle excitations in this region.* On the other hand, it is also possible to treat the particle-line insertions as one of a class of higher-order diagrams (e.g. of third order in $\tilde{t}$, as for Figure 10.2b) that are to be kept together for calculation. The role of particle insertions in convergence of the theory (which is discussed later) is not completely understood.

We return to our original problem of finding the ground-state energy of our uniform Fermi fluid at some stable density $\rho > \rho_0$: We can evaluate the matrix elements of $\tilde{t}$ by solving the integral equation 14 with the Q-operator *and* the self-consistency conditions (Eqs. 32, 33), both of which will depend on ρ (or k_F). The lowest-order energy shift is obtained from Eq. 31, which we now express in a form analogous to Eq. 28:

$$\left(\frac{\Delta E}{\Omega}\right)_{\text{ladder}} = \tfrac{1}{2}\rho^2 \langle \tilde{t} \rangle , \tag{34}$$

where $\langle \tilde{t} \rangle$ is the average t-matrix element over the occupied states, or more descriptively, the average two-body interaction.

For the case of nuclear matter,† the stable density is $\rho_0 \cong 0.17$ nucleons/ fm^3 (for equal numbers of protons and neutrons), or $k_F \cong 1.4$ fm^{-1}. The kinetic energy *per nucleon* is

$$\frac{E_0}{N} \cong 25 \text{ MeV} . \tag{35a}$$

For a "standard" nucleon–nucleon potential (Reid) the energy per nucleon from the ladder series, given by Eq. 34 divided by ρ_0, is about

$$\left(\frac{\Delta E}{N}\right)_{\text{ladder}} \cong -35 \text{ MeV} , \tag{35b}$$

giving a lowest-order energy per nucleon of $\cong -10$ MeV. The value extracted from actual nuclei via the empirical mass formula is $\cong -16$ MeV. The discrepency of $\cong -6$ MeV could be expected to come from higher-order corrections, *if* the Reid potential (or any two-nucleon potential, for that matter) is alone responsible for the binding of nuclei. There are good physical reasons to think that two-body interactions may not be the whole story, but that is not our subject. For the *given* interaction, one wants to know how accurate is the ladder approximation of Eq. 35b. This involves going to higher order.

Higher order in what? We have rearranged the original perturbation series into an expansion in the $\tilde{t}$-matrix, in which the first term is Eq. 34, and

---

*J. P. Jeukenne, A. Lejeune, and C. Mahaux, Phys. Reports **25**, 83 (1976).
†For a review of Brueckner theory as it has been applied to nuclear matter, see B. D. Day, Rev. Mod. Phys. **50**, 495 (1978).

the next terms are of third order in $\tilde{t}$, e.g. Figure 10.2b. It has been shown that convergence in powers of $\tilde{t}$ may be poor. We have already summed one partial series in $\tilde{t}$ by self-consistency: Eqs. 32, 33. A systematic approach to other higher-order terms seems to be that of the *hole-line* expansion (which will not be further developed here; see Day, *op. cit.*). The ladder series of Figure 10.3 sums all diagrams with two independent hole lines, k_1, k_2. Diagrams with three hole lines include all those of third order in $\tilde{t}$ (e.g. Figure 10.2b) as well as a class of diagrams of higher order. The technique for summing this entire group (Bethe–Faddeev method) is related to methods used in three-body scattering theory. The next order has four hole lines, and so on. It emerges that there is a relevant expansion parameter (approximately) given by $\rho\omega$, where ω is a correlation volume in which wave functions are strongly distorted by the potential. For nuclear matter, it has been estimated that $\rho_0\omega \cong 0.15$, so that we may expect that corrections to the ladder-approximation result, Eq. 35b, could be of order

$$(0.15) \times (35\,\text{MeV}) \cong 5\,\text{MeV} , \tag{35c}$$

which is the *order* of the discrepancy between the first-order and the empirical values. In this sense, the Brueckner expansion for nuclear matter is a low-density expansion. Brueckner calculations have been carried out to fourth order; problems remain in reproducing the empirical binding energy *and* the saturation density simultaneously (Day, *op. cit.*; Jackson[*]) with two-body potentials.

For liquid ^{3}He, the expansion parameter $\rho_0\omega$ is not small, and the hole-line expansion is not convergent. Alternative methods have been developed, based on the variational principle for the ground-state energy, which are not limited by density. Results for nuclear matter have been rather similar to the hole-line results of Day.

10.5   ELECTRON GAS: SINGULARITIES AND RING DIAGRAMS

In Section 6.3 we calculated the ground-state energy of the electron gas to second order in e^2, the strength of the Coulomb potential. While the first-order result, Eq. 6.31, was finite, the second-order term showed a singularity, as in Eq. 6.41, which we associated with the small-q (momentum transfer) dependence of the Coulomb potential $V(q)$. (See the discussion following Eq. 6.41.) In this section we shall again apply the method of summation of partial series to obtain a well-behaved expansion for the ground-state energy.

For this discussion we shall find it more convenient to work with Feynman diagrams, first to classify the singular behavior of diagrams, and

[*]A. D. Jackson, Ann. Rev. Nucl. Part. Sci **33**, 105 (1983).

second to perform the formal summation which leads to nonsingular results. We begin by redoing the second-order calculation of Section 6.3.2, for which we used a Rayleigh–Schrödinger (or Goldstone) expression of Eq. 6.16, and the Coulomb potential matrix elements of Eq. 6.34 and 6.36. The appropriate Feynman diagrams are shown in Figure 10.5, corresponding to the expressions given in Eq. 8.42 (and to Figure 8.7), with plane-wave particle and hole states for the present translationally uniform problem.

We first consider the direct term of Figure 10.5a, which we may think of as consisting of two simultaneous particle–hole (p–h) excitations, which begin and end in time through mutual interactions. Let us introduce a function which describes the propagation of one of these p–h excitations (see Figure 10.6) from t_2 to t_1, defined as follows:

$$\Pi^0(\mathbf{q}, t_2 - t_1) = -\frac{i}{(2\pi)^3} \sum_\mu \int d\mathbf{k}\, G^0_\mu(\mathbf{k}, t_2 - t_1) G^0_\mu(\mathbf{k} + \mathbf{q}, t_1 - t_2), \quad (36)$$

where $\mathbf{q} = \mathbf{p} - \mathbf{k}$ is the total momentum of the excitation, and μ is the spin projection of the particle states (which is conserved by the spin-independent Coulomb interaction). This particle–hole propagator is also called the unperturbed *polarization part*, referring to the p–h excitation as a local polarization of the medium by the Coulomb potential. It is clear that for an isotropic system, Π^0 will be independent of the direction of $\mathbf{q}$: $\Pi^0(q, t_2 - t_1)$.

To construct the Feynman diagrams of Figure 10.5a, we rewrite Eq 8.42 as a product of two such propagators (one for each p–h excitation) multiplied by the Coulomb matrix elements in momentum space, given by Eqs. 6.33 and 6.34; the summations over particle and hole labels become integrations over $\mathbf{k}_1$ and $\mathbf{k}_2$. Using momentum conservation, as in Eq. 6.34, and converting the sums to integrals for $\Omega \to \infty$ (see, e.g., Eq. 1.56), the expression for the diagram becomes

$$D(\text{iii})_a = \frac{-\Omega}{2(2\pi)^3} \int d\mathbf{q}\, V(q)\Pi^0(q, t_2 - t_1) V(q)\Pi^0(q, t_1 - t_2), \quad (37)$$

where we have taken the direct part only, corresponding to Eq. 6.38a.

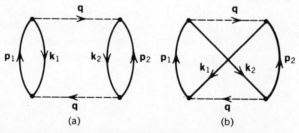

FIGURE 10.5 Feynman diagrams for the second-order correlation energy: (a) direct, (b) exchange. Arrows on interaction lines indicate direction of momentum transfer.

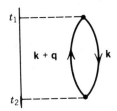

t_1

$\mathbf{k} + \mathbf{q}$ $\mathbf{k}$

t_2

FIGURE 10.6 Diagram for particle–hole propagator, or unperturbed polarization part, defined in Eq. 36.

Finally, to obtain the contribution to the energy, we must integrate Eq. 37 over times, with one time fixed, as in Eq. 9.21, which also brings in a factor of $-i/2$ for $n = 2$. Taking $t_1 = 0$ and $t_2 = t$, we find the energy per particle to be

$$\frac{\Delta E_a^{(2)}}{N} = \frac{i}{4(2\pi)^3 \rho_0} \int_{-\infty}^{\infty} dt \int d\mathbf{q}\, V(q)\Pi^0(q, t)V(q)\Pi^0(q, -t). \quad (38)$$

The time integral can be converted to a frequency or energy integral by introducing the Fourier transforms, as in Section 8.6:

$$\Pi^0(q, \omega) = -i \sum_\mu \int \frac{d\mathbf{k}\, d\omega'}{(2\pi)^4}\, G^0_\mu(\mathbf{k}, \omega')G^0_\mu(\mathbf{k} + \mathbf{q}, \omega + \omega'), \quad (39)$$

so that Eq. 38 may be rewritten in the form

$$\frac{\Delta E_a^{(2)}}{N} = \frac{i}{4\rho_0(2\pi)^4} \int d\omega\, d\mathbf{q}\, V(q)\Pi^0(\mathbf{q}, \omega)V(q)\Pi^0(\mathbf{q}, \omega). \quad (40)$$

The ω'-integral and spin sum (μ is the spin projection) in Eq. 39 are easily performed, using Eq. 7.33b for $G^0_\mu(\mathbf{k}, \omega)$, to yield

$$\Pi^0(\mathbf{q}, \omega) = 2 \int \frac{d\mathbf{k}}{(2\pi)^3}\, \theta(|\mathbf{k} + \mathbf{q}| - k_F)\theta(k_F - k)$$

$$\times \left\{ \frac{1}{\omega + \epsilon(\mathbf{k}) - \epsilon(\mathbf{k} + \mathbf{q}) + i\eta} - \frac{1}{\omega - \epsilon(\mathbf{k}) + \epsilon(\mathbf{k} + \mathbf{q}) - i\eta} \right\}. \quad (41)$$

This function was studied first by Lindhard.*

Now the Feynman expression for the direct second-order energy, Eq. 40, is completely equivalent to the direct part of Eq. 6.16, as we have seen, e.g. in Section 8.5. The singular behavior of this term, for small values of the q-integrand, becomes apparent for Eq. 40 on performing the ω-integration, using Eq. 41, and dropping constant factors:

*J. Lindhard, Kgl. Dan. Vid. Sel. Mat.-Fys. Medd. **28**, no. 8 (1954).

$$\int d\omega \, [\Pi^0(\mathbf{q}, \omega)]^2 \propto \int d\mathbf{k}_1 \, d\mathbf{k}_2 \, [\epsilon(\mathbf{k}_1) + \epsilon(\mathbf{k}_2) - \epsilon(\mathbf{k}_1 + \mathbf{q}) - \epsilon(\mathbf{k}_2 - \mathbf{q})]^{-1}$$

$$\propto \int d\mathbf{k}_1 \, d\mathbf{k}_2 \, [\mathbf{q} \cdot (\mathbf{k}_1 - \mathbf{k}_2 + \mathbf{q})]^{-1} , \qquad (42)$$

where the momentum restrictions at k_F are $k_1, k_2 \leq k_F$ and $|\mathbf{k}_1 + \mathbf{q}|$, $|\mathbf{k}_2 - \mathbf{q}| \leq k_F$. The last integral brings us back to the form of Eq. 6.40, which diverges as $\ln q_0$ (see Eq. 6.41). As we noted in Section 6.3.2, the second-order exchange term, represented by Figure 10.5b, is not singular, giving the contribution of Eq. 6.42.

The divergent behavior of the direct term in second order also appears in all higher orders, as can be seen by considering the two fourth-order diagrams of Figure 10.7 as examples. In Figure 10.7a, all four interactions must carry the same momentum transfer $\mathbf{q}$, by conservation of linear momentum. The Coulomb interaction of Eq. 6.33 will enter as $V^4(q)$, contributing a factor q^{-8} to the integrand. The argument following Eq. 42 can be applied in a similar way here, to show that the $\mathbf{q}$-integral behaves at small q as

$$\frac{\Delta E_a^{(4)}}{N} \propto \int_{q_0} \frac{dq}{q^5} \propto q_0^{-4} , \qquad (43)$$

that is, more singularly than the $\ln q_0$ divergence of $\Delta E_a^{(2)}/N$.

In contrast, the diagram of Figure 10.7b has two interactions with the same value of $\mathbf{q}$; the other momentum transfers $(\mathbf{q}_1, \mathbf{q}_2)$ are independent. The $\mathbf{q}$-integration is like that for $\Delta E_a^{(2)}/N \propto \ln q_0$. The integrations on $\mathbf{q}_1$ and $\mathbf{q}_2$ each contribute a finite factor $\propto r_s$ (as can be seen on dimensional grounds), so that

$$\frac{\Delta E_b^{(4)}}{N} \propto r_s^2 \frac{\Delta E_a^{(2)}}{N} \propto r_s^2 \ln q_0 . \qquad (44)$$

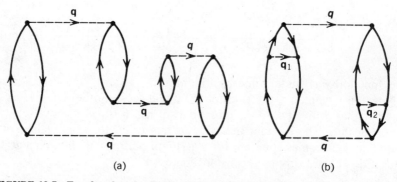

(a)   (b)

FIGURE 10.7 Two fourth-order Feynman diagrams, with different behavior at small q.

It is fairly easy to see the general result from these examples: in each order n, the diagram most singular at small q is that which is constructed entirely out of n particle–hole propagators (Figure 10.6) connected at their ends by n interaction lines, giving $V^n(q)$. For $n = 4$, this most singular diagram is shown in Figure 10.7a. (Remember that for Feynman diagrams, all time orderings of the interactions are included.) These diagrams are commonly called *ring* diagrams.

The ring diagrams are to the electron-gas problem what the ladder diagrams of Section 10.1 were to the hard-sphere problem: they are individually singular, but they form a partial series whose sum may be properly evaluated to yield a well-defined, finite result. For the hard-sphere gas, the method of summation was accurate at low density. In contrast, the evaluation of the ring diagrams gives a good estimate of the electron correlation energy at high densities (i.e. $r_s \ll 1$; see Eq. 6.29b).

10.6 CORRELATION ENERGY: RING APPROXIMATION

First, let us see that if we can sum the ring diagrams, we have the leading contribution to the correlation energy at high density. As an example, compare the contribution of Figure 10.5a, the lowest-order ring diagram, with that of Figure 10.7b, in which two interactions have been added to the diagram, acting within the particle–hole propagators. The ratio of the fourth-order to the second-order diagram is given in Eq. 44 (assuming a cutoff) as $\propto r_s^2$, which is small at high density ($r_s \ll 1$). This property persists to all orders; contributions from diagrams which are made from ring diagrams by adding interaction lines are smaller than those of the original ring diagram by powers of r_s—one for each interaction—at high density. Since all diagrams for $n > 2$ can be so generated, the result is that only the ring diagrams themselves survive in the limit $r_s \rightarrow 0$, with the exception of the $n = 2$ exchange diagram of Figure 10.5b, which gives a density-independent contribution (see Eq. 6.42).

Now let us consider the ring diagram of nth order, illustrated in Figure 10.8a. Following the method leading to Eq. 40, we find that the contribution to the energy can be expressed in the form

$$\frac{\Delta E_{\text{ring}}^{(n)}}{N} = \frac{1}{2\rho_0} \frac{i}{n} \int_{-\infty}^{\infty} \frac{d\omega \, d\mathbf{q}}{(2\pi)^4} \left[\Pi^0(q, \omega) V(q) \right]^n , \qquad (45)$$

where we have made use of the conservation of energy (frequency) ω as well as momentum $\mathbf{q}$, followed around the ring in one direction, as in Eq. 40 for $n = 2$. It is useful to introduce a new representation for this diagram, illustrated in Figure 10.8b, in which the ring consists of one p–h propagator $\Pi^0(q, \omega)$ and an effective interaction $v^{(n)}(q, \omega)$ (of nth order), the latter

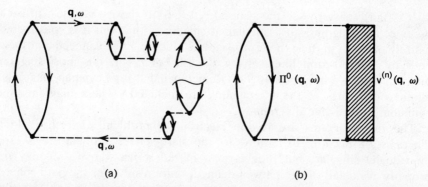

FIGURE 10.8 (a) Form of the *n*th-order ring diagram (Eq. 45); (b) the same diagram represented as polarization to lowest order in the presence of an effective interaction $v^{(n)}$.

representing a chain of n Coulomb interactions $V(q)$ connected by $n-1$ propagators Π^0, expressed as

$$v^{(n)}(q, \omega) = V(q)[\Pi^0(q, \omega)V(q)]^{n-1},\tag{46}$$

with $n \geq 2$. We rewrite Eq. 45 for the *n*th-order energy as

$$\frac{\Delta E_{\text{ring}}^{(n)}}{N} = \frac{i}{2\rho_0 n} \int_{-\infty}^{\infty} \frac{d\omega \, d\mathbf{q}}{(2\pi)^4} \, \Pi^0(q, \omega)v^{(n)}(q, \omega) \,.\tag{47}$$

Next, we define the full effective interaction by the sum

$$v(q, \omega) = \sum_{n=1}^{\infty} v^{(n)}(q, \omega) \,,\tag{48}$$

where we now include a first-order term defined by

$$v^{(1)}(q, \omega) = V(q) \,.\tag{49}$$

The series defined by Eqs. 46 and 48 is of simple geometric structure and may be summed formally to give both a linear equation for v,

$$v(q, \omega) = V(q) + V(q)\Pi^0(q, \omega)v(q, \omega) \,,\tag{50}$$

and its solution

$$v(q, \omega) = V(q)[1 - \Pi^0(q, \omega)V(q)]^{-1} \,.\tag{51}$$

Equations (48) and (50) are represented diagrammatically in Figure 10.9.

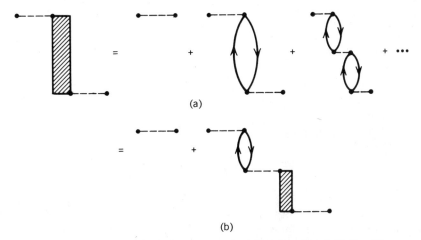

(a)

(b)

FIGURE 10.9   Diagrammatic representation of (a) Eq. 48 and (b) Eq. 50.

We note that in contrast to the ladder series, Eq. 9, whose sum involved solution of a (linear) *integral* equation, in the present case, Eq. 50 is *algebraic*, and Eq. 51 is the solution, of the form of Eq. 1. The formal operation of summing the geometric series by the fractional expression of Eq. 51 is well defined for any $q \neq 0$, by analytic continuation. We shall also find that the contribution to the energy is finite: the singularities which appear in each order of the perturbation expansion have been systematically removed by this procedure.

At this point we face a technical problem: to calculate the sum of the contributions to the energy from Eq. 47, we need the expression for $\sum_{n=2}^{\infty} v^{(n)}/n$, while we have calculated a different sum, not weighted by n^{-1}, in Eq. 48. Here we can use a device of integration over a strength parameter λ, as follows. First we replace the interaction V by λV everywhere: specifically, $v_\lambda^{(n)} \propto \lambda^n$. Then we integrate the parameter in the form $d\lambda/\lambda$, between the limits $[0, 1]$, which brings out a factor of

$$\int_0^1 \frac{d\lambda}{\lambda} \lambda^n = \frac{1}{n} \tag{52}$$

for the nth-order term $v^{(n)}$. Using this method, the sum of Eq. 47 may be evaluated in the form

$$\frac{\Delta E_{\text{ring}}}{N} = \sum_{n=2}^{\infty} \frac{\Delta E_{\text{ring}}^{(n)}}{N}$$

$$= \frac{i}{2\rho_0} \int_0^1 \frac{d\lambda}{\lambda} \int \frac{d\omega \, d\mathbf{q}}{(2\pi)^4} \Pi^0(q, \omega)[v_\lambda(q, \omega) - \lambda V(q)]. \tag{53}$$

The integrations involved in Eq. 53 may be shown to be nonsingular, and have been carried out, yielding

$$\frac{\Delta E_{\text{ring}}}{N} = \frac{2}{\pi^2}(1 - \ln 2)\ln r_s + \text{constant} \tag{54}$$

to leading order in $r_s \ln r_s$; the constant term is found to be -0.142 Ry. The correlation energy, in the high-density approximation, is then given by the sum of Eqs. 6.42 and 54, numerically given as

$$\frac{\Delta E_{\text{cor}}}{N} = [+0.0622 \ln r_s - 0.096]\text{Ry}. \tag{55}$$

The next correction is of order $r_s \ln r_s$. The result Eq. 55 may be compared to the Hartree–Fock energy of Eq. 9. For $2 < r_s < 5$, as for real metals, the correlation energy is dominated by the constant term, which is actually comparable to the net value of Eq. 6.31 in this range. Of course, for real metals, there are many further physical effects due to the nonuniformity of the crystal and electron–phonon interactions, which cannot be neglected.

Why has the singular behavior at $q \to 0$ disappeared? We can see this in part by looking at the factor in brackets in Eq. 53, which we may rewrite, using Eq. 50.

$$v_\lambda(q, \omega) - \lambda V(q) = \lambda V(q)\Pi^0(q, \omega)v_\lambda(q, \omega). \tag{56}$$

Comparing the resulting expression in the integrand of Eq. 53 with that of Eq. 40 for the (divergent) second-order ring diagram, we see that Eq. 53 has one Coulomb-interaction matrix element $V(q)$ of Eq. 40 replaced by $\int_0^1 d\lambda\, v_\lambda(q, \omega)$, the effective interaction integrated over the strength parameter. The singularity in q of Eq. 40 can be traced to the behavior of the integral of Eq. 42 near $\omega = 0$. The effective interaction at low excitation may be obtained from Eq. 51, using the limiting value of the Lindhard function, Eq. 41:

$$\lim_{q \to 0} \Pi^0(q, 0) = -\frac{mk_F}{\pi^2}. \tag{57}$$

Substituting this into Eq. 51, we find that for small energy excitations, the effective interaction develops a range, i.e.

$$v(q, \omega) \cong \frac{4\pi e^2}{q^2 + q_0^2}, \qquad q^2 < q_0^2, \quad \omega \ll q^2, \tag{58}$$

where the inverse range (cutoff) q_0 is given by

$$q_0^2 = \frac{4mk_F e^2}{\pi} = \frac{4k_F}{\pi a_0} = \left(\frac{16}{3\pi^2}\right)^{2/3} r_s k_F^2. \tag{59}$$

The effective potential now looks like a finite-range potential in configuration space,

$$v(r) = e^2 \frac{e^{-q_0 r}}{r} , \tag{60}$$

which is often called the "screened Coulomb" potential. It is as if the longitudinal "photon" of the Coulomb potential had developed a mass, $m = q_0$. The change of form from $V(q) = 4\pi e^2/q^2$ to that of Eq. 58 makes the integrals of Eq. 53 finite, in contrast to those of Eq. 40. (To complete the argument, we must rescale with λ, and integrate:

$$\int d\lambda \, v_\lambda(q, \omega) \cong \int \frac{4\pi e^2 \lambda^2 \, d\lambda}{q^2 + q_0^2 \lambda^2} ,$$

which still leads to a finite result in Eq. 53.) This explains the "taming" of the singularity of the Coulomb interaction, but not the entire correlation, since q is not limited as in Eq. 58, and there are $\omega \neq 0$ contributions to Eq. 53. We extend the discussion of screening in the next section.

10.7   RESPONSE TO EXTERNAL POTENTIAL: SCREENING AND COLLECTIVE EXCITATIONS

There is more information in the theoretical methods just developed than just the ground-state correlation energy; in particular, one may also find the response of the electron gas to a weak, externally applied time-dependent potential. For this discussion it will be useful to generalize some of the functions defined earlier in the chapter. First we introduce the *proper polarization part*, or reduced particle–hole propagator, $\Pi(q, \omega)$, which we define as a series of Feynman diagrams, as illustrated in Figure 10.10a. The shaded block includes any interaction of the particle and hole, *excluding* that which connects two distinct particle–hole loops; i.e., one cannot sever the proper polarization part into disconnected parts by removing any single interaction line. An example of an excluded diagram is given in Figure 10.10b. As we discussed in the previous section, the leading term at high density is the unperturbed propagator $\Pi^0(q, \omega)$, given by the first diagram on the right-hand side of Figure 10.10a (compare Figure 10.7); the other diagrams are smaller by powers of r_s. For later use (in Section 11.6) we also define the full particle–hole propagator $\tilde{\Pi}(q, \omega)$ as the series of Figure 10.10a extended to *include* the interaction of Figure 10.10b. It is easy to show that the two propagators are related by

$$\tilde{\Pi}(q, \omega) = \Pi(q, \omega) + \Pi(q, \omega) V(q) \tilde{\Pi}(q, \omega) . \tag{61}$$

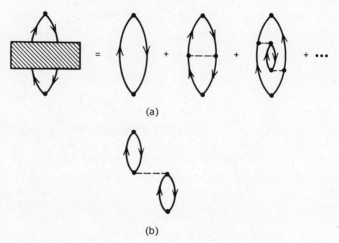

(a)

(b)

FIGURE 10.10 (a) Diagram expansion of the proper polarization part, or reduced particle–hole propagator, Π. (b) Example of a diagram excluded from Π, but included in the full propagator $\tilde{\Pi}$.

Next, we generalize the effective interaction defined in Eqs. 46, 48, and 50, by replacing Π^0 with Π, so that $\tilde{v}$ replaces v, and satisfies the linear equation

$$\tilde{v}(q, \omega) = V(q) + V(q)\Pi(q, \omega)\tilde{v}(q, \omega) \qquad (62a)$$

and has the form (compare Eq. 51)

$$\tilde{v}(q, \omega) = V(q)[1 - \Pi(q, \omega)V(q)]^{-1} . \qquad (62b)$$

It is conventional to introduce a *generalized dielectric function* $\epsilon(q, \omega)$ (a function of momentum and frequency) to relate the effective interaction to the Coulomb interaction:

$$\frac{\tilde{v}(q, \omega)}{V(q)} = \epsilon(q, \omega)^{-1} . \qquad (63)$$

We see from Eq. 62b that

$$\epsilon(q, \omega) = 1 - \Pi(q, \omega)V(q) , \qquad (64a)$$

while from Eq. 62a we may also write

$$\epsilon(q, \omega) = [1 + \Pi(q, \omega)\tilde{v}(q, \omega)]^{-1} . \qquad (64b)$$

In the ring approximation (good at high density)

$$\epsilon(q, \omega)_{\text{ring}} = 1 - \Pi^0(q, \omega)V(q) . \tag{65}$$

For $\omega = 0$ and $q < q_0$, we have the result already quoted in Eq. 58, in the form

$$\epsilon(q, 0)_{\text{ring}} \cong 1 + \frac{q_0^2}{q^2} \tag{66}$$

with q_0 defined in Eq. 59. The variation of the dielectric function from unity is one way of expressing the phenomenon of screening.

Now let us apply these new definitions to the following problem: the electron gas in its ground state is perturbed by a weak external potential, which may vary with time as well as space. We shall work directly with the Fourier transform, $V_{\text{ext}}(q, \omega)$. (A static external potential has $\omega = 0$.) The electron gas will be polarized in the presence of the external field, producing an induced field due to the change in the charge distribution from uniformity, which we call $V_{\text{ind}}(q, \omega)$. We want to determine $V_{\text{ind}}(q, \omega)$ for an arbitrary function $V_{\text{ext}}(q, \omega)$ of small amplitude; the last assumption allows us to make a *linear approximation* for the response. We show V_{ext} and V_{ind} diagrammatically in Figure 10.11a, b, acting on a single electron line. Figure

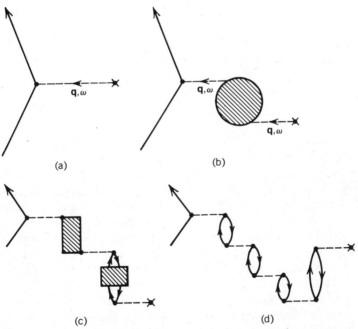

(a)   (b)

(c)   (d)

FIGURE 10.11   (a) External applied potential, V_{ext}; (b) induced potential V_{ind} produced in linear response of system to V_{ext}; (c) division into diagrams for effective interaction (Figure 10.9) and polarization part (Figure 10.10), as in Eq. 67; (d) example of contribution from ring series.

10.11b illustrates the general form of a Feynman diagram in which V_{ind} depends linearly on V_{ext}; clearly the two interactions must be connected for V_{ind} to be *induced* by V_{ext}. (The linked-cluster theorem will remove all unlinked contributions to the response, as it did for the ground-state energy.)

The general diagram of Figure 10.11b can always be redrawn in the form shown in Figure 10.11c; i.e., the "blob" can be divided uniquely into two recognizable parts, namely the effective interaction $\tilde{v}$ and the polarization part Π, as can be shown by examining the definitions of these quantities. As an example, a term in the ring series (actually, a "chain" series here, since the ring has been opened) is shown in Figure 10.11d: the first three links in the chain (from the left) contribute to v, and the last is Π^0 in this case. The division of the diagrams corresponds to the following equation:

$$V_{ind}(q, \omega) = \tilde{v}(q, \omega)\Pi(q, \omega)V_{ext}(q, \omega) \tag{67}$$

which is exact to all orders in the Coulomb interaction among electrons, but only to first order in V_{ext}. So the linear response can be obtained directly from $\tilde{v}$ and Π, or in the ring approximation, from v and Π^0. It is easy to see from the form of Figure 10.11c that Eq. 67 holds for fixed (q, ω).

If we add the induced and external potential, we obtain the total potential seen by each electron, which can then be expressed as follows, using Eq. 67:

$$\begin{aligned} V_{tot}(q, \omega) &= V_{ext}(q, \omega) + V_{ind}(q, \omega) \\ &= [1 + \tilde{v}(q, \omega)\Pi(q, \omega)]V_{ext}(q, \omega) \\ &= \frac{V_{ext}(q, \omega)}{\epsilon(q, \omega)} , \end{aligned} \tag{68}$$

where we have used Eq. 64b to get the last line. This equation shows that the dielectric function which expresses the screening of the Coulomb field in the ground state of the electron gas (see Eq. 63) also gives the dielectric screening of an external field. If the external field is time independent and slowly varying in space, then the static ($\omega = 0$) long-wavelength result given in Eq. 66 applies (at high density). Not surprisingly, this static result can also be obtained in the Thomas–Fermi approximation (see Chapter 4), since in this case the screening effect is given by a readjustment of the Hartree–Fock ground state, and the Thomas–Fermi method is valid for long wavelengths ($q < k_F$).

Study of the response of a system to a weak external probe can be translated into information about the excitation spectrum of the system. For the electron gas, we have characterized the polarizability in the presence of an external field by the dielectric function $\epsilon(q, \omega)$. For fixed nonzero frequency $\omega > 0$, the electron gas absorbs energy from the driving field, exciting those states of energy ω which can be reached from the ground state by a one-body excitation (since both V_{ext} and V_{ind} are one-electron operators, as shown in Figure 10.11). The character and density of states reached at a given ω will vary with the momentum $\mathbf{q}$ transferred to the system, and vice versa.

If the system were of finite size, the excitation spectrum would have discrete energy levels, as in any atom or molecule, for example (plus a continuum if ionization is possible). The response of the system to an external field would be expected to show a *resonance* in ω at each energy level, at which $V_{tot}(\omega)$ or $V_{ind}(\omega)$ is maximum, or equivalently, at which $\epsilon(q, \omega)$ is a minimum, as we see from Eq. 68. For sufficiently narrow resonances, we have

$$\epsilon(q, \omega) \cong 0 , \tag{69a}$$

(i.e., vanishes with the widths of the resonances) so that from Eq. 64a we find

$$V(q)\Pi(q, \omega) \cong 1 \tag{69b}$$

as a condition on $\Pi(q, \omega)$ at each excited state. In Green-function language, the excited states are given by complex poles in ω: when the width of the states vanishes (no damping mechanism), the poles are real. Then Eq. 69 is called a *dispersion* formula for the spectrum. In the present case, the poles of the particle–hole Green function (or propagator) are the zeros of $\epsilon(q, \omega)$; Eq. 69b expresses the existence of a state of narrow width (almost real pole–zero position in ω). For a very large system, the resonant states overlap in a continuous spectrum, but Eq. 69 still applies in the limit of infinite size.

This is illustrated most easily in the ring approximation, for which $\Pi \cong \Pi^0$, and Eq. 69b gives the dispersion formula

$$V(q)\Pi^0(q, \omega) = 1 . \tag{70}$$

This last equation can be shown with some algebra to be equivalent to the random phase approximation (RPA) equations given in Section 4.6, e.g. Eq. 4.84, for the electron gas. For exact correspondence, one must neglect the "direct" parts of p–h matrix elements in the RPA equations, which contribute interactions like that of Figure 10.9b to the polarization diagrams, plus others omitted in the ring approximation. The "exchange" p–h interaction is retained, and contributes the $V(q)$ matrix elements to the ring diagrams. In this approximation there are no damping mechanisms, so $\epsilon(q, \omega)$ has *real* zeros given by Eq. 70. We know from the explicit form of Eq. 41 that $\Pi^0(q, \omega)$ diverges at the unperturbed energy of each particle–hole excitation,

$$\omega = \epsilon(\mathbf{k} + \mathbf{q}) - \epsilon(\mathbf{k}) = \epsilon_p - \epsilon_h . \tag{71}$$

Therefore, the zeros of $\epsilon(\omega)$, given by Eq. 70, will lie between these particle–hole energies. This is easily seen in Figure 10.12, where we plot $V\Pi^0(\omega)$ at fixed q, against ω, given by the solid curves. The singularities in $V\Pi^0(\omega)$ fall at all possible values of $\epsilon_p - \epsilon_h$, given by Eq. 71. (We imagine a periodic system, for the moment, to keep the spectrum discrete.) The actual excitations fall at the values at which Eq. 70 is satisfied, shown as the intersections of $V\Pi^0(\omega)$ with the unit line in the figure. We see that for all but the highest state, the resonant frequencies are close to the nearby particle–hole energies, shifted by less than the spacing of the next highest level. However, the highest state may be shifted considerably further, since it is not constrained by $\epsilon_p - \epsilon_h$ above. This last state is a collective excitation, called the *plasmon*. In the limit $q \to 0$, the frequency of the plasmon approaches the classical frequency for an electron plasma oscillation (with a fixed positive background, in our case), given by

$$\omega_0^2 = \frac{4\pi\rho_0 e^2}{m} , \tag{72a}$$

which shows that this excitation is indeed a collective state. This may be obtained directly from Eq. 70, using the following limiting expression:

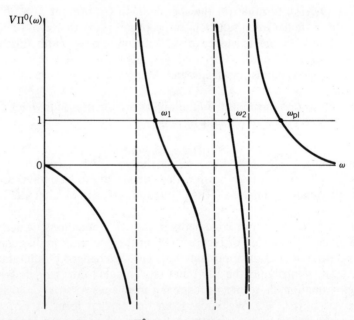

FIGURE 10.12 Schematic plot of $V\Pi^0(\omega)$ for a discrete spectrum. Vertical dashed lines give position of unperturbed p–h states. Solid lines are values of $V\Pi^0(\omega)$ for fixed q. The resonances are at the intersections of these curves with the unit line. The highest excited state is the collective *plasmon* state, marked ω_{pl}. The curves are symmetric in $\omega \to -\omega$.

$$\lim_{q \to 0} \Pi^0(q, \omega) = \frac{q^2 k_F^3}{3\pi^2 m\omega^2} = \frac{q^2 \rho^0}{m\omega^2}, \qquad \omega \neq 0. \qquad (72b)$$

For long wavelengths, the plasmon frequency can be shown to increase with q^2.

The screening of the Coulomb field in the ring approximation can also be understood in terms of Figure 10.12. We saw in Sections 6.3.2 and 10.5 that the second-order ring diagram diverges at $q \to 0$, where $V(q) \propto q^{-2}$ and where the lowest-energy particle-hole states have excitation energy $\epsilon(\mathbf{k} + \mathbf{q}) - \epsilon(k) \to 0$. The effect of summing the ring diagrams is to produce a shifted spectrum for the particle–hole states, which can be seen from Figure 10.12 and Eq. 70 to be always upward. This shift removes the singularity of the diagram; the mechanism can be thought of as a result of coupling the electron-gas ground state to the plasmon mode. This was in fact the original approach to the problem taken by Bohm and Pines* in the random-phase approximation.

10.8 SUMMARY

Let us summarize the applications of perturbation theory we have encountered in this chapter. We consider many-body systems interacting through a strong, short-range, repulsive potential (hard-sphere gases: fermions or bosons), an attractive potential with a repulsive core (nuclear matter), and the long-range repulsive Coulomb potential (electron gas). For reasons we first saw in Chapter 6, these systems all exhibit singularities when considered order by order in a perturbative expansion based on the interaction. The removal of the problem of the singularities required the method of "summation of partial series," which we applied to the ladder series in Sections 10.1–10.3 and the ring-diagram series in Sections 10.5 and 10.6. However, it is good to remember that the results are still in the form of perturbation expansions. The original interactions have been replaced by effective interactions: the t-operator $t(\epsilon)$ for the ladder series (see Eq. 8 or 10), and $v(q, \omega)$ for the ring series (see Eqs. 46–51). Thus in the hard-sphere gas, we could generate a low-density expansion in the scattering length, Eq. 22. For the electron gas, we obtained a high-density expansion in the dimensionless parameter r_s, given in Eq. 55. Neither expansion is analytic, but each does have a pertinent domain of validity.

The situation is analogous to that of quantum electrodynamics, where the original perturbation expansion in the "bare" charge is replaced by a well-behaved (renormalized) perturbation expansion in an "effective charge" defined through the renormalization procedure. The effective charge (in the form of $\alpha = e^2/\hbar c$) is then the small expansion parameter

*D. Bohm and D. Pines, Phys. Rev. **92**, 609 (1953).

actually used in calculation of phenomena such as electron scattering, atomic radiation, and the magnetic moment of the electron.

The analogy with quantum electrodynamics is not accidental. The diagrammatic approach to many-body systems originally developed from similar considerations first applied to quantum field theory (details of which can be found in a number of textbooks). Such methods, in various forms, continue to be extensively used in the quantum field theory of elementary-particle reactions. Some elementary examples are included in the discussion of scattering in the following chapter (Section 11.5).

The application of diagrammatic methods to many-body systems goes considerably further than the examples of this chapter illustrate. In dealing with condensed matter (solids and fluids), one is interested in many properties beyond those of the ground state, including transport of electrical charge or heat, collective excitations, magnetism, effects of impurities, and so on. For many of these properties, the diagrammatic approach to dynamics provides a useful tool for theoretical discussion and calculation.

EXERCISES

10.1 (a) Derive Eqs. 25a and 25b for the first-order energy of a Bose gas.
(b) Evaluate Eq. 25a for the δ-function potential $V(r) = \lambda\delta(r)$, and verify Eq. 25b in this case. (See Exercise 6.1.)

10.2 The screened form of the Coulomb potential given in Eq. 60 may also be derived from the Thomas–Fermi approximation of Section 4.5.2. Starting with Eqs. 4.56–4.59, and modifying the last to include the uniform positive charge density:

$$\nabla^2 \mathscr{V}(r) = -4\pi e^2[\rho(r) - \rho_0] , \qquad (4.59')$$

derive an equation for $\mathscr{V}(r)$. Neglecting all but linear terms in $\mathscr{V}$ in the equation, show that there are solutions of the form of Eq. 60. This may be regarded as the shielded potential due to one electron at the origin.

CHAPTER **11**

Particle Propagation and Scattering

Thus far in this book we have mainly been concerned with ground-state properties of many-particle systems. Excitations were implicit in this domain as well, since we attempted to include fully the dynamics of the system and thus summed over excitations virtually present in intermediate states. In the course of generating the tools to deal with this, we have found ways to characterize single-particle propagation in the many-particle system. These are obviously of great importance for treating system dynamics beyond ground-state properties, and so are developed further here: We shall see how to characterize the overall many-particle effect on one-particle propagation by studying the general single-particle Green function. Approximate methods emerge for dealing with this propagation in terms of Dyson's equation, and it proves to be convenient to describe single-particle dynamics in the many-particle system by means of an equivalent potential for the single particle, called the optical potential. This potential contains an imaginary part to account for the fact that the single particle in interaction with the rest of the particles may remove the overall system from the ground state, and this must be represented as flux lost from the original system state—one manifestation of many-particle effects seen from the perspective of single-particle motion.

In this final chapter we also study the application of the techniques we have developed to the problem of scattering in a many-body system; the power of these techniques clearly allows such an extension. Although it is not our purpose in this book to study scattering problems exhaustively, we do wish to illustrate the connection between methods of treating many-body (ground) states and scattering in field theory, so that when this latter topic is encountered it will be recognized as a straightforward broadening of these methods. Last, we address the question of excitations of the many-particle system in response to a probe that undergoes inelastic scattering upon it. This makes more explicit the ways in which the dynamics of excitations appear in ground-state features of the system and illustrates further the capabilities of the theory for handling fully the dynamics of the many-particle system.

11.1 SINGLE-PARTICLE GREEN FUNCTION

The causal single-particle Green function for an interacting system was introduced in Section 7.2 as an amplitude that describes the propagation in space and time of an excitation of the system by adding (or subtracting) a single particle at a point. (See Eq. 7.26.) The definition made no reference to perturbation theory; here we shall study the expansion of the Green function in Feynman diagrams. In the following section we shall find a partial summation method for the expansion which leads to Dyson's integral equation.

First we define the Green function

$$iG_{\alpha\beta}(t, t') = \langle \Psi_0 | \{ \tilde{a}_\alpha(t) \tilde{a}_\beta^\dagger(t') \}_T | \Psi_0 \rangle \,, \tag{1}$$

where we have rewritten Eq. 7.26 in an orbital representation, with $\tilde{a}_\alpha(t)$ or $\tilde{a}_\beta^\dagger(t')$ the annihilation or creation operator in the Heisenberg picture. These can be reexpressed in the interaction picture by combining Eqs. 7.8, 7.36, and 7.38:

$$\tilde{a}_\alpha(t) = U(0, t) a_\alpha(t) U(t, 0) \,. \tag{2}$$

The ground state Ψ_0 in Eq. 1 is the fully interacting state; we may obtain an expression for it by the formal device of adiabatic switching, as discussed in Section 7.4. The state vector was given in Eq. 7.46, and was defined so that it has unit overlap with the unperturbed ground state Φ_0, as shown in Eq. 7.50. For the Green function of Eq. 1 we must normalize $\langle \Psi_0 | \Psi_0 \rangle = 1$; therefore we rewrite Eq. 7.46 as

$$\Psi_0 = \lim_{\alpha \to 0^+} \frac{U_\alpha(0, -\infty) | \Phi_0 \rangle}{\langle \Phi_0 | U_\alpha(\infty, -\infty) | \Phi_0 \rangle^{1/2}} \,, \tag{3}$$

where we have used $U_\alpha^\dagger(0, -\infty) = U_\alpha(\infty, 0)$ and $U_\alpha(\infty, 0) U_\alpha(0, -\infty) = U_\alpha(\infty, -\infty)$ to obtain the denominator. The last is a special case of the multiplication law for the time-development operators,

$$U(t_1, t_3) = U(t_1, t_2) U(t_2, t_3) \tag{4}$$

which follows directly from the definition, Eq. 7.38. Writing out the time ordering in Eq. 1 explicitly, and incorporating Eqs. 2–4, the Green function becomes

$$iG_{\alpha\beta}(t, t') = \frac{1}{\langle \Phi_0 | U(\infty, -\infty) | \Phi_0 \rangle} [\langle \Phi_0 | U(\infty, t) a_\alpha(t) U(t, t') a_\beta^\dagger(t')$$

$$\times U(t', -\infty) | \Phi_0 \rangle \theta(t - t')$$

$$- \langle \Phi_0 | U(\infty, t') a_\beta^\dagger(t') U(t', t) a_\alpha(t) U(t, -\infty) | \Phi_0 \rangle \theta(t' - t)] , \quad (5)$$

where the adiabatic limit $(\alpha \to 0^+)$ of the ratio is understood.

To make use of the Wick contraction theorem, we need to reexpress Eq. 5 in terms of time-ordered products of a, $a^\dagger$ operators. We already have an expression for the expansion of $U(t, t_0)$ given in Eqs. 8.8, 8.34. This may be used to obtain a multiple expansion of the numerator of Eq. 5, which can be recombined (with some labor) into the useful form

$$iG_{\alpha\beta}(t, t') = \sum_{n=0}^{\infty} \frac{(-i)^n}{n!} \int_{-\infty}^{\infty} dt_1 \cdots \int_{-\infty}^{\infty} dt_n$$

$$\times \frac{\langle \Phi_0 | \{V(t_1) \cdots V(t_n) a_\alpha(t) a_\beta^\dagger(t')\}_T | \Phi_0 \rangle}{\langle \Phi_0 | U(\infty, -\infty) | \Phi_0 \rangle} . \quad (6)$$

The numerator of this expression can now be treated by the Wick method, which will lead to an expansion in Feynman diagrams of the type illustrated in Figure 11.1. The shaded part of the diagram represents interactions connecting the α, β lines to other excitations of the system. There may also be disconnected parts in the shaded part. However, a linked-cluster theorem similar to that of Section 9.1 also applies here: the contribution of the disconnected parts of the numerator is a factor exactly equal to the denominator of Eq. 6. The result is that the Green function may be expanded in linked diagrams only:

$$iG_{\alpha\beta}(t, t') = \sum_{n=0}^{\infty} \frac{(-i)^n}{n!} \int_{-\infty}^{\infty} dt_1 \cdots \int dt_n$$

$$\times \langle \Phi_0 | \{V(t_1) \cdots V(t_n) a_\alpha(t) a_\beta^\dagger(t')\}_T | \Phi_0 \rangle_l , \quad (7)$$

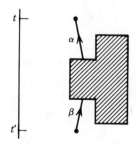

FIGURE 11.1 General form of the Feynman diagram for the numerator of Eq. 6, or for the Green function (Eq. 7) for the case $t > t'$. The lines α, β are shown as particle lines, but could be changed to hole lines by changing interaction times within the rest of the diagram (shaded). For Eq. 7, the diagram must be connected.

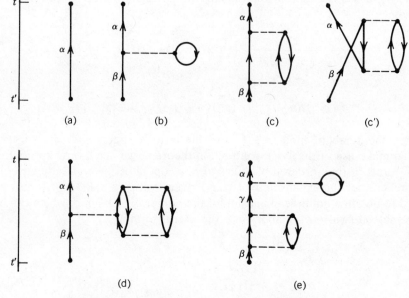

FIGURE 11.2 Examples of Figure. 11.1, through third order.

which are of the form of Figure 11.1, with the shaded part entirely connected to the α, β lines.

Some examples of linked diagrams of this form are shown in Figure 11.2, including some to third order in V. Diagrams (c) and (c') are actually the same Feynman diagram, with different time ordering of the two interactions. In fact, we could also regard the diagrams of Figure 11.2 as *Goldstone* diagrams for the Green function, imagining that we had used the Goldstone method of Section 9.2 to obtain an expansion of Eq. 5. The arguments used there for the ground-state energy shift could also be applied to the definition of diagrams for the Green function, including the elimination of unlinked diagrams. For the Goldstone expansion, Figure 11.2c and c' are distinct diagrams, representing different intermediate states in the two cases.

11.2 DYSON'S EQUATION; QUASIPARTICLES

The diagram expansion of the single-particle Green function is an interesting example for which the method of summing partial series is particularly useful. As in Section 10.1, where we discussed the ladder approximation, the formal summation of the partial series defines an integral equation, whose solution gives the sum. The integral equation for the Green function is called Dyson's equation.

The general form of the Feynman diagrams for $G_{\alpha\beta}(t, t')$ of Eq. 7 was shown in Figure 11.1, where the shaded part represents any connected combination of interaction and particle (hole) lines. One way of further characterizing diagrams of this form is in terms of parts which are connected to each other only by a single particle (hole) line, each part connected to at most two other parts. If these parts cannot be further separated into lesser parts with only single particle-line connections, the parts are called *irreducible*. The decomposition of the Green function into irreducible parts is shown in Figure 11.3, where the rectangular block stands for all connected Feynman diagrams of the type of Figure 11.1, and the oval shapes are irreducible parts. We have also specified that the first interaction of β is at t_1, and the last interaction of α at t_2. As usual in Feynman diagrams, the particle lines may go forward or backward, so that all time orderings of the irreducible parts relative to each other are included in the decomposition. Of the examples shown in Figure 11.2, the unperturbed particle line in (a), representing $G_\alpha^0(t - t')\delta_{\alpha\beta}$, is shown again as Figure 11.3a. The three diagrams of Figure 11.2b, c, d are all irreducible, forming part of Figure 11.3, while Figure 11.2e is reducible, containing two irreducible parts: (b) and (c). The decomposition of the Green function into combinations of irreducible parts is unique. There are still an infinite number of possible Feynman diagrams which are irreducible.

Imagine that we knew how to sum all the irreducible Feynman diagrams, represented by Figure 11.3b. We shall define this sum

$$-i\Sigma_{\alpha\beta}(t_2 - t_1) = \text{sum of irreducible diagrams for } iG_{\alpha\beta}(t - t')\,, \qquad (8)$$

where we set $t = t_2$ and $t' = t_1$. This has the effect of shrinking the particle lines α, β, to points coinciding with the interactions at t_2, t_1. We have here used the time-translation invariance of the Green function to write Eq. 8

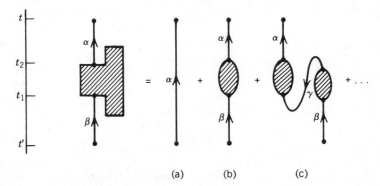

(a)   (b)   (c)

FIGURE 11.3 Decomposition of the Feynman diagram for a Green function into irreducible parts connected only by single particle lines, each part connected to at most two others.

explicitly in terms of relative times. This property can be derived directly from Eqs. 1 and 2 for a time-independent Hamiltonian, or from the properties of the Feynman-diagram expansion, e.g. Eq. 7. The quantity Σ defined in Eq. 8 is sometimes called the *proper self-energy*, from its origin in the theory of the electron self-energy (proper = irreducible).

Now we can write the decomposition represented by Figure 11.3 as a series:

$$iG_{\alpha\beta}(t-t') = iG_\alpha^0(t-t')\delta_{\alpha\beta}$$

$$+ i \iint_{-\infty}^{\infty} dt_1\, dt_2\, G_\alpha^0(t-t_2)\Sigma_{\alpha\beta}(t_2-t_1)G_\beta^0(t_1-t')$$

$$+ i \sum_\gamma \iiiint_{-\infty}^{\infty} dt_1\, dt_1'\, dt_2'\, dt_2\, G_\alpha^0(t-t_2)\Sigma_{\alpha\gamma}(t_2-t_2')$$

$$\times\, G_\gamma^0(t_2'-t_1')\Sigma_{\gamma\beta}(t_1'-t_1)G_\beta^0(t_1-t') + \cdots \quad (9)$$

where the first three terms are represented by (a), (b), and (c). The propagation between irreducible parts is given by the unperturbed G^0. The infinite series of Eq. 9 is formally equivalent to a linear (matrix) integral equation for G, which is Dyson's equation:

$$G_{\alpha\beta}(t-t') = G_\alpha^0(t-t')\delta_{\alpha\beta}$$

$$+ \sum_\gamma \iint_{-\infty}^{\infty} dt_1\, dt_2\, G_\alpha^0(t-t_2)\Sigma_{\alpha\gamma}(t_2-t_1)G_{\gamma\beta}(t_1-t')\,.$$

$$(10)$$

Clearly, Eq. 9 is the series solution for Eq. 10, in orders of the proper self-energy Σ, as can be checked by solving Eq. 10 iteratively. The integral equation could be solved for $G_{\alpha\beta}(t-t')$, if we indeed knew Σ, as assumed. Of course, we have actually replaced the problem of summing the original Feynman expansion of G with the reduced, but still infinite expansion of Σ in Eq. 8. We return shortly to the question of summing Σ; for the moment we continue to consider it as known.

We may find a formal solution to Eq. 10 by taking the Fourier transforms from relative time $\tau = t - t'$ to ω of each function G, G^0, and Σ, as in Eq. 8.50. The transformed equation is

$$G_{\alpha\beta}(\omega) = \delta_{\alpha\beta}G_\alpha^0(\omega) + \sum_\gamma G_\alpha^0(\omega)\Sigma_{\alpha\gamma}(\omega)G_{\gamma\beta}(\omega)\,. \quad (11)$$

Note that the energy variable ω is the same in all factors; this is an example of the conservation of excitation energy for an isolated system, as discussed in Section 8.6. For fixed ω, Eq. 11 is a matrix equation on the single-particle orbital labels, which can be solved by matrix inversion. First multiply Eq. 11 on the left by the matrix $[G^0]^{-1}$, rewriting the result as

$$\sum_{\gamma} \{\delta_{\alpha\gamma}[G_{\alpha}^{0}(\omega)]^{-1} - \Sigma_{\alpha\gamma}(\omega)\} G_{\gamma\beta}(\omega) = \delta_{\alpha\beta} , \tag{12}$$

the solution to which can be written immediately as the inverse matrix

$$G_{\alpha\beta}(\omega) = \{[G^{0}(\omega)]^{-1} - \Sigma(\omega)\}_{\alpha\beta}^{-1} . \tag{13}$$

In special cases, the matrix Σ may be put into diagonal form simultaneously with the unperturbed Green function $G^{0}(\omega)$. Generally this is the result of a symmetry of the system, such as translational invariance for a uniform medium, for which the linear momentum $\mathbf{k}$ may be chosen as a diagonal quantum number. Whenever Σ and G^{0} are diagonal together, Eq. 13 gives a diagonal solution for the Green function, using the inverse of Eq. 8.51:

$$G_{\alpha\beta}(\omega) = \delta_{\alpha\beta} G_{\alpha}(\omega) , \tag{14a}$$

$$G_{\alpha}(\omega) = [\omega - (\epsilon_{\alpha} - \epsilon_{F}) - \Sigma_{\alpha}(\omega) \pm i\eta]^{-1} \tag{14b}$$

with $\pm i\eta$ for particle or hole orbits $\epsilon_{\alpha} \gtrless \epsilon_{F}$).

What has been accomplished by rearranging the Feynman perturbation series for the Green function in terms of the Dyson integral equation? As mentioned earlier, the problem of calculating the proper self-energy Σ still remains. However, several things have been gained. The first is formal: the expressions in Eqs. 13 and 14 for the Fourier-transformed Green function display the ω, or energy, dependence of $G(\omega)$ directly in terms of $\Sigma(\omega)$. These relations can be continued to complex ω, and lead to useful analytic relations, in what is called the *spectral representation* of the Green function. The analytic properties obtained are actually more general than the perturbation theory in which we have been working. The theory goes beyond the limits of this book; the interested reader may refer to a standard work on Green functions.*

The second gain is suggestive: the proper self-energy $\Sigma(\omega)$ enters the equations very much as a single-particle potential would; e.g. in Eq. 14b $\Sigma_{\alpha}(\omega)$ is added to the unperturbed orbital energy, as if it represented a potential imposed on the orbit α, due to interactions of a particle in that orbit with the rest of the many-body system. The idea is partly correct, and is reminiscent of the Hartree–Fock approximation of Chapter 4. The fact that $\Sigma(\omega)$ is in general a function of the excitation energy of the system is important, however, and alters the simplicity of the picture of Σ as an average potential. The dependence on ω is related to the time dependence of $\Sigma(t - t')$ through the Fourier transform. It is easy to see that if $\Sigma_{\alpha}(\omega) = U_{\alpha}$, independent of ω, then the time transform is

*For example, A. A. Abrikosov, L. P. Gorkov, and I. E. Dzyaloshinski, *Methods of Quantum Field Theory in Statistical Physics* (Prentice-Hall, Englewood Cliffs, N.J., 1963).

$$\Sigma_\alpha(t - t') = \int_{-\infty}^{\infty} \frac{d\omega}{2\pi} \, e^{-i\omega(t-t')} U_\alpha$$

$$= U_\alpha \delta(t - t') \,, \tag{15}$$

which acts as an instantaneous and time-independent operator. With any ω-dependence of $\Sigma_\alpha(\omega)$, $\Sigma_\alpha(t - t')$ exhibits a dependence on the time *retardation* $(t - t')$. This is a dynamical effect of the interaction of the single particle in orbit α with the rest of the system. Even with these qualifications, it is still useful to think of the proper self-energy as a *potential-like* operator on the single particle. This is useful for developing models in which single-particle behavior is central, as we shall see in Section 11.3 for the optical model for scattering.

The third accomplishment of the Dyson-equation formulation is that it makes approximation methods more powerful. This comes about as follows: if it is possible to find a good approximation to Σ (say) in low order of the Feynman-diagram expansion, then it is clear that the Green function calculated from the integral equation 11, or from Eqs. 13, 14, is at least as accurate, in terms of perturbation theory, as it would be to approximate G to the same order as Σ. But in fact, one may have done very much better than the same order by summing the partial series of Feynman diagrams implied by finding G from Σ. This is because a good approximation to $\Sigma(\omega)$ makes the functional dependence of $G(\omega)$ approximately correct. For example, if the ω-dependence of $\Sigma(\omega)$ is very weak, then Eq. 15 is a good approximation, and Eq. 14b tells us that $G(\omega)$ has its poles shifted from $\epsilon_\alpha - \epsilon_F$ to $\epsilon_\alpha - \epsilon_F - U_\alpha$. This shift could not be well represented by a direct perturbation expansion of $G(\omega)$. Since the location of the poles of $G(\omega)$ has important effects on the propagation of particles, the use of Dyson reordering would be very important in such a case.

The simplest approximation to $\Sigma(\omega)$ is the ω-independent approximation

$$\Sigma_{\alpha\beta}(\omega) = \sum_{\gamma \leq F} \langle \alpha\gamma | v | \beta\gamma \rangle_A \,, \tag{16}$$

which is equivalent to the Hartree–Fock approximation (see Eq. 9.35), without assuming Σ to be diagonal in the unperturbed representation. This is equivalent to approximating $-i\Sigma$ by the diagram of Figure 11.2b, plus its exchange diagram, as in Figure 11.4a. One can include higher-order contributions to $-i\Sigma$, e.g. from Figure 11.2c–d. Further, one can sum a partial series to evaluate Σ, e.g. using the ladder approximation of Section 10.1, with

$$\Sigma_{\alpha\beta}(\omega) \cong \sum_{\gamma < F} \langle \alpha\gamma | \tilde{t}(\omega) | \beta\gamma \rangle \,. \tag{17}$$

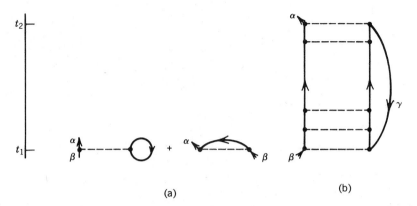

(a)

(b)

FIGURE 11.4 Approximations to Eq. 8: (a) Hartree–Fock, (b) ladder series (typical term).

The diagrams summed are shown in Figure 11.4b. (Exchange ladder diagrams can also be included, as in Figure 10.1b, which we would denote by $\langle\ \rangle_A$ in Eq. 17.)

Diagrammatic approximations to $\Sigma(\omega)$ beyond first order (Eq. 16) are invariably ω-dependent; this leads to a noninstantaneous self-energy, as we have already discussed. Under some circumstances, particularly when only very high virtual excitation energies are involved (as for short-range interactions), the ω-dependence may be weak, even though higher orders of perturbation theory contribute. An example is the ladder approximation for a short-range interaction. In such cases, the *quasiparticle* approximation may be valid, namely (in a diagonal representation)

$$\Sigma_\alpha(\omega) \cong \Delta_\alpha \mp i\gamma_\alpha\,, \qquad \gamma_\alpha > 0 \qquad (18)$$

with $-(+)$ for particles (holes). This gives an instantaneous, but *complex* single-particle potential. The imaginary part γ_α represents damping of the excitation in the orbit α; with the $\mp$ and positive γ_α, the excitation damps as time moves forward. The quasiparticle Green function is then

$$G_\alpha(\omega) \cong [\omega - (\epsilon_\alpha - \epsilon_F) - \Delta \pm i\gamma_\alpha]^{-1}\,. \qquad (19)$$

The pole of $G_\alpha(\omega)$ is shifted off the real ω-axis by $\mp i\gamma_\alpha$: this is necessarily an approximation, since one can show generally that the singularities of $G(\omega)$ all have real ω. In fact, there is a singularity for $\omega = E_n - E_0$ for every excited state n of the system that is reached by adding or removing a single particle. The pole approximation gives the average over these singularities. The position of the pole is shifted by the real part Δ_α in either direction along the real ω-axis.

The use of $\mp i\gamma_\alpha$ in Eq. 18 represents a process in which probability flux is lost from the excitation of α at energy ω: this requires that there be a continuum of states at that energy. This can happen for an infinite system, e.g. a uniform medium, with continuous plane-wave states, $\alpha = \mathbf{k}$. Even for a system with discrete excitation α, like an atom, there may be unbound states of the same energy to which the excitation α may couple; e.g. a 2p–1h state of a bound atom may be degenerate with a single-electron state in the continuum; this describes the Auger process. In these cases, there are real damping processes involved, which are represented by the imaginary part of Σ. In a discrete system with closely spaced levels, there is no actual damping, but the *mixing* of nearby states may be well approximated by the same device of a complex potential. We shall return to a discussion of the quasiparticle approximation in the next section, in the context of the theory of the optical potential.

Last, we should note that the approximation of Eq. 18 is used widely in the theory of the low-energy excitations of infinite Fermi systems, e.g. for the theory of metals.

This concludes our introduction to Green functions. Clearly, one of the uses of the Green function, calculated in any reasonable approximation, is to sum partial series of Feynman diagrams, by replacing G^0 with G in any diagram under consideration. If we do this using the first-order approximation, Eq. 16, it is equivalent to the calculation of Goldstone diagrams with the use of *self-consistent* or Hartree–Fock insertions, which we discussed in Section 9.4, in connection with Eq. 9.35 and Figure 9.9.

11.3 OPTICAL POTENTIAL

We now turn to a scattering problem often encountered in the context of systems of many degrees of freedom, namely, elastic scattering of a projectile from a complex target, such as an atom, molecule, nucleus, or a hadron composed of more-elementary constituents (quarks). The scattering projectile could be an elementary particle (electron, proton, photon, etc.) or could itself be a complex system. Elastic scattering, in which by definition no energy is transferred between projectile and target, always takes place, if there is any scattering whatsoever. Since the quantum state of the target (and of the projectile, if it is complex) is not altered in the process, there is often a considerable degree of coherence in the scattering amplitude, reflecting the contribution, in phase, of each of the elements of the complex target (projectile).

The common experimental situation for an isolated target is that it is initially in its ground state, and is left in this state by elastic scattering. So we may think of the description of this process as a characterization of some property or properties of the many-body ground state itself. For a bulk-matter target (gas, liquid, crystal, etc.) the initial state is usually that of equilibrium at a fixed temperature, so that elastic scattering can be related analogously to an equilibrium property of the material.

A particularly useful characterization of this ground state (or equilibrium) property is through an *optical potential*. The basic notion is that elastic scattering can always be described by a Schrödinger equation with an appropriately defined potential interaction between projectile and target. The potential may be nonlocal, and may also depend on the energy of the scattering, but can have as kinematic variables only the relative distance and momentum of the projectile and target, treated as point particles, since the internal state is the ground state of each system. (For degenerate ground states, e.g. of nonzero spin, the ground-state spin variable also appears in the Schrödinger equation.) The Schrödinger equation with these restricted variables does not, of course, completely describe all the dynamics of the complicated scattering process, but it does have just sufficient degrees of freedom to describe the projectile and target well separated in space. This is all that is required to give a complete characterization of the scattering amplitude (or S-matrix: see Sections 7.5, 11.4), which represents only asymptotic properties of the scattering wave. A potential which will produce the correct asymptotic wave for elastic scattering at a given energy is called an *optical potential*; as suggested by the phrasing, this requirement does not uniquely define the potential.

Under most circumstances, elastic scattering on a complex target is not flux conserving: The possibility of excitation of many states other than the ground state removes probability flux from the elastic wave—less comes out than went in. (The exceptional case is when the energy available is less than the first inelastic energy.) A well-known result of elementary quantum mechanics is that a Schrödinger wave equation with a Hermitian Hamiltonian conserves flux. Therefore, in general an optical potential will be non-Hermitian, to allow for this flux loss.

If we now limit our considerations to a projectile which is itself elementary, we may connect the notion of an optical potential to that of the single-particle Green function or propagator, as discussed in Sections 11.1 and 11.2. The connection is most direct if the projectile is identical with (some of) the constituents of the target (e.g., in electron–atom scattering), in which case the scattering channel (projectile plus ground-state target) is that reached by exciting one particle out of the ground state of the combined system of projectile plus target particles. This is the situation for which the Green functions of Eq. 7.26 or Eq. 11.1 were introduced. But only a slight generalization is required to use the same notion (and notation) for the case of a distinct projectile (e.g., in electron–nucleus scattering). Let us see how the optical potential may be related to the single-particle Green function.

11.3.1 Connection of the Potential to the Green Function

We first establish the connection of the elastically scattered wave to the single-particle Green function, and second, using Dyson's equation of

Section 11.2, find that the optical potential may be identified with Dyson's proper self-energy function Σ. We follow a line of argument originally given by Bell and Squires.*

We start by constructing the initial scattering state of Eq. 7.52a that represents the projectile and target in a noninteracting state in the infinite past. As we discussed in Section 7.5, this state may be reached by separating projectile and target into nonoverlapping packets, or by the mathematically equivalent device† of switching on the scattering interaction adiabatically in the distant past. With the latter choice, the initial state may be constructed by adding the projectile in momentum state $\mathbf{k}$ to the target in its ground state Ψ_0 as follows:

$$|\Phi_i\rangle = |\mathbf{k}, \Psi_0\rangle = a_{\mathbf{k}}^{\dagger}|\Psi_0\rangle , \qquad (20)$$

giving a time-independent state, as required in Eq. 7.52a. We may alternatively rewrite this state, using the Heisenberg operator $\tilde{a}_{\mathbf{k}}(t')$ (see Eq. 1), with the time label in the distant past $(t' \to -\infty)$, at which time the interaction has not yet been switched on. Without interaction, Eq. 7.11b gives the operator relation

$$a_{\mathbf{k}}^{\dagger} = \tilde{a}_{\mathbf{k}}^{\dagger}(t')e^{-i\epsilon(k)t'} \qquad (21)$$

with $\epsilon(k) = k^2/2m$, the kinetic energy. If this substitution is made in Eq. 20, a Heisenberg-picture state is produced; to stay in the interaction picture we transform with $U^{-1}(0, t') = U(t', 0)$ as in Eq. 2, so that Eq. 20 becomes

$$|\Phi_i\rangle = \lim_{t' \to -\infty} e^{-i\epsilon(k)t'}U(t', 0)\tilde{a}_{\mathbf{k}}^{\dagger}(t')|\Psi_0\rangle . \qquad (22)$$

The scattered wave in the interaction picture evolves as in Eq. 7.53,

$$|\Phi(t)\rangle = U(t, -\infty)|\Phi_i\rangle , \qquad (23)$$

now generating all possible excitations of the target system present at time t. We determine the component of this complicated state, which has the target in its ground state and the projectile in momentum state $\mathbf{k}'$ at time t, from the overlap of Eq. 23 with the state $\langle\Psi_0|\tilde{a}_{\mathbf{k}'}(t)$, again transformed to the interaction picture

$$\langle\Phi_0|\tilde{a}_{\mathbf{k}'}(t)U(0, t)|\Phi(t)\rangle = \psi(\mathbf{k}', t) . \qquad (24)$$

We have introduced the symbol $\psi(\mathbf{k}', t)$ for this amplitude, which can be

---

*J. S. Bell and E. J. Squires, Phys. Rev. Lett. **3**, 96 (1959).
†This device presents problems when applied to cases for which the projectile is identical with target particles; it should then be understood as representing the wave-packet construction.

thought of as the wave function for the *elastic wave* in a momentum representation, for the initial wave given by Eq. 20. Combining Eqs. 22–24 and 1, we can relate this wave function to the single-particle Green function, also in momentum representation, by

$$\psi(\mathbf{k}', t) = i \lim_{t' \to -\infty} G(\mathbf{k}', \mathbf{k}; t, t') e^{-i\epsilon(k)t'} , \tag{25}$$

where we write the momentum indices of Eq. 1 as variables in Eq. 25.

We shall now show that this wave function does obey a Schrödinger-like equation with an energy-dependent potential, which plays the role of the optical potential. First we make use of Dyson's equation (Eq. 10) for the Green function to rewrite Eq. 25:

$$\psi(\mathbf{k}', t) = i \lim_{t' \to -\infty} e^{-i\epsilon(k)t'} \left[G^0(\mathbf{k}', t - t') \delta(\mathbf{k}' - \mathbf{k}) \right.$$

$$+ \int d\mathbf{k}'' \iint dt_1 \, dt_2 \, G^0(\mathbf{k}', t - t_2) \Sigma(\mathbf{k}', \mathbf{k}''; t_2 - t_1)$$

$$\left. \times G(\mathbf{k}'', \mathbf{k}; t_1 - t') \right]$$

$$= \delta(\mathbf{k}' - \mathbf{k}) e^{-i\epsilon(k)t}$$

$$+ \int d\mathbf{k}'' \iint dt_1 \, dt_2 \, G^0(\mathbf{k}, t - t_2) \Sigma(\mathbf{k}', \mathbf{k}''; t_2 - t_1) \psi(\mathbf{k}'', t_1) , \tag{26}$$

which is now in the form of an equation of motion for the wave function $\psi(\mathbf{k}, t)$. [We have used Eq. 7.29, setting $\epsilon_F = 0$, which is appropriate for the present case, to evaluate $G^0(\mathbf{k}', t - t')$ in the first term of Eq. 26.] This may be put into a more familiar form by use of the following result:

$$\left(i \frac{\partial}{\partial t} - \epsilon(k) \right) G^0(\mathbf{k}, t - t') = \delta(t - t') , \tag{27}$$

which was derived in Exercise 7.2. Now, if we differentiate both sides of Eq. 26 as on the left side of Eq. 27, we find (dropping primes)

$$\left(i \frac{\partial}{\partial t} - \epsilon(k) \right) \psi(\mathbf{k}, t) = \int d\mathbf{k}'' \int dt_1 \, \Sigma(\mathbf{k}, \mathbf{k}''; t - t_1) \psi(\mathbf{k}'', t_1) . \tag{28}$$

This last result is similar to the form of the Schrödinger equation in momentum space, which can be written, for a wave function ϕ,

$$\left(i \frac{\partial}{\partial t} - \epsilon(k) \right) \phi(k, t) = \int d\mathbf{k}'' \, v(\mathbf{k}, \mathbf{k}'') \phi(\mathbf{k}'', t) \tag{29}$$

with a potential v. The difference between Eqs. 28 and 29 is in the extra time integration in Eq. 28: $\Sigma(t - t_1)$, unlike an ordinary potential, acts over time, as we have already discussed in Section 2 (see Eq. 15 and remarks following), where we found that the time dependence of $\Sigma(t - t_1)$ is connected to the energy, or ω, dependence of $\Sigma(\omega)$ by Fourier transformation.

If we perform the t-to-ω transformation of Eq. 28 (see Eq. 8.50), we obtain

$$[\omega - \epsilon(k)]\psi(\mathbf{k}, \omega) = \int d\mathbf{k}'' \, \Sigma(\mathbf{k}, \mathbf{k}''; \omega)\psi(\mathbf{k}'', \omega) \tag{30a}$$

with

$$\psi(\mathbf{k}, \omega) = \int_{-\infty}^{\infty} dt \, e^{i\omega t}\psi(\mathbf{k}, t) , \tag{30b}$$

while the transformation of Eq. 29 gives the time-independent Schrödinger equation

$$[\omega - \epsilon(k)]\phi(\mathbf{k}, \omega) = \int d\mathbf{k}'' \, v(\mathbf{k}, \mathbf{k}'')\phi(\mathbf{k}'', \omega) , \tag{31}$$

where $\phi(\mathbf{k}, \omega)$ is defined as in Eq. 30b. This form is appropriate for a wave-packet description of scattering, for which the energy (or frequency) ω has a spread; energy eigenstates have singular amplitudes in momentum space [e.g., the plane wave momentum $\mathbf{k}_0$ becomes $\phi_0(\mathbf{k}) = \delta(\mathbf{k} - \mathbf{k}_0)$].

Comparison of Eqs. 30 and 31 clearly shows that the elastically scattered wave ψ can be calculated from a Schrödinger equation with an energy-dependent potential function $\Sigma(\mathbf{k}, \mathbf{k}''; \omega)$, which we identify as the optical potential. We have already discussed in Section 2 that $\Sigma(\omega)$ is in general non-Hermitian, reflecting the loss of flux involved in single-particle propagation. In a diagonal representation, $\Sigma(\omega)$ is therefore *complex*, with a negative imaginary part, as in Eq. 18. (For scattering, we have only particle states at large distance from the target.) It is easy to show that this sign corresponds to *loss* of flux (Exercise 11.2). The spatial form of the optical potential is obtained by a double Fourier transform

$$v_{\text{opt}}(\mathbf{r}, \mathbf{r}'; \omega) = (2\pi)^{-6} \int d\mathbf{k} \, d\mathbf{k}' \, \Sigma(\mathbf{k}, \mathbf{k}'; \omega)\exp i(\mathbf{k}' \cdot \mathbf{r}' - \mathbf{k} \cdot \mathbf{r}) , \tag{32}$$

and is generally nonlocal, so that the wave equation takes the form

$$\left(\omega + \frac{\nabla^2}{2m}\right)\psi(\mathbf{r}, \omega) = \int d\mathbf{r}' \, v_{\text{opt}}(\mathbf{r}, \mathbf{r}'; \omega)\psi(\mathbf{r}', \omega) . \tag{33}$$

If the optical potential is known, either by phenomenology or by calculation (as in the following subsection), then the wave equation 33 may be solved by standard techniques of scattering theory, for fixed ω. The elastic scattering amplitude and differential cross section are then obtained directly from ψ or from the S-matrix (as given in Section 11.4).

11.3.2 Expansion of the Optical Potential in Diagrams

Since we have found that the optical potential may be obtained directly from Dyson's proper self-energy function for a single particle (the projectile) propagating through the target, we may now use the diagrammatic techniques of Section 2 to calculate the optical potential directly. The calculation of the scattered wave from the potential, by solving Eq. 33, is completely equivalent to solving Dyson's equation (Eq. 11) for the Green function.

We remember that the self-energy function Σ was defined in Eq. 8 to include all *irreducible* diagrams for single-particle propagation, as illustrated in Figure 11.3. Two simple examples were shown in Figure 11.4: (a) the Hartree–Fock insertion diagrams, direct and exchange, and (b) the ladder diagrams. The first represent (see Section 9.4) a single interaction of the projectile with an individual particle in the target, without excitation of the target particle from its ground-state orbit (in the Hartree–Fock approximation for the target). This is just the Born approximation for scattering from the Hartree–Fock ground state. The exchange diagram is present only if the projectile is identical with some particles in the target, so that they can "change places" and still lead to elastic scattering.

The ladder diagrams are an infinite series which was formally summed in Section 10.1, in connection with the hard-sphere gas. In the present context, it would provide a useful first approximation to the optical potential for short-range, strong interaction v between projectile and target particle. With very little modification, we can adopt Eqs. 10.9 and 10.10 to our present needs. One particle in the "ladder" is our projectile, and the other, a target particle (to be summed over all occupied orbits: γ in Figure 11.4b). For scattering we can no longer take an infinite medium with orbits diagonal in momentum. The self-energy function can be expressed (compare Eq. 10.9) as

$$\Sigma(\mathbf{k}, \mathbf{k}'; \omega) = \sum_{\gamma} \langle \mathbf{k}, \gamma | \tilde{t}(\omega + \epsilon_{\gamma}) | \mathbf{k}', \gamma \rangle_A , \qquad (34)$$

where $\tilde{t}(\epsilon)$ was defined in Eq. 10.10. If exchange of projectile and target particles is possible, then the exchange-ladder diagrams are included in Eq. 34, and the projectile may not occupy intermediate states which are occupied in the target [see the discussion of the operator Q (Eq. 10.2) in Section 10.1].

For high-energy scattering, these exchange effects are usually small, and may be neglected. For similar reasons, we neglect the energy of the occupied orbit relative to the scattering energy: $|\epsilon_{\gamma}| \ll \omega$ in Eq. 34. The difference between $\tilde{t}(\omega)$ and the free t-matrix $t(\omega)$ of Eq. 10.21 is also small at high energy: $\tilde{t}(\omega) \cong t(\omega)$ is called the *impulse* approximation in scattering theory. (Recall that ω is the projectile energy, as is made clear in Eq. 30.) With these assumptions, Eq. 34 can be simplified to give

$$\Sigma(\mathbf{k}, \mathbf{k}'; \omega) \cong \sum_{\gamma} \langle \mathbf{k}, \gamma | t(\omega) | \mathbf{k}' \gamma \rangle$$

$$= (2\pi)^{-6} \int d\mathbf{p} \, d\mathbf{p}' \, \langle \mathbf{k}, \mathbf{p} | t(\omega) | \mathbf{k}', \mathbf{p}' \rangle \rho(\mathbf{p}', \mathbf{p}) \,, \qquad (35)$$

where we have expanded the orbital wave functions in a momentum representation, using a one-body density matrix

$$\rho(\mathbf{p}', \mathbf{p}) = \sum_{\gamma} \langle \mathbf{p}' | \gamma \rangle \langle \gamma | \mathbf{p} \rangle = \sum_{\gamma} \phi_{\gamma}(\mathbf{p}) \phi_{\gamma}^{*}(\mathbf{p}') \,. \qquad (36)$$

If we can further neglect the p-dependence of the t-matrix (p, $p' \ll k$, k' for projectiles moving much faster than target particles), Eq. 35 factors in the form

$$\Sigma(\mathbf{k}, \mathbf{k}'; \omega) \cong \langle \mathbf{k} | t(\omega) | \mathbf{k}' \rangle F(\mathbf{q}) \,, \qquad (37)$$

where we have used Eq. 10.16a, and defined the momentum transfer $\mathbf{q} = \mathbf{k} - \mathbf{k}' = \mathbf{p}' - \mathbf{p}$ and the *form factor*

$$F(\mathbf{q}) = (2\pi)^{-3} \int d\mathbf{p} \, \rho(\mathbf{p} + \mathbf{q}, \mathbf{p}) \,. \qquad (38a)$$

[Note that we have neglected $\mathbf{p}$, $\mathbf{p}'$ in the t-matrix of Eq. 37, but not in $F(\mathbf{q})$.] Fourier transformation of the orbital wave functions allows us to write Eq. 38a also in terms of the particle density $\rho(\mathbf{r})$ as follows:

$$F(\mathbf{q}) = \int d\mathbf{r} \, e^{i\mathbf{q} \cdot \mathbf{r}} \, \rho(\mathbf{r}) \qquad (38b)$$

with

$$\rho(\mathbf{r}) = \sum_{\gamma} |\phi_{\gamma}(\mathbf{r})|^{2} \,. \qquad (38c)$$

A last approximation: at high energy, again since k, $k' \gg p$, p', the scattering *angle* between $\mathbf{k}$ and $\mathbf{k}'$ is restricted by the form factor (Eq. 38a) to be small, so we may take the *forward-scattering* approximation for t: $\langle \mathbf{k} | t(\omega) | \mathbf{k}' \rangle \cong \langle \mathbf{k} | t(\omega) | \mathbf{k} \rangle = t(\omega)$ [but *not* for $F(\mathbf{q}) = F(\mathbf{k} - \mathbf{k}')$], to write Eq. 37 as

$$\Sigma(\mathbf{k}, \mathbf{k}'; \omega) \cong t(\omega) F(\mathbf{q}) \,, \qquad (39)$$

where $t(\omega)$ is the forward-scattering amplitude at (laboratory) energy ω. This can now be transformed to position space, as in Eq. 32, to produce a *local* (but energy-dependent) optical potential of the form

$$v_{\text{opt}}(\mathbf{r}, \mathbf{r}'; \omega) = v(\mathbf{r}, \omega)\delta(\mathbf{r} - \mathbf{r}') , \tag{40a}$$

$$v(\mathbf{r}, \omega) = t(\omega)\rho(\mathbf{r}) ; \tag{40b}$$

in Eq. 40b we have inverted Eq. 38b. This last approximation to the optical potential is obtained in many ways in multiple scattering theory[*] and is called the $t\rho$ (or factored impulse) approximation. With the t-matrix obtained from scattering of the projectile on free target particles, Eq. 40b often gives a good approximation to the optical potential for (a) high-energy scattering, and (b) a short-range scattering interaction.

11.4 THE FORMULATION OF SCATTERING PROBLEMS

We now turn to scattering processes in general, with several interests in mind. First, we want to establish the relationship between the main theoretical formulation of the preceeding chapters and the usual expressions of scattering theory. Second, we can now fill in some background for applications of scattering theory we have already made earlier, for example, in the ladder approximation of Section 10.1. Third, we shall show explicitly how to handle scattering processes with many degrees of freedom. In Section 11.5 we apply these techniques to scattering problems involving the Yukawa interaction of Section 2.3. Then in Section 11.6 we consider inelastic scattering of a particle on a many-body target.

The main tools for formulating scattering problems in terms of diagrammatic perturbation theory are already at our disposal and are here used to calculate scattering cross sections. One of these tools is the time development of the time-evolution operator $U(t, t_0)$, discussed in Section 7.3, and the second is the scattering operator (or S-matrix) of Eqs. 7.54–7.56,

$$S = U(\infty, -\infty) , \tag{41}$$

which describes how the system evolves from the distant past, when preparations are made for the scattering event, to the distinct future, when scattering has taken place and the results are detected. A matrix element $S_{f0} = \langle \Phi_f | S | \Phi_0 \rangle$ of this operator, as in Eq. 7.56, is then the probability amplitude for the system to have developed, through scattering, from the state $|\Phi_0\rangle$ to the state $|\Phi_f\rangle$.

---

[*]See, e.g. J. M. Eisenberg and D. S. Koltun, *Theory of Meson Interactions with Nuclei* (Wiley-Interscience, New York, 1980), Chapter 4.

The perturbation expansion for U, developed in Section 8.1 for use in the many-body ground state, is, of course, just as applicable to the scattering problem, so we may take over the result of Eq. 8.8, as applied to the scattering operator,

$$S = \sum_{n=0}^{\infty} \frac{(-i)^n}{n!} \int_{-\infty}^{\infty} dt_1 \cdots dt_n \, \{V(t_1) \cdots V(t_n)\}_T \,, \tag{42}$$

where the label T means that the operators $V(t_i)$ are to be ordered from right to left according to increasing time, that is, if $t_i > t_j$ then $V(t_i)$ stands to the left of $V(t_j)$.

The dynamical ingredient in the calculation of a scattering cross section – at least to the degree that the interaction is weak and perturbation theory is applicable – is the evaluation of S in Eq. 42. Once this is done, the calculation of the transition rate or a cross section is aided by considering the transition operator

$$\mathcal{T} = 1 - S \,, \tag{43}$$

which is constructed in such a way that there is a cancellation of terms for which no transition occurs: $U(\infty, -\infty) = 1$. Matrix elements of S between an initial state labeled by 0 and a final state f are given by

$$S_{f0} = \delta_{f0} - \langle f | \mathcal{T} | 0 \rangle = \delta_{f0} - 2\pi i \delta(E_f - E_0) \langle f | T | 0 \rangle \,, \tag{44}$$

where the last form embodies conservation of total system energy E in the scattering event and introduces the T-matrix with a particular normalization.

The transition probability is not given directly by the square of S_{f0},

$$|S_{f0}|^2 = |\mathcal{T}_{f0}|^2 = [2\pi \delta(E_f - E_0)]^2 |\langle f | T | 0 \rangle|^2 \,. \tag{45}$$

This contains a factor $2\pi\delta(0)$ which at the moment is without well-defined meaning. The problem is in the order of taking time limits. First we write the matrix element

$$\langle f | U(t_2, t_1) | 0 \rangle = -\langle f | T | 0 \rangle \int_{t_1}^{t_2} dt \, e^{i(E_f - E_0)t} \,, \tag{46}$$

corresponding to Eq. 44 (for $f \neq 0$) before taking limits. The time integration gives

$$\frac{\exp i\omega t_2 - \exp i\omega t_1}{i\omega} = \frac{\sin[\omega(t_2 - t_1)/2]}{\omega/2} \exp i\omega \frac{t_2 + t_1}{2} \,, \tag{47}$$

where we set $\omega = E_f - E_0$. The transition probability $(0 \rightarrow f)$ is given by

$$dW_{f0}(t_2, t_1) = |\langle f|U(t_2, t_1)|0\rangle|^2 \, dn_f$$

$$= |\langle f|T|0\rangle|^2 \, \frac{\sin^2[\omega(t_2 - t_1)/2]}{(\omega/2)^2} \, dn_f \,,$$
(48)

where the differential refers to variables (denoted by dn_f, e.g. solid angles) not integrated. The transition *rate* at time t_2 is then given by the derivative d/dt_2 of Eq. 48, which, after some manipulation, may be written

$$dw_{f0}(t_2, t_1) = |\langle f|T|0\rangle|^2 \left(\frac{2}{\omega}\right) \sin \omega(t_2 - t_1) \, dn_f \,.$$
(49)

We now take the limits $t_2 \rightarrow \infty$, $t_1 \rightarrow -\infty$, and using the representation of the δ-function

$$\lim_{t \rightarrow \infty} \frac{\sin \omega t}{\omega} = \pi\delta(\omega) \,,$$
(50)

obtaining the (time-independent) rate formula

$$dw_{f0} = 2\pi\delta(E_f - E_0)|\langle f|T|0\rangle|^2 \, dn_f \,,$$
(51a)

where, as in Eq. (44), the conservation of energy is given by the δ-function. The result is given *symbolically* by dividing Eq. 45 by $2\pi\delta(E_f - E_0)$:

$$dw_{f0} = \text{"}|S_{f0}|^2 \, dn_f/2\pi\delta(E_f - E_0)\text{"} \,,$$
(51b)

as if $2\pi\delta(0)$ represented the transition time. The cross section is then generated from this transition rate, as usual, by dividing by the flux I of incident particles:

$$d\sigma_{f0} = \frac{1}{I} \, dw_{f0} \,,$$
(52)

where I is the number of particles impinging on the target per unit time and per unit area.

The form in which we have developed the scattering problem is particularly appropriate to the use of diagrammatic methods for many-body systems, since we have cast the scattering operator S of Eq. 42 in a form to which Wick's algebraic methods directly apply. Before we give some explicit examples of such cases (in the following section), it is useful to recover a few of the standard results of potential-scattering theory, for which one particle scatters from a fixed potential, given by the operator V in the Schrödinger picture. In this case, Eq. 42 represents the usual Born series for the scattering amplitude, given in the interaction picture. For example, the first Born approximation is obtained from the $n = 1$ term,

$$\langle f|S^{(1)}|0\rangle = -i \int_{-\infty}^{\infty} dt \, \langle f|V(t)|0\rangle$$

$$= -i \int_{-\infty}^{\infty} dt \, e^{i(E_f - E_0)t} \langle f|V|0\rangle$$

$$= -2\pi i\delta(E_f - E_0)\langle f|V|0\rangle , \tag{53}$$

where we use Eqs. 8.9 for $V(t)$, with H_0 the kinetic energy of the projectile, and Eq. 46 for the time integration, taking the time limits indicated. Lastly, from Eq. 44 we write the first Born approximation in T-matrix form

$$\langle f|T|0\rangle_{\text{BA}} = \langle f|V|0\rangle . \tag{54}$$

The amplitude for potential scattering may also be generated in a time-independent form, using an operator $t(E)$ which obeys the following linear equation (derived in the Appendix):

$$t(E) = V + V(E - H_0 + i\eta)^{-1}t(E) , \tag{55}$$

(often called the Lippmann–Schwinger equation for t). The operator is defined so that its (energy-conserving) matrix elements give the t-matrix of Eqs. 44, 45, i.e.

$$\langle f|T|0\rangle = \langle f|t(E)|0\rangle , \tag{56}$$

with $E = E_0 = E_f$. Note that potential scattering only produces elastic waves: $E_f = E_0$.

The connection of the t-matrix to the ordinary scattering amplitude for the elastic wave is given in most textbooks for quantum mechanics as

$$f(E_0, \theta) = -\frac{m}{2\pi} \langle \mathbf{k}_f|t(E_0)|\mathbf{k}_0\rangle , \tag{57}$$

where the matrix element is taken between plane-wave states $\mathbf{k}_0$, $\mathbf{k}_f$ of energy $E_0 = k_0^2/2m = k_f^2/2m$, and $\cos\theta = \mathbf{k}_0 \cdot \mathbf{k}_f$ gives the c.m. scattering angle. Then the general cross-section formula, Eq. 52, reduces to the elementary result for potential (elastic) scattering

$$\frac{d\sigma}{d\Omega}(\theta) = |f(\theta, E_0)|^2 . \tag{58}$$

For the scattering of two particles by a mutual potential, we may also use the operator $t(E)$, now defined as a two-body operator, with H_0 the sum of the two kinetic energies and V the two-body interaction. This is the form of Eq. 55 used in Chapter 10 (Eq. 10.15) in connection with the ladder approximation.

11.5 APPLICATIONS OF SCATTERING TECHNIQUES TO THE YUKAWA INTERACTION

We now illustrate the application to scattering problems of some of the methods developed in this book. We start with a case that is not a scattering situation, but rather involves decay through particle emission. The reaction in question involves a change in the number of bosons by one, and thus complements the general points raised in this regard in Chapters 2 and 3. We then discuss boson–fermion and fermion–fermion scattering for the scalar Yukawa coupling; although there is no overall change of boson number in these cases, they are paradigms of the consequences of the instantaneous change in boson number for fermion systems in interaction with bosons.

We consider the decay in which a fermion emits a boson and converts to a fermion of lower energy through the mechanism of the Yukawa interaction introduced in Section 2.3. In the interaction picture (see Section 7.3) the Yukawa interaction of Eqs. 2.43 and 2.44 is

$$V(t) = g \sum_{\alpha\beta\sigma} \int_\Omega d\mathbf{r}\, u_\beta^\dagger(\mathbf{r}) u_\alpha(\mathbf{r}) e^{i(\epsilon_\beta - \epsilon_\alpha)t}$$

$$\times [w_\sigma(\mathbf{r}) b_\sigma e^{-i\omega_\sigma t} + w_\sigma^\dagger(\mathbf{r}) b_\sigma^\dagger e^{i\omega_\sigma t}] a_\beta^\dagger a_\alpha , \qquad (59)$$

where we have now shown the full time dependence of the field operators (suppressed in Eq. 2.44) in the interaction picture, which involves, in the operators, only the unperturbed energies. The interaction in Eq. 59 usually refers to fermions [entering here in wave functions $u_\gamma(\mathbf{r})$, creation and annihilation operators $a_\gamma^\dagger, a_\gamma$, and energies ϵ_γ] and bosons [here with wave functions w_σ, creation and annihilation operators $b_\sigma^\dagger, b_\sigma$, and energies ω_σ]. In the case of plane-wave states, for example, we take the fermion wave functions with conventional nonrelativistic normalization,

$$u_\gamma(\mathbf{r}) = \frac{1}{\sqrt{\Omega}}\, e^{i\mathbf{p}_\gamma \cdot \mathbf{r}} , \qquad (60)$$

where Ω the normalization volume. The boson wave functions carry a further factor in their normalization, as for the photon case in Eq. 3.33,

$$w_\sigma(\mathbf{r}) = \frac{1}{\sqrt{2\omega_{\mathbf{k}_\sigma}\Omega}}\, e^{i\mathbf{k}_\sigma \cdot \mathbf{r}} , \qquad (61)$$

ensuring that the calculation of the energy from the Hamiltonian density $H = \frac{1}{2}\dot{\phi}^2 + \frac{1}{2}(\nabla\phi)^2 + \frac{1}{2}m^2\phi^2$ indeed yields $\omega_{\mathbf{k}_\sigma}$ for each $\mathbf{k}_\sigma$ (see Eq. 3.45). The Yukawa interaction is constructed so as not to change the number of fermions, but to increase or decrease the boson population by unity, as illustrated in Figure 2.1.

Consider the absorption or emission of a boson (Figure 2.1) further: As a specific example one might think of the decay of a Δ-isobar through pion emission, leaving a nucleon N along with the π-meson in the final state. In order for the interaction of Eq. 59 to be applicable here, one must suppose that the fermion in question has internal degrees of freedom that change appropriately to describe that Δ or the N, and one must generalize Eq. 59 to deal with these internal degrees of freedom and with the fact that the pion is a pseudoscalar—rather than a scalar—boson. We ignore these features here, since our concern is to see how the general form of the Yukawa interaction induces transitions; the labels Δ, N, and π are thus merely convenient ways to distinguish initial and final fermion states and the state of the emitted boson. (The discussion of boson absorption proceeds along very similar lines, though in the "opposite direction"; see Exercise 11.3.)

Under the assumption—not valid for $\Delta \to N\pi$, but we only wish to illustrate the method here—that the interaction is weak, the S-matrix of Eq. 42 may be taken in lowest order,

$$S^{(1)} = -i \int_{-\infty}^{\infty} V(t)\,dt$$

$$= -ig \sum_{\alpha\beta\sigma} \int_{-\infty}^{\infty} dt \int_{\Omega} d\mathbf{r}\, u_\beta^\dagger(\mathbf{r})u_\alpha(\mathbf{r})e^{i(\epsilon_\beta - \epsilon_\alpha)t}$$

$$\times [w_\sigma(\mathbf{r})b_\sigma e^{-i\omega_\sigma t} + w_\sigma^\dagger(\mathbf{r})b_\sigma^\dagger e^{i\omega_\sigma t}]a_\beta^\dagger a_\alpha\,. \tag{62}$$

The S-matrix element for boson emission has the structure

$$\langle \pi N|S^{(1)}|\Delta\rangle = -ig \sum_{\alpha\beta\sigma} \int_{-\infty}^{\infty} dt \int_{\Omega} d\mathbf{r}\, u_\beta^\dagger(\mathbf{r})u_\alpha(\mathbf{r})w_\sigma^\dagger(\mathbf{r})$$

$$\times e^{i(\epsilon_\beta - \epsilon_\alpha + \omega_\sigma)t}\langle \pi N|b_\sigma^\dagger a_\beta^\dagger a_\alpha|\Delta\rangle$$

$$= -ig \int_{-\infty}^{\infty} dt \int_{\Omega} d\mathbf{r}\, u_N^\dagger u_\Delta(\mathbf{r})w_\pi^\dagger(\mathbf{r})e^{i(\epsilon_N + \omega_\pi - \epsilon_\Delta)t}\,, \tag{63}$$

where we have simply used the basic properties of creation and annihilation operators. If we further assume plane-wave states for all three particles involved (momenta $\mathbf{p}_\Delta$, $\mathbf{p}_N$, and $\mathbf{k}_\pi$ for the corresponding particle states), the matrix element is

$$\langle \pi N | S^{(1)} | \Delta \rangle = \frac{-ig}{\sqrt{2\omega_\pi \Omega^3}} \int_{-\infty}^{\infty} dt \int_\Omega d\mathbf{r} \, e^{i(\epsilon_N + \omega_\pi - \epsilon_\Delta)t} e^{-i(\mathbf{p}_N + \mathbf{k}_\pi - \mathbf{p}_\Delta)\cdot\mathbf{r}}$$

$$\times (\chi_N, \phi_\pi^\dagger \chi_\Delta),$$

$$\qquad\qquad\qquad (64)$$

where χ_N and χ_Δ are fermion spinors, which include all internal quantum numbers for these states, and ϕ_π describes the internal quantum numbers of the scalar boson of our illustrative example (e.g., isospin). Note that the exponentials in Eq. 64 involve a covariant combination $Et - \mathbf{P}\cdot\mathbf{r}$, where E is the overall energy difference and $\mathbf{P}$ is the overall three-momentum difference. The integration $\int dt \int d\mathbf{r}$ may also be handled in ways that make manifest the covariance of the S-matrix element, so that the scheme that emerges here easily lends itself to covariant treatment. (We do not pursue this point further here, but it is of great importance in relativistic field theory.)

The time integration in Eq. 64 is easily performed, leading to a δ-function for energy conservation, as we anticipated in the discussion of Eqs. 45–51:

$$\langle \pi N | S^{(1)} | \Delta \rangle = -2\pi i g \delta(\epsilon_N + \omega_\pi - \epsilon_\Delta)(2\omega_\pi \Omega^3)^{-1/2}$$

$$\times \int_\Omega d\mathbf{r} \, e^{-i(\mathbf{p}_N + \mathbf{k}_\pi - \mathbf{p}_\Delta)\cdot\mathbf{r}}(\chi_N, \phi_\pi^\dagger \chi_\Delta),$$

$$\qquad\qquad (65)$$

so that the boson emission rate is given by Eq. 51,

$$dw_{N\pi\leftarrow\Delta} = 2\pi\delta(\epsilon_N + \omega_\pi - \epsilon_\Delta)g^2(2\omega_\pi \Omega^3)^{-1}|(\chi_N, \phi_\pi^\dagger \chi_\Delta)|^2$$

$$\times \left| \int_\Omega d\mathbf{r} \, e^{-i(\mathbf{p}_N + \mathbf{k}_\pi - \mathbf{p}_\Delta)\cdot\mathbf{r}} \right|^2 .$$

$$\qquad\qquad (66)$$

In the limit of large quantization volume Ω, the spatial integral yields a factor of overall momentum conservation,

$$(2\pi)^3 \delta(\mathbf{p}_N + \mathbf{k}_\pi - \mathbf{p}_\Delta),$$

again as was to be expected for a free system. This factor is squared and thus echoes the problem we encountered with the δ-function for energy conservation. In the present case the resolution of this difficulty is fairly immediate, since it is clear that with one factor to ensure momentum conservation, the other integral is simply the volume,

$$\int_\Omega 1 \, d\mathbf{r} = \Omega,$$

$$\qquad\qquad (67)$$

whence

$$dw_{N\pi\leftarrow\Delta} = (2\pi)^4 \delta(\epsilon_N + \omega_\pi - \epsilon_\Delta)\delta(\mathbf{p}_N + \mathbf{k}_\pi - \mathbf{p}_\Delta)g^2(2\omega_\pi\Omega^2)^{-1}|(\chi_N, \phi_\pi^\dagger\chi_\Delta)|^2.$$

(68)

For a total transition rate this must be integrated over the various final-state contributions,

$$w_{N\pi\leftarrow\Delta} = \int dw_{N\pi\leftarrow\Delta} = (2\pi)^4 \int \delta(\epsilon_N + \omega_\pi - \epsilon_\Delta)$$

$$\times \delta(\mathbf{p}_N + \mathbf{k}_\pi - \mathbf{p}_\Delta)g^2(2\omega_\pi\Omega^2)^{-1}|(\chi_N, \phi_\pi^\dagger\chi_\Delta)|^2$$

$$\times \frac{\Omega \, d\mathbf{k}_\pi}{(2\pi)^3} \frac{\Omega \, d\mathbf{p}_N}{(2\pi)^3} ;$$

(69)

note that the quantization volume now cancels, as it should. We shall take a system in which the initial Δ is at rest: $\mathbf{p}_\Delta = 0$. Then the momentum integration yields

$$w_{N\pi\leftarrow\Delta} = g^2|(\chi_N, \phi_\pi^\dagger\chi_\Delta)|^2(2\omega_\pi)^{-1}\, 2\pi$$

$$\times \int \delta[\epsilon_N(-\mathbf{k}_\pi) + \omega_\pi(\mathbf{k}_\pi) - \epsilon_\Delta(0)] \frac{d\mathbf{k}_\pi}{(2\pi)^3} ,$$

(70)

which embodies the obvious kinematic fact that when a body at rest undergoes two-particle decay, the products come out with momenta of equal magnitudes and opposite directions. For relativistic kinematics

$$\epsilon_N = (\mathbf{p}_N^2 + m_N^2)^{1/2} \Rightarrow (\mathbf{k}_\pi^2 + m_N^2)^{1/2} ,$$

(71a)

$$\omega_\pi = (\mathbf{k}_\pi^2 + m_\pi^2)^{1/2} ,$$

(71b)

with total final energy

$$E_f = \epsilon_N(-\mathbf{k}_\pi) + \omega_\pi(\mathbf{k}_\pi) ,$$

(71c)

this last integration is

$$w_{N\pi\leftarrow\Delta} = g^2|(\chi_N, \phi_\pi^\dagger\chi_\Delta)|^2(2\omega_\pi)^{-1} \frac{4\pi}{(2\pi)^3} \int_0^\infty 2\pi\delta(\epsilon_N + \omega_\pi - \epsilon_\Delta)k_\pi^2 \, dk_\pi$$

$$= g^2|(\chi_N, \phi_\pi^\dagger\chi_\Delta)|^2 \frac{1}{\pi \cdot 2\omega_\pi} \int_0^\infty \delta[E_f - \epsilon_\Delta(0)]k_\pi^2 \frac{dk_\pi}{dE_f} \, dE_f$$

$$= \frac{g^2}{\pi \cdot 2\omega_\pi} |(\chi_N, \phi_\pi^\dagger\chi_\Delta)|^2 \left[\frac{k_\pi^2}{dE_f/dk_\pi}\right]_{E_f=\epsilon_\Delta(0)}$$

$$= \frac{g^2}{2\pi} |(\chi_N, \phi_\pi^\dagger\chi_\Delta)|^2 \frac{k_\pi\epsilon_N}{\epsilon_\Delta}\bigg|_{\mathbf{p}_N=-\mathbf{k}_\pi,\, \epsilon_N+\omega_\pi=\epsilon_\Delta} .$$

(72)

This form shows clearly the phase-space factor $k_\pi \epsilon_N / \epsilon_\Delta(0)$ arising from the integration over the density of states restricted by energy and momentum conservation. For this very simple, scalar case, the dynamics produced a factor g^2 independent of kinematics. The internal degrees of freedom have been left untreated in the factor $|(\chi_N, \phi_\pi^\dagger \chi_\Delta)|^2$, and thus the result is appropriate to a case where a Δ of given spin polarization (and given values for all other internal quantum numbers as well) decays into a nucleon of given polarization (again with definite internal quantum numbers in the final state). Otherwise summation and averaging over internal quantum numbers are in order.

As we have noted, the specific physical case of $\Delta \to N\pi$ is not well suited to the treatment here, both because of a coupling which is inappropriate to the internal symmetries involved and because the strong interaction in the decay does not lend itself to lowest-order treatment (except perhaps as a phenomenological study of effective coupling strength, say). Had we considered atomic decay with photoemission, for example, lowest-order results would have sufficed, but the Yukawa form would only be pertinent in the general features of the $a_\alpha^\dagger a_\beta b_\sigma^\dagger$ structure. The detailed coupling would require construction of the electromagnetic current in place of the simple $u_\alpha^\dagger(\mathbf{r}) u_\beta(\mathbf{r})$, and the transverse nature of the photon would have to be incorporated in $w_\sigma^\dagger(\mathbf{r})$ and in the sum over final polarization states.

Next we consider the scattering of a scalar boson on a fermion to lowest order, using the present formalism. As is seen, say with the aid of Figure 2.2, the lowest order that yields scattering is second order, since the Yukawa interaction must act once to absorb the boson and once to emit it (in either time order). Thus we require

$$S^{(2)} = -\frac{1}{2} \int_{-\infty}^{\infty} dt \int_{-\infty}^{\infty} dt' \, \{V(t)V(t')\}_T$$

$$= -\frac{1}{2} \int_{-\infty}^{\infty} dt \int_{-\infty}^{t} dt' \, V(t)V(t') - \frac{1}{2} \int_{-\infty}^{\infty} dt \int_{t}^{\infty} dt' V(t')V(t)$$

$$= -\int_{-\infty}^{\infty} dt \int_{-\infty}^{t} dt' V(t)V(t') , \tag{73}$$

where the last step follows from reversing the dummy variables and the order of integration in the second integral. The scattering matrix element is thus

$$\langle b'f'|S^{(2)}|bf\rangle = -g^2 \sum_{\alpha\beta\sigma} \sum_{\gamma\delta\tau} \int_\Omega d\mathbf{r} \int_\Omega d\mathbf{r}' \, u_\beta^\dagger(\mathbf{r}) u_\alpha(\mathbf{r}) u_\delta^\dagger(\mathbf{r}') u_\gamma(\mathbf{r}')$$

$$\times \int_{-\infty}^{\infty} dt \int_{-\infty}^{t} dt' \, e^{i(\epsilon_\beta - \epsilon_\alpha)t} e^{i(\epsilon_\delta - \epsilon_\gamma)t'}$$

$$\times \langle b'f'|a_\beta^\dagger a_\alpha [w_\sigma(\mathbf{r})b_\sigma e^{-i\omega_\sigma t} + w_\sigma^\dagger(\mathbf{r})b_\sigma^\dagger e^{i\omega_\sigma t}] a_\delta^\dagger a_\gamma$$

$$\times [w_\tau(\mathbf{r}')b_\tau e^{-i\omega_\tau t'} + w_\tau^\dagger(\mathbf{r}')b_\tau^\dagger e^{i\omega_\tau t'}]|bf\rangle , \tag{74}$$

where the boson–fermion scattering is labeled as $bf \rightarrow b'f'$. Corresponding to the two diagrams in Figure 2.2, there are two ways to obtain nonvanishing matrix elements in the Fock space of the meson, namely, from the combination $\langle b'|b_\sigma^\dagger b_\tau|b\rangle = \delta_{\tau b}\delta_{\sigma b'}$ and from $\langle b'|b_\sigma b_\tau^\dagger|b\rangle = \delta_{\tau b'}\delta_{\sigma b}$. In the fermion space one has $\langle f'|a_\beta^\dagger a_\alpha a_\delta^\dagger a_\gamma|f\rangle = \delta_{\gamma f}\delta_{\beta f'}\delta_{\alpha\delta}$. Thus

$$\langle b'f'|S^{(2)}|bf\rangle = -g^2 \int_\Omega d\mathbf{r} \int_\Omega d\mathbf{r}'\, u_{f'}^\dagger(r)u_f(r') \int_{-\infty}^{\infty} dt \int_{-\infty}^{t} dt'$$

$$\times\, e^{i\epsilon_{f'}t - i\epsilon_f t'} \left\{ \sum_\alpha e^{i\epsilon_\alpha(t'-t)} u_\alpha(\mathbf{r})u_\alpha^\dagger(\mathbf{r}') \right\}$$

$$\times\, [w_{b'}^\dagger(\mathbf{r})w_b(\mathbf{r}')e^{i\omega_b t - i\omega_b t'}$$

$$+\, w_b(\mathbf{r})w_{b'}^\dagger(\mathbf{r}')e^{-i\omega_b t + i\omega_{b'} t'}] , \tag{75}$$

where the quantity in braces plays the role of the unperturbed single-particle propagator, or Green function (compare Section 7.2, noting that we have not introduced hole states here). For a systematic and thoroughgoing development of scattering methods we would use Wick's theorem (see Section 8.3) and construct diagrammatic methods in terms of these propagators; for our present illustrative purposes it suffices to carry out the time integrations as they are explicitly shown in Eq. 75. Then we require

$$\int_{-\infty}^{\infty} dt\, e^{i(\epsilon_{f'} - \epsilon_\alpha + \omega_{b'})t} \int_{-\infty}^{t} dt'\, e^{i(\epsilon_\alpha - \epsilon_f - \omega_b)t'}$$

and the same quantity with $\omega_b \leftrightarrow -\omega_{b'}$ (the so-called crossed term, Figure 2.2b). Using, as before, an adiabatic switching factor $e^{i\epsilon t'}$, with $\epsilon \rightarrow 0^+$, this gives

$$-2\pi i\delta(\epsilon_{f'} + \omega_{b'} - \epsilon_f - \omega_b)\, \frac{1}{\epsilon_\alpha - \epsilon_f - \omega_b} .$$

The δ-function expresses energy conservation, as expected, and is unchanged, of course, for $\omega_b \leftrightarrow -\omega_{b'}$; the second factor is the energy denominator of ordinary perturbation theory (see Exercise 11.6a).

The S-matrix element for $bf \rightarrow b'f'$ for free particles (plane waves for $u(\mathbf{r})$, $w(\mathbf{r})$ is

$$\langle b'f'|S^{(2)}|bf\rangle = +ig^2(2\pi)^4\delta(\epsilon_{f'} + \omega_{b'} - \epsilon_f - \omega_b)$$

$$\times \delta(\mathbf{p}_{f'} + \mathbf{k}_{b'} - \mathbf{p}_f - \mathbf{k}_b)(4\omega_b\omega_{b'})^{-1/2}\Omega^{-2}$$

$$\times \left[\frac{1}{\epsilon(\mathbf{p}_f + \mathbf{k}_b) - \epsilon(\mathbf{p}_f) - \omega(\mathbf{k}_b)} \right.$$

$$\left. + \frac{1}{\epsilon(\mathbf{p}_f - \mathbf{k}_{b'}) - \epsilon(\mathbf{p}_f) + \omega(\mathbf{k}_{b'})} \right] \tag{76}$$

where we have suppressed the internal quantum numbers of the particles. The differential scattering cross section is obtained from this using Eqs. 51 and 52 with the same consideration for removing the "extra" δ-function of momentum conservation as entered in Eqs. 66–68, whence

$$d\sigma(b'f' \leftarrow bf) = \frac{1}{I} (2\pi)^4 \delta(\epsilon_{f'} + \omega_{b'} - \epsilon_f - \omega_b)$$

$$\times \delta(\mathbf{p}_{f'} + \mathbf{k}_{b'} - \mathbf{p}_f - \mathbf{k}_b) g^4 (4\omega_b \omega_{b'} \Omega^3)^{-1}$$

$$\times \left[\frac{1}{\epsilon(\mathbf{p}_f + \mathbf{k}_b) - \epsilon(\mathbf{p}_f) - \omega(\mathbf{k}_b)} \right.$$

$$\left. + \frac{1}{\epsilon(\mathbf{p}_f - \mathbf{k}_{b'}) - \epsilon(\mathbf{p}_f) + \omega(\mathbf{k}_{b'})} \right]^2. \tag{77}$$

If we consider the scattering in a frame in which the initial fermion is at rest, the incident boson flux is $I = v_b/\Omega = k_b/(\omega_b\Omega)$, where v_b is the incident boson velocity. Inserting this, together with density-of-states factors, yields

$$d\sigma(b'f' \leftarrow bf) = \frac{\omega_b \Omega}{k_b} (2\pi)^4 \int \delta(\epsilon_{f'} + \omega_{b'} - M - \omega_b)\delta(\mathbf{p}_{f'} + \mathbf{k}_{b'} - \mathbf{k}_b)$$

$$\times g^4 (4\omega_b \omega_{b'} \Omega^3)^{-1} \left[\frac{1}{\epsilon(\mathbf{k}_b) - M - \omega(\mathbf{k}_b)} \right.$$

$$\left. + \frac{1}{\epsilon(-\mathbf{k}_{b'}) - M + \omega(\mathbf{k}_{b'})} \right]^2$$

$$\times \frac{\Omega \, d\mathbf{p}_{f'}}{(2\pi)^3} \frac{\Omega \, d\mathbf{k}_{b'}}{(2\pi)^3}, \tag{78}$$

where M is the fermion mass. Note that the factors of the normalization volume Ω again cancel. The integration over the fermion recoil momentum is easily carried out by exploiting the momentum conservation δ-function,

$$d\sigma(b'f' \leftarrow bf) = \frac{1}{4k_b \omega_b'} 2\pi \int \delta(\epsilon_{f'} + \omega_{b'} - M - \omega_b) \frac{d\mathbf{k}_{b'}}{(2\pi)^3}$$

$$\times g^4 \left[\frac{1}{\epsilon(\mathbf{k}_b) - M - \omega(\mathbf{k}_b)} + \frac{1}{\epsilon(-\mathbf{k}_{b'}) - M + \omega(\mathbf{k}_{b'})} \right]^2$$

$$= \frac{g^4}{(2\pi)^2} \frac{D^2 \, d\hat{\mathbf{k}}_{b'}}{4k_b \omega_{b'}} k_{b'}^2 \left\{ \frac{d}{dk_{b'}} [\epsilon(\mathbf{p}_{f'} + \mathbf{k}_{b'}) + \omega(\mathbf{k}_{b'}] \right\}^{-1}, \tag{79a}$$

where

$$D = \left[\frac{1}{\epsilon(\mathbf{k}_b) - M - \omega(\mathbf{k}_b)} + \frac{1}{\epsilon(-\mathbf{k}_{b'}) - M + \omega(\mathbf{k}_{b'})} \right]. \tag{79b}$$

Using relativistic kinematics, the differential cross section is

$$\frac{d^2\sigma(b'f' \leftarrow bf)}{d \cos \theta_{b'} \, d\phi_{b'}} = \frac{g^4}{(2\pi)^2} \frac{k_b^2 D^2}{4k_b \omega_{b'}} \left[\frac{k_{b'} + p_{f'} \cos (\hat{\mathbf{k}}_{b'}, \hat{\mathbf{p}}_{f'})}{\epsilon_{f'}} + \frac{k_{b'}}{\omega_{b'}} \right]^{-1}. \tag{80}$$

If the mass of the fermion is much greater than the momenta, then the kinematical factor simplifies and we have

$$\frac{d^2\sigma(b'f' \leftarrow bf)}{d \cos \theta_{b'} \, d\phi_{b'}} = \frac{g^4}{16\pi^2} \frac{k_{b'}}{k_b} D^2, \tag{81}$$

for which there is no angular dependence, and so the total cross section is easily evaluated as

$$\sigma(b'f' \leftarrow bf) = \frac{g^4}{4\pi} \frac{k_{b'}}{k_b} D^2. \tag{82a}$$

The limit of negligible recoil is dangerous, however, because for the scalar Yukawa interaction this limit, if applied to the dynamical factor as well, leads to a vanishing cross section. That is, for $M \rightarrow \infty$, we have $\mathbf{k}_{b'} = \mathbf{k}_b$ and the energy-denominator factor D in Eq. 82a is zero. (The cross section would not necessarily vanish in the no-recoil limit if internal quantum numbers were involved in the Yukawa interaction, i.e. if the interaction were not scalar.) From the form of the Yukawa interaction, Eq. 59, the coupling constant g is seen to be dimensionless. This is confirmed by the result for the decay rate, Eq. 72, which then has dimensions $[w] \sim$ [energy] $\sim$ [time]$^{-1}$, as it should. Our cross-section result then has units $[\sigma] \sim$ [energy]$^{-2} \sim$ [length]^2, again as expected.

Note also that the result of Eq. 81 clearly refers to s-wave scattering, since it has no angular dependence. Scattering in this single partial wave should be limited by unitarity so that the total cross section of Eq. 82 fulfils

$$\sigma(b'f' \leftarrow bf) = \frac{4\pi}{k_{\text{c.m.}}^2} \sin^2\delta_0 \leq \frac{4\pi}{k_{\text{c.m.}}^2}, \tag{82b}$$

where $k_{\text{c.m.}}$ is the center-of-mass momentum and δ_0 is the (real) s-wave phase shift. The result of Eqs. 81 and 82 obviously may violate this condition as a consequence of the fact that it is valid only in lowest-order scattering (Born approximation).

Last, we consider the interaction induced between two fermions by the exchange of a boson—the purpose for which Yukawa introduced the interaction named for him. The case differs from the previous ones in that the boson that is emitted is reabsorbed, so that no real boson appears in either the initial or the final state. The pictorial representation of this process is in Figure 2.3; again the second-order scattering matrix of Eq. 73 is involved, this time taken between two-fermion states, $f_1 f_2 \rightarrow f_1' f_2'$:

$$
\langle f_1' f_2' | S^{(2)} | f_1 f_2 \rangle = -g^2 \sum_{\alpha\beta\sigma} \sum_{\gamma\delta\tau} \int_\Omega d\mathbf{r} \int_\Omega d\mathbf{r}' \, u_\beta^\dagger(\mathbf{r}) u_\alpha(\mathbf{r}) u_\delta^\dagger(\mathbf{r}') u_\gamma(\mathbf{r}')
$$

$$
\times \int_{-\infty}^\infty dt \int_{-\infty}^t dt' \, e^{i(\epsilon_\beta - \epsilon_\alpha)t} e^{i(\epsilon_\delta - \epsilon_\gamma)t'}
$$

$$
\times \langle f_1' f_2' | a_\beta^\dagger a_\alpha [w_\sigma(\mathbf{r}) b_\sigma e^{-i\omega_\sigma t} + w_\sigma^\dagger(\mathbf{r}) b_\sigma^\dagger e^{i\omega_\sigma t}]
$$

$$
\times a_\delta^\dagger a_\gamma [w_\tau(\mathbf{r}') b_\tau e^{-i\omega_\tau t'} + w_\tau^\dagger(\mathbf{r}') b_\tau^\dagger e^{i\omega_\tau t'}] | f_1 f_2 \rangle
$$

$$
= -g^2 \int_\Omega d\mathbf{r} \int_\Omega d\mathbf{r}' \, u_{f_1}^\dagger(\mathbf{r}) u_{f_2}^\dagger(\mathbf{r}') u_{f_1}(\mathbf{r}) u_{f_2}(\mathbf{r}') \int_{-\infty}^\infty dt \int_{-\infty}^t dt'
$$

$$
\times e^{i(\epsilon_{f_1} - \epsilon_{f_1})t} e^{i(\epsilon_{f_2'} - \epsilon_{f_2})t'} \sum_\sigma w_\sigma(\mathbf{r}) w_\sigma^\dagger(\mathbf{r}') e^{i\omega_\sigma(t'-t)}
$$

$$
+ \text{antisymmetrizing terms} , \tag{83}
$$

where we have, for the sake of simplicity, suppressed the explicit antisymmetrization induced by $a_\alpha a_\gamma | f_1 f_2 \rangle$ and $\langle f_1' f_2' | a_\beta^\dagger a_\delta^\dagger$.

Carrying out the time integration as before, and introducing plane-wave states for the particles, we have, for $f_1(\mathbf{p}_1, \epsilon_1) + f_2(\mathbf{p}_2, \epsilon_2) \rightarrow f_1'(\mathbf{p}_1', \epsilon_1') + f_2'(\mathbf{p}_2', \epsilon_2')$,

$$
\langle f_1' f_2' | S^{(2)} | f_1 f_2 \rangle = +ig^2 \int_\Omega d\mathbf{r} \int_\Omega d\mathbf{r}' e^{i(\mathbf{p}_1 - \mathbf{p}_1')\cdot\mathbf{r}} e^{i(\mathbf{p}_2 - \mathbf{p}_2')\cdot\mathbf{r}}
$$

$$
\times 2\pi\delta(\epsilon_1' + \epsilon_2' - \epsilon_1 - \epsilon_2) \sum_k \frac{e^{i\mathbf{k}\cdot(\mathbf{r}-\mathbf{r}')}}{2\omega^2(\mathbf{k})}
$$

$$
= i \frac{g^2}{2\omega^2(\mathbf{p}_1 - \mathbf{p}_1')} (2\pi)^4 \delta(\epsilon_1' + \epsilon_2' - \epsilon_1 - \epsilon_2)
$$

$$
\times \delta(\mathbf{p}_1' + \mathbf{p}_2' - \mathbf{p}_1 - \mathbf{p}_2) . \tag{84}
$$

In addition to the energy- and momentum-conserving δ-functions, whose treatment we have already discussed, this expression contains the Yukawa, or one-boson-exchange, form

$$
\frac{g^2}{2\omega^2(\mathbf{p}_1 - \mathbf{p}_1')} = \frac{g^2}{2} \frac{1}{(\mathbf{p}_1 - \mathbf{p}_1')^2 + m^2} \tag{85}
$$

for the fermion–fermion scattering matrix element. The Fourier transform of this is proportional to the configuration-space Yukawa form

$$- g^2 \frac{e^{-mr}}{mr} ,$$

and the lowest-order T-matrix element of Eq. 85 measures the interaction energy introduced into the system by the Yukawa coupling, so that this is the fermion–fermion (static) potential implied by that coupling.

The introduction of the Yukawa coupling thus leads automatically to relationships between several kinds of processes: boson emission and absorption vertices, boson–fermion scattering, and fermion–fermion scattering as mediated by boson exchange. The resulting transition matrix elements contain appropriate factors for conservation of energy and momentum, and the squares of these elements are easily related to transition rates and cross sections through standard manipulations of essentially kinematic character; two fairly common cases have been shown here for illustration.

Our intent has been to show as explicitly as possible the ligature between the diagrammatic formalisms we have developed in the second half of this book and the corresponding formalisms for scattering problems. A number of specialized features are quite central for the further exploration of the use of these techniques in scattering. These include the full treatment of system symmetries and internal quantum numbers (which change the detailed form of the Yukawa interaction, but not its basic $a^{\dagger}ab$-plus-$a^{\dagger}ab^{\dagger}$ structure), covariance, renormalization formalism, and detailed computational application to electrodynamics, as well as more complete handling of scattering features as such. We do not examine these further here, since a great number of texts at various levels and with assorted ranges of topics already exist in these areas.

11.6 INELASTIC SCATTERING AND LINEAR RESPONSE

Our treatment of the many-body problem in this book—especially in the context of diagrammatic methods—has by and large dealt with the system ground state. Excitations of the system have, of course, always been implicitly present in that we handle the dynamics with some degree or other of completeness and thus obtain the more exact ground state in terms of the unperturbed ground state together with the admixture of excitations. In the present subsection we consider the inelastic scattering of a relatively weakly interacting probe on a complex target of fermions, restricting ourselves to the simple situation in which the inelastic transition takes place as a one-step process insofar as probe interaction with target particles is concerned. As we shall see, the inclusive cross section for the scattering (i.e. the cross section summed over all excitations) then relates explicitly to the polarization properties of the target ground state discussed in Chapter 10, especially in the context of linear response to external potentials (Section 10.7).

Assume that the incident probing particle has momentum $\mathbf{k}_0$ and is scattered by interaction with the complex target system into momentum $\mathbf{k}$. The inclusive cross section can be calculated by the methods of Section 11.4; the inclusivity is contained in the fact that we sum over all target-system final states that are consistent with the probe kinematics from the point of view of energy conservation. The cross section is

$$\frac{d^3\sigma}{d\mathbf{k}} = \frac{1}{(2\pi)^2 v_0} \sum_f |\langle \mathbf{k}, f|V|\mathbf{k}_0, 0\rangle|^2 \delta(E_0 + \epsilon_0 - E_f - \epsilon_f)\,, \tag{86}$$

where

$$V = \sum_{\alpha=1}^{N} V_\alpha \tag{87}$$

is the interaction energy between the probe and the target, and is given, by virtue of our assumptions, as a sum over interactions with the N individual particles comprising the target system. The incident velocity v_0 here equals the incident flux factor I of Eq. 52, since we now use a unit normalization volume Ω (having seen in Section 11.4 that physical results are always independent of Ω). We ignore recoil in the kinematics of Eq. 86. The initial and final states for the target are 0 and f, with energies E_0 and E_f, while those for the probe have energies $\epsilon_0 = \epsilon(k_0)$ and $\epsilon_f = \epsilon(k_f)$. We have integrated over the time dependence of the lowest-order interaction of Eq. 86, using the methods described in Section 11.4, to obtain the energy-conserving δ-function exhibited here.

The transition matrix element in Eq. 86 is, more explicitly,

$$\langle \mathbf{k}, f|V|\mathbf{k}_0, 0\rangle = \int d\mathbf{r}\, e^{i(\mathbf{k}_0 - \mathbf{k})\cdot\mathbf{r}} \langle f| \sum_{\alpha=1}^{N} V\delta(\mathbf{r} - \mathbf{r}_\alpha)|0\rangle\,,$$

$$= V \int d\mathbf{r}\, e^{i\mathbf{q}\cdot\mathbf{r}} \rho_{f0}(\mathbf{r})$$

$$= V F_{f0}(\mathbf{q})\,, \tag{88}$$

where we have taken plane waves in the space of the probe, introducing the momentum transfer $\mathbf{q} = \mathbf{k}_0 - \mathbf{k}$, and defined, for the space of the target system, the transition density $\rho_{f0}(\mathbf{r})$ and its Fourier transform, the transition form factor $F_{f0}(\mathbf{q})$ (see Eqs. 37–40). For this purpose we have assumed point interactions between the probe and the target particles, with equal interaction coefficients V for all particles and no reference to internal quantum numbers in V. These latter, technical restrictions are made only for the sake of simplicity; they do no damage to the basic physical features we shall illustrate. (The earlier assumptions of a weak interaction V and of plane waves for the probe can be relaxed for appropriate systems by using the impulse approximation to replace the interaction V with the T-matrix for scattering between probe and target particle, and using the approximation of distorted waves to represent the overall effect of elastic scattering of the probe by the target on its way to and from the single "hard" scattering that causes the system excitation. These are rather specialized techniques whose validity must be studied—often with considerable computational tedium—in the context of the specific system under consideration; their consideration does not change the basic underlying features we explore here.)

If we were to deal with a single, point-like, nonrecoiling particle placed at the origin in an elastic scattering event, then we would have a cross section $\sigma_0(\mathbf{k}_0, \mathbf{q})$ for this basic process given by Eqs. 86 and 88, but with $\rho_{f0}(\mathbf{r}) = \rho_{00}(\mathbf{r}) = \delta(\mathbf{r})$ and thus $F_{f0}(\mathbf{q}) = F_{00}(\mathbf{q}) = 1$, i.e. $\sigma_0(\mathbf{k}_0, \mathbf{q}) = |V|^2/(2\pi)^2 v_0$. Thus the inclusive cross section in Eq. 86 for scattering on the many-particle target can be written as

$$\frac{d^3\sigma}{d\mathbf{k}} = \sigma_0^{\text{eq}}(\mathbf{k}_0, \mathbf{q}) \sum_f \delta(E_0 + \epsilon_0 - E_f - \epsilon_f)|F_{f0}(\mathbf{q})|^2$$

$$= \sigma_0^{\text{eq}}(\mathbf{k}_0, \mathbf{q})S(\mathbf{q}, \omega) , \tag{89}$$

where $\omega = \epsilon_0 - \epsilon_f$ is the energy transfer, and the structure factor $S(\mathbf{q}, \omega)$ will be discussed at length in a moment. The basic cross section σ_0 is labeled σ_0^{eq} in Eq. 89 because the different kinematical conditions of elastic scattering on a single particle for σ_0 from those of inelastic scattering on a complex target for $d^3\sigma/d\mathbf{k}$ require the selection of equivalent kinematics for $\sigma_0(\mathbf{k}_0, \mathbf{q})$ in order that it parallel the situation for $d^3\sigma/d\mathbf{k}$. [Thus if, in practice, σ_0 is measured in the center-of-mass system, the momentum transfer for this elastic $(k_f = k_0)$ case is given by

$$q = [2k_0^2(1 - \cos\theta)]^{1/2} = 2k_0 \sin\tfrac{1}{2}\theta , \tag{90a}$$

where θ is the scattering angle, whereas for the inelastic situation $(k_f \neq k_0)$

$$q = [k_0^2 - 2k_0 k_f \cos\theta + k_f^2]^{1/2} , \tag{90b}$$

so that an equivalent incident momentum must be chosen to achieve the same q for given scattering angle θ, say, if the cross section is characterized by q and θ. Moreover, the density of final states will carry a factor involving k_f, with different k_f in the two cases; and when recoil is included, a further factor enters in the phase-space term (see Eq. 79a, for example). All of these issues are primarily kinematic in nature, and thus are not essential complications for our purpose here.]

The central many-body quantity in Eq. 89 is the structure factor

$$S(\mathbf{q}, \omega) = \sum_f \delta(E_0 + \omega - E_f)|F_{f0}(\mathbf{q})|^2 , \qquad (91)$$

upon which we now focus. The inelastic form factor that appears here can be written in Fock space in terms of a nondiagonal single-particle density operator as

$$F_{f0}(\mathbf{q}) = \int d\mathbf{r}\, e^{i\mathbf{q}\cdot\mathbf{r}} \rho_{f0}(\mathbf{r}) = \int d\mathbf{r}\, e^{i\mathbf{q}\cdot\mathbf{r}} \langle f|\psi^\dagger(\mathbf{r})\psi(\mathbf{r})|0\rangle$$

$$= \int d\mathbf{r}\, e^{i\mathbf{q}\cdot\mathbf{r}} \int \frac{d\mathbf{p}\, d\mathbf{p}'}{(2\pi)^6} \langle f|e^{-i\mathbf{p}'\cdot\mathbf{r}} a^\dagger(\mathbf{p}') e^{i\mathbf{p}\cdot\mathbf{r}} a(\mathbf{p})|0\rangle$$

$$= \int \frac{d\mathbf{p}}{(2\pi)^3} \langle f|a^\dagger(\mathbf{p}+\mathbf{q}) a(\mathbf{p})|0\rangle , \qquad (92)$$

where $\psi(\mathbf{r})$ is the field operator for particles in the target system. Using this form in Eq. 91, we can write, using closure,

$$S(\mathbf{q}, \omega) = \sum_f \delta(E_0 + \omega - E_f) \int \frac{d\mathbf{p}'\, d\mathbf{p}}{(2\pi)^6} \langle 0|a^\dagger(\mathbf{p}') a(\mathbf{p}'+\mathbf{q})|f\rangle$$

$$\times \langle f|a^\dagger(\mathbf{p}+\mathbf{q}) a(\mathbf{p})|0\rangle$$

$$= \int \frac{d\mathbf{p}\, d\mathbf{p}'}{(2\pi)^6} \langle 0|a^\dagger(\mathbf{p}') a(\mathbf{p}'+\mathbf{q}) \delta(E_0 + \omega - H) a^\dagger(\mathbf{p}+\mathbf{q}) a(\mathbf{p})|0\rangle , \qquad (93)$$

where H is the target-system Hamiltonian, $H|f\rangle = E_f|f\rangle$, and the structure factor is expressed in terms of an expectation value based on the ground state of the target system.

If we consider a sum-rule result based on all values of the energy transfer different from zero (so as to exclude the elastic case, in which no energy is transferred and the target system remains in the ground state), we require

$$\int_\eta^\infty S(\mathbf{q}, \omega)\, d\omega = \int \frac{d\mathbf{p}'\, d\mathbf{p}}{(2\pi)^6} \langle 0|a^\dagger(\mathbf{p}') a(\mathbf{p}'+\mathbf{q}) a^\dagger(\mathbf{p}+\mathbf{q}) a(\mathbf{p})|0\rangle$$

$$- |F_{00}(\mathbf{q})|^2 . \qquad (94)$$

Here η is a number greater than zero and smaller than the energy of the first excited state of the target system, and is chosen in this way to exclude the target ground-state contribution, which is explicitly subtracted off on the right-hand side of the relationship. Since the first term on the right-hand side of Eq. 94 involves a two-particle operator, it is to be expected that it contains information on two-particle correlations in the target ground state.

This may be made more explicit in configuration space: reversing the operations that took us from the forms of Eq. 88 to those of Eq. 92, we have

$$\int_\eta^\infty S(\mathbf{q}, \omega)\, d\omega = \int d\mathbf{r}'\, d\mathbf{r}\, e^{i\mathbf{q}\cdot(\mathbf{r}-\mathbf{r}')} \langle 0|\psi^\dagger(\mathbf{r}')\psi(\mathbf{r}')\psi^\dagger(\mathbf{r})\psi(\mathbf{r})|0\rangle$$

$$- \left| \int d\mathbf{r}\, e^{i\mathbf{q}\cdot\mathbf{r}} \langle 0|\psi^\dagger(\mathbf{r})\psi(\mathbf{r})|0\rangle \right|^2$$

$$= \int d\mathbf{r}\, \langle 0|\psi^\dagger(\mathbf{r})\psi(\mathbf{r})|0\rangle + \int d\mathbf{r}'\, d\mathbf{r}\, e^{i\mathbf{q}\cdot(\mathbf{r}-\mathbf{r}')}$$

$$\times \langle 0|\psi^\dagger(\mathbf{r}')\psi^\dagger(\mathbf{r})\psi(\mathbf{r})\psi(\mathbf{r}')|0\rangle$$

$$- \left| \int d\mathbf{r}\, e^{i\mathbf{q}\cdot\mathbf{r}} \langle 0|\psi^\dagger(\mathbf{r})\psi(\mathbf{r})|0\rangle \right|^2$$

$$= N + N(N-1) \int d\mathbf{r}'\, d\mathbf{r}\, e^{i\mathbf{q}\cdot(\mathbf{r}-\mathbf{r}')} P(\mathbf{r}', \mathbf{r}) - |F_{00}(\mathbf{q})|^2 , \qquad (95)$$

where we have used the field anticommutators, and defined

$$P(\mathbf{r}', \mathbf{r}) = \frac{1}{N(N-1)} \langle 0|\psi^\dagger(\mathbf{r}')\psi^\dagger(\mathbf{r})\psi(\mathbf{r})\psi(\mathbf{r}')|0\rangle , \qquad (96)$$

the two-particle *correlation function*. This represents the probability of finding one particle in the target system at $\mathbf{r}$ and a second at $\mathbf{r}'$, with the possibility of both particles being at the same point, $\mathbf{r} = \mathbf{r}'$, excluded for the target fermions by the chosen juxtapositions $\psi(\mathbf{r})\psi(\mathbf{r}')$ and $\psi^\dagger(\mathbf{r}')\psi^\dagger(\mathbf{r})$, which are each zero when $\mathbf{r} = \mathbf{r}'$. The correlation function is normalized to unity,

$$\int d\mathbf{r}'\, d\mathbf{r}\, P(\mathbf{r}', \mathbf{r}) = 1 . \qquad (97)$$

Equation 95 shows explicitly that the sum-rule result for the structure factor yields the Fourier transform of the two-particle correlation function, which can thus be measured using inclusive inelastic scattering.

The structure factor $S(\mathbf{q}, \omega)$ may be related to the many-body system polarization operator $\Pi(\mathbf{q}', \mathbf{q}; \omega)$ discussed in Sections 10.5–10.7 (where infinite, uniform media were considered, so we had always $\mathbf{q}' = \mathbf{q}$). The general methods developed there for calculating Π thus apply for the structure factor as well. We start from the definition of the polarization operator for this case, which will be the full particle–hole propagator (see Eq. 10.61)

$$i\tilde{\Pi}(\mathbf{q}', \mathbf{q}; t' - t) = \int \frac{d\mathbf{p}'\, d\mathbf{p}}{(2\pi)^6} \langle 0|[\tilde{a}^\dagger(\mathbf{p}'; t')\tilde{a}(\mathbf{p}' + \mathbf{q}'; t')$$

$$\times \tilde{a}^\dagger(\mathbf{p} + \mathbf{q}; t)\tilde{a}(\mathbf{p}; t)]_T|0\rangle , \qquad (98)$$

where we use operators in the Heisenberg picture (see Section 7.1) and the operator product is time ordered. This form suggests a kind of two-particle Green function describing time propagation of a particle–hole excitation at time t and of net momentum $\mathbf{q}$ to a similar excitation at time t' with momentum $\mathbf{q}'$. The Fourier transform of the quantity in Eq. 98 is given by

$$i\tilde{\Pi}(\mathbf{q}', \mathbf{q}; \omega) = \int_{-\infty}^{\infty} d\tau \, e^{i\omega\tau} \tilde{\Pi}(\mathbf{q}', \mathbf{q}; \tau)$$

$$= \int_{-\infty}^{\infty} d\tau \, e^{i\omega\tau} \int \frac{d\mathbf{p}' \, d\mathbf{p}}{(2\pi)^6}$$

$$\times \langle 0|[e^{iH\tau}a^{\dagger}(\mathbf{p}')e^{-iH\tau}e^{iH\tau}a(\mathbf{p}' + \mathbf{q}')e^{-iH\tau}a^{\dagger}(\mathbf{p} + \mathbf{q})a(\mathbf{p})]_T|0\rangle \ , \tag{99}$$

where we have used Eq. 7.8 and measured time from $\tau = 0$. Then (introducing an adiabatic switching factor $e^{-\eta|\tau|}$, $\eta \to 0^+$, for the time integration)

$$\tilde{\Pi}(\mathbf{q}', \mathbf{q}; \omega) = -i \int \frac{d\mathbf{p}' \, d\mathbf{p}}{(2\pi)^6} \left\{ \int_0^{\infty} d\tau \, e^{i\omega\tau} e^{-\eta\tau} \right.$$

$$\times \langle 0|e^{iH\tau}a^{\dagger}(\mathbf{p}')a(\mathbf{p}' + \mathbf{q}')e^{-iH\tau}a^{\dagger}(\mathbf{p} + \mathbf{q})a(\mathbf{p})|0\rangle$$

$$\left. + \int_{-\infty}^{0} d\tau \, e^{i\omega\tau} e^{\eta\tau} \langle 0|a^{\dagger}(\mathbf{p} + \mathbf{q})a(\mathbf{p})e^{iH\tau}a^{\dagger}(\mathbf{p}')a(\mathbf{p}' + \mathbf{q}')e^{-iH\tau}|0\rangle \right\}$$

$$= \int \frac{d\mathbf{p}' \, d\mathbf{p}}{(2\pi)^6} \langle 0|\left[a^{\dagger}(\mathbf{p}')a(\mathbf{p}' + \mathbf{q}') \frac{1}{\omega + E_0 - H + i\eta} a^{\dagger}(\mathbf{p} + \mathbf{q})a(\mathbf{p}) \right.$$

$$\left. - a^{\dagger}(\mathbf{p} + \mathbf{q})a(\mathbf{p}) \frac{1}{\omega - E_0 + H - i\eta} a^{\dagger}(\mathbf{p}')a(\mathbf{p}' + \mathbf{q}') \right]|0\rangle \ , \tag{100}$$

whence

$$\operatorname{Im} \tilde{\Pi}(\mathbf{q}, \mathbf{q}; \omega) = -\pi \int \frac{d\mathbf{p}' \, d\mathbf{p}}{(2\pi)^6} [\langle 0|a^{\dagger}(\mathbf{p}')a(\mathbf{p}' + \mathbf{q})\delta(\omega + E_0 - H)$$

$$\times a^{\dagger}(\mathbf{p} + \mathbf{q})a(\mathbf{p})|0\rangle$$

$$+ \langle 0|a^{\dagger}(\mathbf{p} + \mathbf{q})a(\mathbf{p})\delta(\omega - E_0 + H)a^{\dagger}(\mathbf{p}')a(\mathbf{p}' + \mathbf{q})|0\rangle]$$

$$= -\pi S(\mathbf{q}, \omega) \ . \tag{101}$$

The last step here follows from comparing with Eq. 93 and noting that for $\omega > 0$ the second δ-function cannot have vanishing argument.

Thus the inclusive inelastic cross section of Eq. 89 may be written as

$$\frac{d^3\sigma}{d\mathbf{k}} = -\frac{1}{\pi}\sigma_0^{eq}(\mathbf{k}_0, \mathbf{q})\mathrm{Im}\tilde{\Pi}(\mathbf{q}, \mathbf{q}; \omega)\,. \tag{102}$$

This dynamic quantity may then be calculated using techniques developed in Chapter 10 for the treatment of the ground-state properties of the many-particle system, thus illustrating the ready applicability of the methods developed here to scattering problems as well as to the properties of bound systems.

EXERCISES

11.1 Show that the Green function based on the quasiparticle approximation (Eq. 18) shows damping in time for either particle or hole excitations, i.e. that $G_\alpha(t' - t)$ decreases exponentially with increasing $t' - t$. (See Eq. 8.50 for the relation of the Green function in time to the Fourier transform, Eq. 14b.)

11.2 The imaginary part of the optical potential of Eq. 32 is negative, corresponding to loss of flux from the elastically scattered wave.

    (a) Show directly from the continuity equation of quantum mechanics that total probability is not conserved for a particle moving in a complex potential $V(\mathbf{r})$, and therefore changes in time. Further, show that Im $V(\mathbf{r}) < 0$ corresponds to decreasing probability.

    (b) Show how this result can be applied to scattering of a beam of particles from a complex potential. For simplicity, consider a central potential, and show that in each partial wave, the outgoing flux is less than the incoming flux if Im $V(\mathbf{r}) < 0$.

11.3 Calculate the transition rate for boson absorption, the process inverse to that treated in Section 11.5.

11.4 Using methods based on Section 11.5, calculate the photoemission rate for the $2P \to 1S$ decay of the hydrogen atom.

11.5 **(a)** Calculate the density of final states for a particle decaying into two particles, one of which is relativistic and massive, and the other of which is massless (e.g., $\pi \to \mu\nu$).

    (b) Calculate the density of final states for a particle decaying into three nonrelativistic particles of equal mass (e.g., $K \to 3\pi$).

    (c) Assuming the interaction matrix elements for the decays in (a) and (b) are approximately constant, what will be the decay rates?

11.6 **(a)** Using conventional ("old-fashioned") perturbation theory, calculate the diagrams in Figure 2.2 and compare with the result in Section 11.5.

    (b) Calculate the graphs of Figure 2.4.

11.7 Using Wick's theorem, evaluate $\langle b'f'|S^{(2)}|bf\rangle$ of Section 11.5.

11.8 For a scalar, isovector boson having a Yukawa interaction of the structure

$$V = g \int_\Omega d\mathbf{r}\, [\psi^\dagger \boldsymbol{\tau}\psi \cdot \boldsymbol{\phi}]\,,$$

where the fermion field has isospin $\frac{1}{2}$ and $\boldsymbol{\tau}$ is the triad of isospin Pauli matrices,

(a) explain how isospin is conserved, and

(b) calculate $\langle b'f'|S^{(2)}|bf\rangle$ in the limit of an infinitely massive fermion.

11.9 (a) Using conventional perturbation theory for the second-order energy, and fermions at fixed positions separated by distance R, show that the Yukawa interaction leads to the static potential of the expression following Eq. 85.

(b) Show that the Born approximation amplitude for fermion–fermion scattering with such a potential is the same as Eqs. 84 and 85.

Appendix

A. WICK'S THEOREM: EQUATION 8.31

The decomposition of the time-ordered product $\{rstuv\cdots\}_T$ given by Eq. 8.31 is already proven for two operators, directly from Eq. 8.32, by interchanging terms:

$$\{rs\}_T = \{rs\}_N + \langle rs \rangle . \tag{A1}$$

Now we proceed inductively, first assuming the theorem to be true for a product of n operators $(rstuv\cdots z)$. (We have assumed n to be even.) Next we introduce one more operator, d, which we take to have a time *earlier* than all n operators $r,\dots,z$, so that

$$\{rs\cdots z\}_T d = \{rs\cdots zd\}_T . \tag{A2}$$

On the other hand, for a normal-ordered product of m operators $w_1,\dots,w_m$ (a subset of $r,\dots,z$) we may obtain

$$\{w_1\cdots w_m\}_N d = \{w_1\cdots w_m d\}_N \pm \sum_{i=1}^{m} \{w_1 w_2 \cdots w_{i-1}w_{i+1}\cdots w_m\}_N \langle w_i d \rangle , \tag{A3}$$

where the factor w_i which appears in the contraction in the summation is missing from the normal-ordered product which multiplies it. This result is clearly correct for the case that d is an annihilation operator, since then the first term is already normal-ordered, and the summation is then zero, because every contraction $\langle w_i d \rangle = 0$. This last obtains because d occurs at a time earlier than all other operators: see Eqs. 8.25–8.30. If d is a creation operator, then Eq. A3 is obtained by commuting d through all the operators in the set $\{w_i\}$, obtaining one term in the sum from each commutation, and a ($\pm$) sign for even or odd i.

Then, using Eq. 8.31 for the left side of Eq. A2 (by assumption), and applying Eq. A3 repeatedly, we find

$$\{rs \cdots zd\}_T = \{rs \cdots zd\}_N$$
$$+ \{tuv \cdots\}_N \langle rs \rangle + \{rsv \cdots\}_N \langle tu \rangle - \{s \cdots z\}_N \langle rd \rangle$$
$$\pm \text{ other pairs (from } r \cdots zd)$$
$$+ \cdots +$$
$$+ \langle rs \rangle \langle tu \rangle \cdots \langle yz \rangle d \pm \text{permutations} . \tag{A4}$$

This is the Wick result for $n + 1$ operators. For $n + 2$ operators, we again obtain a result of the form of Eq. 8.31. That completes the proof.

B. LINKED-CLUSTER THEOREM: EQUATION 9.3

The theorem concerns the factoring of a given matrix element $\langle m | U(t, t_0) | n \rangle$ of the time-evolution operator between two states of the noninteracting system (m, n). This matrix element can be represented by an infinite series of Feynman diagrams, using the Wick analysis, as in Section 8.4. We have explained in Section 9.1 that diagrams may be grouped into those which are linked, and those which are not, meaning they contain disconnected parts, as in Figure 9.1. We discuss the theorem in terms of Figure 9.2, which shows the general form of a diagram with disconnected parts, in this case for an initial state with one particle and one hole, and a final state with two particles and two holes, which will serve as our examples of (m, n). The first property of interest for such a Feynman diagram is that the expression for $\langle m | U | n \rangle$ factors as

$$\langle m | U(t, t_0) | n \rangle_a = \langle m | U(t, t_0) | n \rangle_b \langle 0 | U(t, t_0) | 0 \rangle_c , \tag{B1}$$

where a, b, c refer to the three parts of Figure 9.2. This is an example of Eq. 9.3, in which the first factor is the expression for the linked part of the diagram, and the second for the vacuum–vacuum part.

The factoring follows from the Wick method of constructing Feynman diagrams, as discussed in Section 8.4. Every particle or hole line connecting two interactions in the diagram corresponds to the Wick contraction of a pair of operators contained in those interactions. A diagram with two disconnected parts has no pair contractions connecting the interactions in the two parts. The relevant variables in the expression for a specific Feynman diagram are the times t_i of interaction, which are integrated (from t_0 to t), and the labels α, β, ... of the particle and hole lines, which are summed over all states. For the unlinked diagram under discussion, each interaction time and each particle (hole) label refers to one or the other disconnected part, not both. Thus the integrations and summations for the two parts are independent, and the expression for the unlinked diagram factors.

The linked part will contribute a factor which is independent of the vacuum–vacuum part, corresponding to the expression for Figure 9.2b alone. Similarly, the vacuum–vacuum part will contribute a factor corresponding to Figure 9.2c, to which the linked part does not contribute. It easily follows that the expression for the original disconnected diagram of Figure 9.2a may be factored, as in Eq. B1.

We complete the theorem in two steps. First, we consider a sequence of Feynman diagrams, all with the same linked part (b), but with any possible vacuum–vacuum diagrams as the unlinked part. Summing the series of expression for this sequence of diagrams may be written

$$\langle m|U(t, t_0)|n\rangle_a = \langle m|U(t, t_0)|n\rangle_b \langle 0|U(t, t_0)|0\rangle , \qquad \text{(B2)}$$

where the last factor now stands for the sum of *all* vacuum–vacuum diagrams (including the diagram of zero order: an empty diagram, corresponding to $U^{(0)} \equiv 1$). If we now sum over all possible *linked* diagrams (b) corresponding to Figure 9.2b, we obtain the general factored expression

$$\langle m|U(t, t_0)|n\rangle = \langle m|U(t, t_0)|n\rangle_l \langle 0|U(t, t_0)|0\rangle , \qquad \text{(B3)}$$

which is the general theorem of Eq. 9.3.

C. EXPONENTIAL FORMULA: EQUATION 9.4

The vacuum expectation value

$$\langle 0|U(t, t_0)|0\rangle \qquad \text{(C1)}$$

can be expanded, using the factoring arguments of Section B above. The infinite series of all linked diagrams (only) contributes a value $\langle 0|U(t; t_0)|0\rangle_l$. For any given diagram (a) with two (and only two) unlinked parts (b and c), the contribution to Eq. C1 may be written in the factored form

$$\langle 0|U(t, t_0)|0\rangle_a = \langle 0|U(t, t_0)|0\rangle_b \langle 0|U(t, t_0)|0\rangle_c , \qquad \text{(C2)}$$

following Eq. B1, applied to the special case $m = n = 0$. In this case, there are no external lines in the diagrams of Figure 9.2, so that (b) and (c) are each a linked, vacuum–vacuum diagram. Then, summing over all possible diagrams of the factored form of (a), we obtain the same result as summing (b) and (c) separately over all possible linked diagrams, but divided by two, since each specific combination of factors (b, c) occurs twice in the sum: i.e.,

$$\sum_a \langle 0|U(t, t_0)|0\rangle_a^{(2)} = \tfrac{1}{2}[\langle 0|U(t, t_0)|0\rangle_l]^2 . \tag{C3}$$

For any given diagram with n unlinked parts, we shall obtain n factors in an expression analogous to Eq. C2:

$$\langle 0|U(t_1, t_0)|0\rangle_a^{(n)} = \langle 0|U(t, t_0)|0\rangle_{b_1} \cdots \langle 0|U(t, t_0)|0\rangle_{b_n} . \tag{C4}$$

Then summing over all n-part diagrams will yield

$$\sum_a \langle 0|U(t_1, t_0)|0\rangle_a^{(n)} = \frac{1}{n!} [\langle 0|U(t, t_0)|0\rangle_l]^n . \tag{C5}$$

The sum over all n gives the complete expression for Eq. C1 in the form

$$\langle 0|U(t, t_0)|0\rangle = \exp \langle 0|U(t_i, t_0)|0\rangle_l , \tag{C6}$$

which is Eq. 9.4.

D. PAULI EXCLUSION PRINCIPLE

One consequence of the elimination of unlinked diagrams from the perturbation expression is the requirement that diagrams which *appear* to violate the Pauli principle must be included for consistency. That has been discussed in detail in Section 9.1. If we approach the expansion starting with the Wick method, defining each Feynman diagram from a particular arrangement of pair contractions, we find immediately that such Pauli-principle-violating diagrams naturally occur, as in Figure 9.1, with $\rho_1 = \rho_3$. What we want to prove is the assertion that in each order of the perturbation expansion, the sum of all Pauli-violating diagrams gives no net contribution.

For this we turn to the expression for the nth-order contribution to the matrix element $\langle m|U^{(n)}(t, t_0)|a\rangle$, given in Eq. 8.12, where complete sets of intermediate states $\Sigma_b|b\rangle\langle b|, \ldots$ have been inserted between interactions $V(t_i)$. The intermediate states are eigenstates of the unperturbed Hamiltonian, as in Eq. 8.10:

$$H_0|b\rangle = \mathscr{E}_b|b\rangle . \tag{D1}$$

Since we are considering a fermion system, the states b are antisymmetric, and automatically obey the Pauli principle. Specifically, each particle or hole state ρ, λ (see Eq. 8.11) is occupied or unoccupied.

Suppose we now consider a system of distinguishable particles, with the same single-particle Hamiltonian H_0, but no restriction of symmetry. The eigenstates for this problem may also be specified by putting each particle in a given orbit (multiple occupancy now permitted), and the energies additive, as in Eq. 8.11b. We write

$$H_0|\tilde{b}\rangle = \mathscr{E}_{\tilde{b}}|\tilde{b}\rangle \tag{D2}$$

for these states, of mixed symmetry with regard to permutations. The complete set of states $\Sigma_{\tilde{b}}|\tilde{b}\rangle\langle\tilde{b}|$ will include the entire antisymmetric space $\Sigma_b|b\rangle\langle b|$, and much more besides.

Now if we use this enlarged space of noninteracting states in the development of Goldstone diagrams from Eq. 8.12 on, the only change will be the introduction of diagrams which appear to violate the Pauli principle. Yet they have no net effect in each order, which we can deduce, still considering Eq. 8.12. First we obtain

$$\sum_{\tilde{b}} |\tilde{b}\rangle\langle\tilde{b}|V(t_n)|a\rangle = \sum_b |b\rangle\langle b|V(t_n)|a\rangle \tag{D3}$$

from the following symmetry argument. The state $|a\rangle$ is antisymmetric, by assumption. The interaction $V(t_n)$ is symmetric for identical particles, and therefore the (ket) vector $|V(t_n)a\rangle$ remains antisymmetric. Therefore this vector has no overlap with the nonantisymmetric parts of $\Sigma_{\tilde{b}}|\tilde{b}\rangle\langle\tilde{b}|$. So it makes no difference whether we use $\Sigma_{\tilde{b}}|\tilde{b}\rangle\langle\tilde{b}|$ or $\Sigma_b|b\rangle\langle b|$ in Eq. 8.12, as shown in Eq. D3. Next we may apply the argument to $\Sigma_c|c\rangle\langle c|$, and so on, to the last complete set $\Sigma_l|l\rangle\langle l|$: it makes no difference to the result whether we use the complete antisymmetric states or the larger complete set of states of all symmetries. Therefore the Pauli-violating diagrams have no net effect, providing they are all summed in each order n.

E. DERIVATION OF THE LIPPMANN–SCHWINGER EQUATION

To complete the background for the discussions in Chapter 10 of Dyson or ladder equations and in Chapter 11 of scattering descriptions, we wish to present here briefly a derivation of the Lippmann–Schwinger equation. This is a formal integral equation that expresses the T-matrix for scattering in terms of the potential V and a particle propagator involving the unperturbed Hamiltonian H_0. It provides the complete formal solution to the problem of how to work beyond the Born approximation for potential scattering.

We have encountered, in Eq. (11.43), the $\mathscr{T}$-operator,

$$\mathscr{T} = 1 - S , \tag{E1}$$

defined so as to vanish for the case of no scattering, $S = U(\infty, -\infty) = 1$. Its matrix elements,

$$\mathcal{T}_{\alpha\beta} = \langle \Phi_\beta | \mathcal{T} | \Phi_\alpha \rangle = \delta_{\beta\alpha} - \langle \Phi_\beta | U(\infty, -\infty) | \Phi_\alpha \rangle , \qquad (E2)$$

are directly related to the cross section for scattering from the state α to the state β, as we shall see below. To establish the Lippmann–Schwinger equation we first derive an integral equation for $\mathcal{T}_{\beta\alpha}$. Since the time-development operator of Eq. (7.38) satisfies

$$U(t, -\infty) = 1 - i \int_{-\infty}^{t} H'(t') U(t', -\infty) \, dt' , \qquad (E3)$$

where $H'(t) = e^{iH_0 t} V e^{-iH_0 t}$ is the potential V transformed to the interaction picture, we have

$$\mathcal{T}_{\beta\alpha} = i \int_{-\infty}^{\infty} \langle \Phi_\beta | H'(t) U(t, -\infty) | \Phi_\alpha \rangle \, dt$$

$$= i \langle \Phi_\beta | V | \Psi_\alpha^{(+)}(E_\beta) \rangle , \qquad (E4)$$

where

$$|\Psi_\alpha^{(+)}(E)\rangle = \int_{-\infty}^{\infty} dt \, e^{i(E-H_0)t} e^{-\epsilon|t|} U(t, -\infty) | \Phi_\alpha \rangle$$

$$= \int_{-\infty}^{\infty} dt \, e^{i(E-E_\alpha)t} e^{-\epsilon|t|} | \Phi_\alpha \rangle$$

$$- i \int_{-\infty}^{\infty} dt \, e^{i(E-H_0)t} e^{-\epsilon|t|}$$

$$\times \int_{-\infty}^{t} dt' \, e^{iH_0 t'} V e^{-iH_0 t'} e^{-\epsilon|t'|} U(t', -\infty) | \Phi_\alpha \rangle ; \qquad (E5)$$

here $\epsilon \to 0^+$ is the adiabatic switching parameter [see Eq. (7.41)], and the second form follows from the integral equation E3.

In second term of Eq. E5, we introduce the variable $\tau = t' - t$, whence the last term is

$$- i \int_{-\infty}^{\infty} dt \, e^{i(E-H_0)t} e^{-\epsilon|t|} \int_{-\infty}^{0} d\tau \, e^{iH_0(\tau+t)} V e^{-iH_0(\tau+t)} e^{-\epsilon|\tau+t|} U(\tau+t, -\infty) | \Phi_\alpha \rangle$$

$$= -i \int_{-\infty}^{0} d\tau \, e^{-iE\tau} e^{iH_0 \tau} V e^{-\epsilon|\tau|} | \Psi_\alpha^{(+)}(E) \rangle , \qquad (E6)$$

where we have performed the t-integration holding τ fixed, and identified $|\Psi_\alpha^{(+)}(E)\rangle$ through its definition in Eq. E5. Carrying out the remaining integrals, we have

$$|\Psi_\alpha^{(+)}(E)\rangle = 2\pi\delta(E - E_\alpha)|\Phi_\alpha\rangle + \frac{1}{E - H_0 + i\epsilon} V|\Psi_\alpha^{(+)}(E)\rangle , \quad (E7)$$

where $(E - H_0 + i\epsilon)^{-1}$ is the system propagator or Green function.

We then introduce a second (so-called "off-shell") wave function $|\psi_\alpha^{(+)}(E)\rangle$, so that we may consider the consequences of the scattering even in situations where energy is not conserved (by virtue, say, of interaction with some external system). We take $|\psi_\alpha^+(E)\rangle$ to satisfy the *Lippmann–Schwinger equation* for the state vector,

$$|\psi_\alpha^{(+)}(E)\rangle = |\Phi_\alpha\rangle + \frac{1}{E - H_0 + i\epsilon} V|\psi_\alpha^{(+)}(E)\rangle , \quad (E8)$$

a formal way of writing an integral equation for the scattering state. Then, the two state-vectors are connected by the "on-energy-shell" relation

$$|\Psi_\alpha^{(+)}(E)\rangle = 2\pi\delta(E - E_\alpha)|\psi_\alpha^{(+)}(E_\alpha)\rangle . \quad (E9)$$

Substituting Eqs. E7–E9 back into Eq. E4 for the *t*-matrix element and introducing $t_{\beta\alpha}$ through

$$\mathcal{T}_{\beta\alpha} = 2\pi i\delta(E_\beta - E_\alpha)t_{\beta\alpha}(E_a) , \quad (E10)$$

we find a formal integral equation for the *t*-matrix element,

$$t_{\beta\alpha}(E) = V_{\beta\alpha} + \langle\Phi_\beta|V \frac{1}{E - H_0 + i\epsilon} V|\psi_\alpha^{(+)}\rangle$$

$$= V_{\beta\alpha} + \langle\Phi_\beta|V \frac{1}{E - H_0 + i\epsilon} t|\Phi_\alpha\rangle , \quad (E11)$$

or, stripping off the unperturbed states for a pure operator relation,

$$t(E) = V + V \frac{1}{E - H_0 + i\epsilon} t(E) . \quad (E12)$$

This is the form shown in Eq. 11.55. It is clear that at the first stage of iteration we have

$$t(E) \cong V , \quad (E13)$$

which yields the Born approximation for the potential scattering, as in Eqs. (11.53) and (11.54). Equation E12 then provides the prescription for solving—at least formally—for the exact quantity, namely $t_{\beta\alpha}(E_\alpha)$, that must replace the approximate $V_{\beta\alpha}$ for the calculation of scattering cross sections, as in Eqs. (11.56)–(11.58).

Index